Tobacco: Its History, Varieties, Culture, Manufacture and Commerce

by E.R. Billings

with an introduction by Roger Chambers

This work contains material that was originally published in 1875.

This publication was created and published for the public benefit, utilizing public funding and is within the Public Domain.

This edition is reprinted for educational purposes and in accordance with all applicable Federal Laws.

Introduction Copyright 2018 by Roger Chambers

IMPORTANT NOTE & DISCLAIMER

IMPORTANT NOTE :
As with all reprinted books of this age that are intended to perfectly reproduce the original edition, considerable pains and effort had to be undertaken to correct fading and sometimes outright damage to existing proofs of this title.

At times, this task can be quite monumental, requiring an almost total rebuilding of some pages from digital proofs of multiple copies. Despite this, imperfections still sometimes exist in the final proof and may detract slightly from the visual appearance of the text.

Some images may suffer from reduced quality due to anomalies in the original scan.

DISCLAIMER :
Due to the age of this book, some methods or practices may have been deemed unsafe or unacceptable in the interim years. In utilizing the information herein, you do so at your own risk.

We republish antiquarian books with no judgment or revisionism, solely for their historical and cultural importance, and for educational purposes.

Self Reliance Books

Get more historic titles on animal and stock breeding, gardening and old fashioned skills by visiting us at:

http://selfreliancebooks.blogspot.com/

Disclaimer

This book was written in an age when little was known about the ill effects of tobacco.

The material presented herein is intended to be strictly for educational purposes with the purpose of enlightening readers about the historical uses of tobacco. Publication of the material is neither an endorsement, nor a criticism of its contents. This book is presented as part of large series of educational material on the history and cultivation of tobacco.

As the reader, please consider it your duty to consult with a medical doctor before utilizing tobacco. It is also the reader's duty to become familiar with local, state, provincial and federal laws relating to the growing of tobacco.

As the author, publisher and retailer cannot control how the reader utilizes the historical information presented in the pages herein, they hereby disclaim any liability to any party for any loss, damage, disruption, death or other liability that may be incurred by the reader's misuse of this material.

introduction

Here at **Self-Reliance Books** we are dedicated to bringing you the best in *dusty-old-book-knowledge* to help you in your quest for self-sufficiency and independence.

We're so pleased to bring you this old title on the culture of Tobacco. These old books on agricultural and horticultural subjects are extremely popular.

This special edition of **Tobacco : Its History, Varieties, Culture, Manufacture and Commerce** was written by E.R. Billings, and first published in 1875, making it over 140 years old.

The book features chapters on *The Tobacco Plant, Tobacco – Its Discovery, Tobacco in America, Tobacco in Europe, Tobacco Pipes, Smoking, and Smokers, Cigars, Tobacco Planters and Plantations, Varieties,* and more.

An incredible old book and an absolute must-read for anybody interested in Tobacco and its history, both culturally and commercially.

~ *Roger Chambers*
State of Jefferson, March 2018

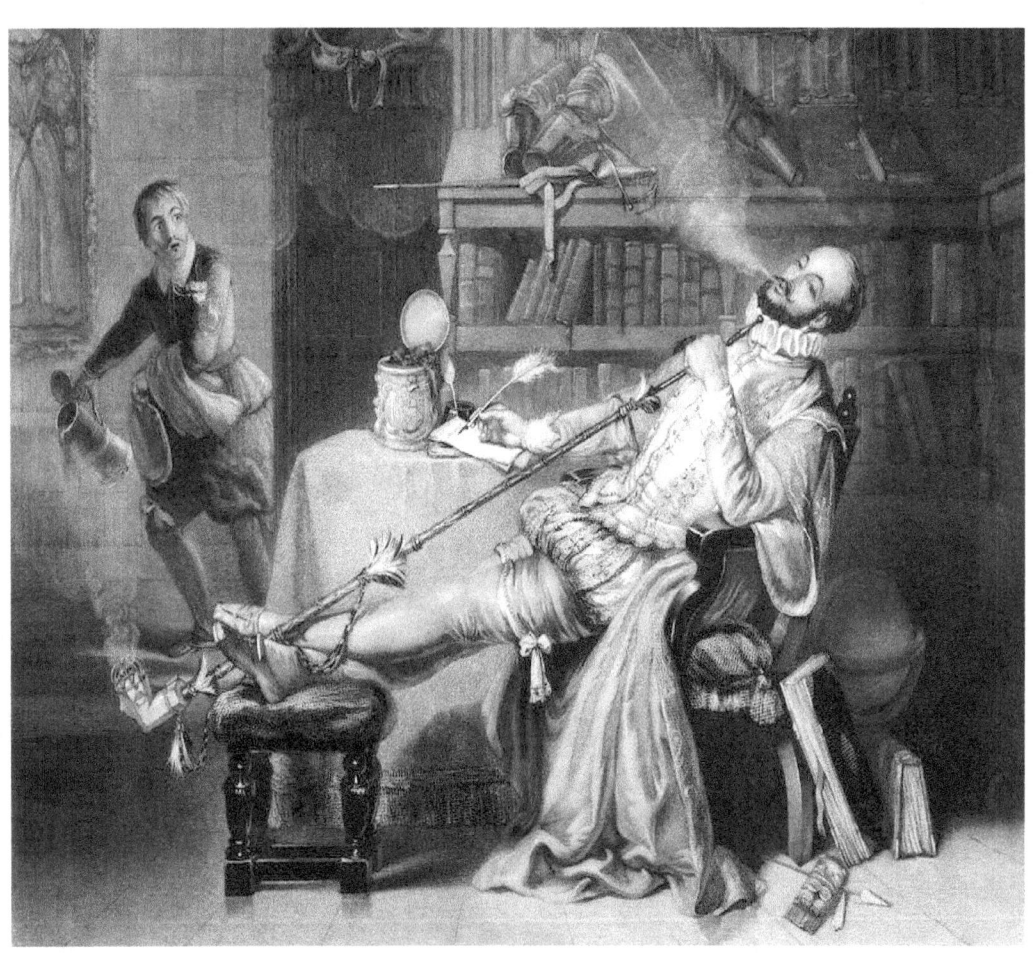

Is it not wondrous strange that there should be
Such different tempers twixt my friend and me?
I burn with heat when I tobacco take,
But he on th' other side with cold doth shake:
To both 'tis physick, and like physick works,
The cause o' th' various operation lurks
Not in tobacco, which is still the same,
But in the difference of our bodies frame:
What's meat to this man, poison is to that,
And what makes this man lean, makes that man fat;
What quenches one's thirst, makes another dry;
And what makes this man wel, makes that man dye.
        THOMAS WASHBOURNE, D. D.

Thy quiet spirit lulls the lab'ring brain,
Lures back to thought the flights of vacant mirth,
Consoles the mourner, soothes the couch of pain,
And wreathes contentment round the humble hearth;
While savage warriors, soften'd by thy breath,
Unbind the captive, hate had doomed to death.
        Rev. WALTER COLTON.

Whate'er I do, where'er I be,
My social box attends on me;
It warms my nose in winter's snow,
Refreshes midst midsummer's glow;
Of hunger sharp it blunts the edge,
And softens grief as some alledge.
Thus, eased of care or any stir,
I broach my freshest canister;
And freed from trouble, grief, or panic,
I pinch away in snuff balsamic.
For rich or poor, in peace or strife,
It smooths the rugged path of life.
        Rev. WILLIAM KING.

HAIL! Indian plant, to ancient times unknown—
A modern truly thou, and all our own!
Thou dear concomitant of nappy ale,
Thou sweet prolonger of an old man's tale.
Or, if thou'rt pulverized in smart rappee,
And reach Sir Fopling's brain (if brain there be),
He shines in dedications, poems, plays,
Soars in Pindarics, and asserts the bays;
Thus dost thou every taste and genius hit—
In smoke thou'rt wisdom, and in snuff thou'rt wit.
        Rev. MR. PRIOR.

TO

CHARLES DUDLEY WARNER,

Whose rare, good gifts have endeared him to all lovers of the English tongue, this volume, historically and practically treating of one of the greatest of plants, as well as the rarest of luxuries, is respectfully dedicated by

THE AUTHOR.

PREFACE.

Ever since the discovery of tobacco it has been the favorite theme of many writers, who have endeavored to shed new light on the origin and early history of this singular plant. Upwards of three hundred volumes have been written, embracing works in nearly all of the languages of Europe, concerning the herb and the various methods of using it. Most writers have confined themselves to the commercial history of the plant; while others have written upon its medicinal properties and the various modes of preparing it for use. For this volume the Author only claims that it is at least a more comprehensive treatise on the varieties and cultivation of the plant than any work now extant. A full account of its cultivation is given, not only in America, but also in nearly all of the great tobacco-producing countries of the world. The history of the plant has been carefully and faithfully compiled from the earliest authorities, that portion which relates to its early culture in Virginia being drawn from hitherto unpublished sources. Materials for such a work have not been found lacking. European authors abound with allusions to tobacco; more especially is it true of English writers, who have celebrated its virtues in poetry and song. All along the highways and by-paths of our literature we encounter much that pertains to this "queen of plants." Considered in what light it may, tobacco must be regarded as the most astonishing of the productions of nature, since it has, in the short period of nearly four centuries,

PREFACE.

dominated not one particular nation, but the whole world, both Christian and Pagan. Ushered into the Old World from the New by the great colonizers—Spain, England, and France—it attracted at once the attention of the authors of the period as a fit subject for their marvel-loving pens. It has been the aim of the writer to give as much as possible of the existing material to be had concerning the early persecution waged against it, whether by Church or State. These accounts, while they invest with additional interest its early use and introduction, serve as well to show its triumph over all its foes and its vast importance to the commerce of the world. This work has been prepared and arranged, not only for the instruction and entertainment of the users of tobacco, but for the benefit of the cultivators and manufacturers as well. As such it is now presented to the public for whatever meed of praise or censure it is found to deserve.

HARTFORD, CONN., 1875.

ILLUSTRATIONS.

	PAGE
1. FRONTISPIECE..	—
2. TOBACCO STALKS...	22
3. TOBACCO LEAVES...	24
4. BUD AND FLOWERS..	25
5. CAPSULES. (FRUIT BUD.)..	27
6. SUCKERS...	28
7. PRIMITIVE PIPE..	33
8. NATIVE SMOKING...	35
9. OLD ENGRAVING..	40
10. THE CONTRAST...	44
11. JOHN ROLFE...	48
12. VIRGINIA TOBACCO FIELD, 1620...................................	51
13. BUYING WIVES...	57
14. GROWING TOBACCO IN THE STREETS...........................	64
15. NATIVES GROWING TOBACCO.......................................	66
16. DESTROYING SUCKERS...	69
17. CARRYING TOBACCO TO MARKET..................................	73
18. ENRICHING PLANT-BED..	75
19. SHIPPING TOBACCO...	78
20. OLD ENGRAVING OF TOBACCO.....................................	86
21. SIR WALTER RALEIGH...	89
22. ENGLISH GALLANTS...	90
23. SMOKING IN THE 17TH CENTURY..................................	94
24. EXHALING THROUGH THE NOSE....................................	97
25. OLD LONDON ALE-HOUSE...	101
26. PUNISHMENT FOR SNUFF TAKING.................................	104
27. SILVER SPITTOONS...	106
28. THE NEGRO IMAGE...	108
29. TOBACCO AND THEOLOGY..	112
30. WEIGHING SMOKE..	117
31. INDIAN PIPE...	126
32. SCULPTURED PIPE..	128
33. PIPE OF PEACE...	130
34. A MODEL CIGAR...	132
35. SOUTH AMERICANS SMOKING......................................	135
36. A WAR PIPE...	139
37. PEACE PIPE..	140

ILLUSTRATIONS.

	PAGE
38. A Tchuktchi Pipe	143
39. Turk Smoking	145
40. Old English Pipes	148
41. French Pipes	149
42. Pipe Colorer	152
43. German Porcelain Pipes	153
44. A Persian Water Pipe	156
45. Searching for Amber	160
46. Fancy Pipes	162
47. Clay and Reed Pipes	164
48. Fairy Pipes	166
49. Female Smoking in Algiers	168
50. African Pipe	170
51. Egyptian Pipes	172
52. Japanese Pipes	173
53. Engraved Boxes	177
54. Tobacco Jars	179
55. Tobacco Stoppers	181
56. Lord and Lackey	185
57. The Strange Youth	190
58. Smokers Reading Epigrams	193
59. The Explosion	195
60. Theory against Experience	199
61. A Faithful Attendant	205
62. Newton and his Pipe	207
63. Tennyson, Smoking	209
64. Modern Smokers	212
65. The Artist	215
66. The Yankee Smoker	216
67. A Tobacco Grater	220
68. Demi-Journeers	222
69. James Gillespie	224
70. Fops Taking Snuff. (From an old print.)	226
71. Horn Snuff-boxes	227
72. Scotch Snuff-mills	232
73. Sweeping from the Pulpit	235
74. Snuff-mill, a Century ago	240
75. Perfuming Snuff	242
76. Fuegian Snuff-Takers	244
77. Snuff-Dipping	247
78. Snuffers	248
79. Fancy Snuff-boxes	251
80. Curing a Headache	255
81. Highlanders	257
82. Cigars	260
83. Cigar-holders	262
84. Life in Mexico	266
85. Cuban Cigar Shop	268
86. Tobacco Leaf	271
87. Wenches Smoking	274
88. A Moonlight Reverie in Havana	275
89. By the Sea	277
90. An American Smoker	279
91. "Light, Sir?"	282
92. Bringing a Light	285
93. Making Cigars	288
94. Havanas	301
95. Yara Cigars	303
96. Manilla Cigar and Cheroot	304
97. Swiss Cigars	305
98. Paraguay Cigars	306
99. Connecticut Tobacco Field	312

ILLUSTRATIONS.

		PAGE.
100.	HOME OF THE CONNECTICUT PLANTER	315
101.	NEGRO QUARTERS	317
102.	THE PLANTER'S HOME	318
103.	"BURNING THE PATCH."	322
104.	STRINGING THE PRIMINGS	323
105.	WORMING	325
106.	OHIO TOBACCO FIELD	329
107.	TOBACCO WAREHOUSE	331
108.	KENTUCKY TOBACCO PLANTATION	332
109.	THE KENTUCKY PLANTER	334
110.	FLORIDA TOBACCO PLANTATION	336
111.	LOUISIANA TOBACCO PLANTATION	338
112.	MEXICAN TOBACCO PLANTATION	342
113.	ST. DOMINGO TOBACCO FIELD, 1535	345
114.	A CUBAN *vega*	346
115.	KILLING BUGS BY NIGHT	348
116.	GOING TO MARKET	349
117.	GERMAN TOBACCO FIELD	351
118.	DUTCH PLANTERS	355
119.	SUCCESS TO VON TROMP	358
120.	TOBACCO FIELD IN ALGIERS	360
121.	TOBACCO FIELD IN AFRICA	361
122.	TOBACCO FIELD IN SYRIA	363
123.	TOBACCO FIELD IN INDIA	365
124.	TURKISH TOBACCO GOING TO MARKET	370
125.	JAPAN TOBACCO FIELD	371
126.	TRANSPLANTING	372
127.	CHINESE TOBACCO FIELD	373
128.	TOBACCO FIELD IN PERSIA	374
129.	GROWING TOBACCO ON THE PHILIPPINE ISLANDS	377
130.	TOBACCO PLOW	378
131.	SPANISH PLANTERS	380
132.	MEXICAN DWARF TOBACCO	384
133.	CONNECTICUT SEED LEAF	385
134.	HAVANA TOBACCO	387
135.	VIRGINIA TOBACCO	388
136.	OHIO WHITE TOBACCO	389
137.	LATAKIA TOBACCO (SYRIA)	393
138.	ORINOCO TOBACCO (VENEZUELA)	397
139.	SHIRAZ TOBACCO (PERSIA)	398
140.	SPANISH TOBACCO	400
141.	JAPAN TOBACCO	402
142.	OLD CONNECTICUT TOBACCO SHED	406
143.	MODERN CONNECTICUT TOBACCO SHED	407
144.	STRIPPING ROOM	408
145.	MODERN VIRGINIA SHED	409
146.	VIRGINIA SHED, 150 YEARS AGO	410
147.	OHIO TOBACCO SHED	412
148.	PERSIAN TOBACCO SHED	414
149.	MAKING THE PLANT BED IN CONNECTICUT	418
150.	COVERING PLANT BED	424
151.	A TOBACCO RIDGER	430
152.	DRAWING THE DIRT AROUND THE FOOT	432
153.	TRANSPLANTING	433
154.	TRANSPLANTING	434
155.	AMERICAN TRANSPLANTER	437
156.	THE WORMS	438
157.	WORMING TOBACCO	439
158.	TOPPING	442
159.	SUCKERING	445
160.	CUTTING THE PLANTS	446
161.	PUTTING ON LATH	447

ILLUSTRATIONS.

	PAGE
162. Carrying to the Shed	448
163. Stripping	456
164. Hands	457
165. Stemming	460
166. Packing	461
167. Prizing in Olden Times	464
168. Tobacco Press	467
169. Firing	470
170. Spanish Seed Tobacco	473

CHAPTER I.

THE TOBACCO PLANT.

Botanical Description—Ancient Plant-Bed—Description of the Leaves—Color of Leaves—Blossoms—The Capsules and Seed—Selection for Seed—Suckers—Nicotine Qualities—Medicinal Properties—Improvement in Plants.. 17

CHAPTER II.

TOBACCO. ITS DISCOVERY.

Early Use—Origin of its Name—Early Snuff-Taking—Tobacco in Mexico—Comparative Qualities of Tobacco—Origin of the Plant—Early Mammoth Cigars—Sacredness of the Pipe—Early Cultivation—Proportions of the Tobacco Trade—Variety of Kinds—Tobacco and Commerce—Original Culture..................... 32

CHAPTER III.

TOBACCO IN AMERICA.

First General Planter—State of the Colony—Conditions of Raising Tobacco—Tobacco Fields, 1620—Increase of Tobacco-Growing—Restriction of Tobacco-Growing—Tobacco used as Money—King James opposes Tobacco-Growing—Buying Wives with Tobacco—Foreign Tobacco Prohibited—King Charles on Tobacco—King Charles as a Tobacco Merchant—Tobacco Taxed—Planting in Maryland—Negro Labor—Competition—Growing Suckers—Virginia Lands—Picture of Early Planters—Large Plantations—Getting to Market—Virginia Plant-Bed—Maryland Plant-Bed—Tobacco Growing in New York and Louisiana—New England Tobacco—Commercial Value of Tobacco—Tobacco a Blessing.............. 47

CHAPTER IV.

TOBACCO IN EUROPE.

Introduction—The Original Importer—Wonderful Cures—How the Herb grew in Reputation—Difference of Opinion—A Smoker's Rhapsody

CONTENTS.

—Old Smokers—The Queen Herb—Drinking Tobacco—Tobacco on the Stage—Shakespeare on Tobacco—Smoking Taught—Ben Jonson on the Weed—Curative Qualities—Modes of Use—Held up to Ridicule—Tirades against Tobacco—Tobacco Selling—Royal Haters of Tobacco—Old Customs—A Racy Poem—A Smoking Divine.. 80

CHAPTER V.

TOBACCO IN EUROPE.—Continued.

Popular use of Tobacco—Tobacco Glorified—Weight of Smoke—Anecdotes—Triumph of Tobacco—A Government Monopoly—Tobacco a Blessing... 111

CHAPTER VI.

TOBACCO PIPES, SMOKING AND SMOKERS.

Indian Pipes—Material for Pipes—Legend of the Red Pipe—Chippewa Pipes—Making the Peace Pipes—South American Pipes—Cigarettes—Tobacco on the Amazon River—Brazilian Tobacco—Patagonians as Smokers—Form and Material—Pipe of the Bobeen Indians—The War Pipe—Pipe Sculpture—Smoking in Alaska—Smoking in Russia—Smoking in Peru—Smoking in Turkey—Moderate Smoking—Female Smoking—Early Manufacture of Pipes—French Pipes. 124

CHAPTER VII.

PIPES AND SMOKERS.—continued.

Meerschaum Pipes—Coloring Meerschaums—The City of Smokers—Hudson as a smoker—Persian Water Pipes—Turkish Pipes—Amber Mouth Pieces—Obtaining Amber—Its Value—Variety of Pipes—History of Pipes—Ancient Habit of Smoking—Buried Pipes—Jasmine Pipes—Smoking in Algiers—Smoking in Africa—Defence of Smoking—Tea and Tobacco—Chinese Pipes—Smoking in Japan—Tobacco Boxes—Tobacco Jars—Musings over a Pipe—Sad Fate of a Chewer—Triumph of the Anti's—The Smoker's Calendar—Doctor Parr as a Smoker—Smoking on the Battle-Field—Literary Smokers—Doctor Clarke on Tobacco—Noted Smokers—Pleasant Pipe—A Tobacco World—Cruelty of Smokers—Men like Pipes—Universal Use.. 150

CHAPTER VIII.

SNUFF, SNUFF-BOXES AND SNUFF-TAKERS.

Its Introduction—Boxes and Graters—Mode of Preparation—Snuff-Boxes—A Celebrated Manufacturer—The Snuffing Period—The Monk and his Snuff-Box—A Pinch of Snuff—Pleasures of Smelling—Frederick the Great—Eminent Snuff-Takers—The Story in Verse—"Come to my Nose"—Snuff Manufacture—Preparation of Tobacco—Grinding the Leaves—Flavoring the Snuff—Profits Made—Love of Tobacco—Chewing and Dipping—Advantages of Dipping—The First Snuffers—Famous Snuff-Takers—Snuff as a Pacificator—A National Stimulant—Different Tastes—Rise and Progress of Snuff-Taking.. 218

CONTENTS.

CHAPTER IX.

CIGARS.

New York Cigars—Havana Cigars—Quality of Havana Cigars—Relative Value and Size—Cigar-Makers—Cuban Cigars—Cigar Manufactories—Preparation of the Tobacco—Sorting the Leaves—Sales, etc.—Large Factories—Universal Smoking—Cigar Etiquette—Reveries—Summer-Day Thoughts—American Smokers—At Home—Sentiment—Ode to a Cigar—Cigar-Lighters—Smoking an Art—Science of Lighting—Age of Fusees—"Home-Made Cigars"—Female Cigar-Makers—A Spicy Article—How to Smoke—Smoking Christians—Lamb's Poem—Tobacco Compliment—Cigarette Smoking—Thomas Hood's Cigar—Lord Byron's Opinion—Kinds of Cigars—Selecting Cigars—Yara Cigars—Manilla Cigars—Swiss Cigars—Paraguay Cigars—Brazilian Cigars—American Cigars—Connecticut Seed Leaf and Havana Cigars—The Exile's Comfort............. 259

CHAPTER X.

TOBACCO PLANTERS AND PLANTATIONS.

The Connecticut Planter—Intelligence of Tobacco Growers—Best Connecticut Seed Leaf—Love for the Plant—Virginia Planters—A Virginia Plantation—The Plant-Patch—Planting, Topping and Priming—Suckering—Crop-Gathering—Curing and Sorting—Tobacco Markets—Ohio Tobacco—Mode of Cure—Kentucky Tobacco-Growing—The Kentucky Planter—Florida Tobacco—Florida Plantation—Tobacco in Lousiana—California Tobacco Lands—Mexican Tobacco—Plants around Vera Cruz—Tobacco in St Domingo—Cuba Plantations—Mode of Working—Soil and Climate—Tobacco-Growing in Germany—Method of Culture—Extent of Culture—Tobacco-Raising in Prussia—Tobacco in Holland—Dutch Planters—A Plea for Tobacco—Tobacco Culture in Australia—Arabian Plantations—Tobacco in Africa—Syrian Tobacco—Latakia Tobacco—Growing Tobacco in India—Curing Tobacco in India—Turks Cultivating Tobacco—Japanese Tobacco—Persian Tobacco—Tobacco Culture, Philippine Islands—Climate of the Islands—Fragrant Manillas—Tropical Tobacco................................ 311

CHAPTER XI.

VARIETIES.

Kinds used for Cigars—Dwarf Tobacco—Havana Tobacco—Yara and Virginia Tobacco—James River Tobacco—Ohio Tobacco—South American Tobacco—Celebrated Brands of Tobacco—Russian Tobacco—Columbian Tobacco—Tobacco of Brazil—The Orinoco Tobacco—Persian Tobacco—French Tobacco—Spanish Tobacco—Japanese Tobacco—Manilla Tobacco......................... 382

CHAPTER XII.

TOBACCO HOUSES.

Tobacco Sheds—Stripping Houses—Virginia Tobacco Sheds—Ordinary Sheds—Superior Sheds—Ohio Sheds—Kentucky and Tennessee Sheds—Foreign Tobacco Sheds............................. 405

CONTENTS.

CHAPTER XIII.

TOBACCO CULTURE.

Hot Beds—Virginia Plant Patch—Tennessee Plant Bed—Cuban Plant Bed—Covering Plant Bed—Selection of Soil—The Soil Affecting Color—Preparing the Soil—Virginia Methods—Burning Brush—Implements—Transplanting Plants—Setting—Seasons in Mexico and Persia—The American Transplanter—Pests—Worming—Backward Plants—Topping—Suckers—Maturation—The Harvest—Cutting—Hanging—Cutting time in Cuba—Harvesting in Virginia—The Season in other Places—Curing—Curing by Smoke—Yellow Tobacco—Stripping—Assorting—Shading—Stemming—Packing—Casing—Old Style—Resistance to Dampness—Prizing—Marking—Baling—Certificates—Firing—White Rust—Seed Plants—Maturing of Seed—Second Growth................................... 415

CHAPTER XIV.

THE PRODUCTION, COMMERCE AND MANUFACTURE OF TOBACCO.

Early History of Tobacco—Cultivation by Spaniards at St. Domingo—Annual Product of Cuba—Amount of Land under Cultivation in U. S.—Cultivation in the South—Annual Product of Europe, Asia and Africa—Government Monopoly—Source of Revenue—Manufacture of Cigarettes—Increase of Tobacco Culture............. 478

CHAPTER I.

THE TOBACCO PLANT.

TOBACCO is a hardy flowering annual* plant, growing freely in a moist fertile soil and requiring the most thorough culture in order to secure the finest form and quality of leaf. It is a native of the tropics and under the intense rays of a vertical sun develops its finest and most remarkable flavor which far surpasses the varieties grown in a temperate region. It however readily adapts itself to soil and climate growing through a wide range of temperature from the Equator to Moscow in Russia in latitude 56°, and through all the intervening range of climate †.

The plant varies in height according to species and locality; the largest varieties reaching an altitude of ten or twelve feet, in others not growing more than two or three feet from the ground. Botanists have enumerated between forty and fifty varieties of the tobacco plant who class them all among the narcotic poisons. When properly cultivated the plant ripens in a few weeks growing with a rapidity hardly equaled by any product either temperate or tropical. Of the large number of varieties cultivated scarcely more than one-half are grown to any great extent while many of them are hardly known outside of the limit of cultivation. Tobacco is a strong growing plant resisting heat and drought to a far

---

* The greater number of the species are annual plants; but two at least are perennial; the *Nicotiana fruticosa*, which is a shrub, a native of the Cape of Good Hope, and of China; and *N. urens*, a native of South America.

† Tatham says that the tobacco plant is peculiarly adapted for an agricultural comparison of climates.

greater extent than most plants. It is a native of America, the discovery of the continent and the plant occurring almost simultaneously. It succeeds best in a deep rich loam in a climate ranging from forty to fifty degrees of latitude. After having been introduced and cultivated in nearly all parts of the world, America enjoys the reputation of growing the finest varieties known to commerce. European tobacco is lacking in flavor and is less powerful than the tobacco of America.

The botanical account of tobacco is as follows:—

"Nicotiana, the tobacco plant is a genus of plants of the order of Monogynia, belonging to the pentandria class, order 1, of class V. It bears a tubular 5-cleft calyx; a funnel-formed corolla, with a plaited 5-cleft border; the stamina inclined; the stigma capitate; the capsule 2-celled, and 2 to 4 valved."

A more general description of the plant is given by an American writer:—

"The tobacco plant is an annual growing from eighteen inches (dwarf tobacco) to seven or eight feet in height*. It bears numerous leaves of a pale green color sessile, ovate lanceolate and pointed in form, which come out alternately from two to three inches apart. The flowers grow in loose panicles at the extremity of the stalks, and the calyx is bell-shaped, and divided at its summit into five pointed segments. The tube of the corolla expands at the top into an oblong cup terminating in a 5-lobed plaited rose-colored border. The pistil consists of an oval germ, a slender style longer than the stamen, and a cleft stigma. The flowers are succeeded by capsules of 2 cells opening at the summit and containing numerous kidney-shaped seeds."

Two of the finest varieties of Nicotiana Tobacum that are cultivated are the Oronoco and the Sweet Scented; they differ only in the form of the leaves, those of the latter variety being shorter and broader than the other. They are annual herbaceous plants, rising with strong erect stems to the height of from six to nine feet, with fine handsome foliage. The stalk near the root is often an inch or more in diameter, and

---

* An old English writer in describing tobacco says:—"When at its just height, it is as tall as an ordinary sized man."

surrounded by a hairy clammy substance, of a greenish yellow color. The leaves are of a light green; they grow alternately, at intervals of two or three inches on the stalk; they are oblong and spear-shaped; those lowest on the stalk are about twenty inches in length, and they decrease as they ascend.

The young leaves when about six inches, are of a deep green color and rather smooth, and as they approach maturity they become yellowish and rougher on the surface. The flowers grow in clusters from the extremities of the stalk; they are yellow externally and of a delicate red within. They are succeeded by kidney shaped capsules of a brown color.

Thompson in his "Notices relative to Tobacco" describes the tobacco plant as follows:—

"The species of Nicotiana which was first known, and which still furnishes the greatest supply of Tobacco, is the N. tobacum, an annual plant, a native of South America, but naturalized to our climate. It is a tall, not inelegant plant, rising to the height of about six feet, with a strong, round, villous, slightly viscid stem, furnished with alternate leaves, which are sessile, or clasp the stems; and are decurrent, lanceolate, entire; of a full green on the upper surface, and pale on the under.

"In a vigorous plant, the lower leaves are about twenty inches in length, and from three to five in breadth, decreasing as they ascend. The inflorescence, or flowering part of the stem, is terminal, loosely branching in that form which botanists term a panicle, with long, linear floral leaves or bractes at the origin of each division.

"The flowers, which bloom in July and August, are of a pale pink or rose color: the calyx, or flower-cup, is bell-shaped, obscurely pentangular, villous, slightly viscid, and presenting at the margin five acute, erect segments. The corolla is twice the length of the calyx, viscid, tubular below, swelling above into an oblong cup, and expanding at the lip into five somewhat plaited, pointed segments; the seed vessel is an oblong or ovate capsule, containing numerous reniform seeds, which are ripe in September and October; and if not collected, are shed by the capsule opening at the apex."

In Stevens and Liebault's Maison Rustique, or the Country Farm, (London, 1606), is found the following curious account of the tobacco plant:—

"This herbe resembleth in figure fashion, and qualities, the great comfrey in such sort as that a man woulde deeme it to be a kinde of great comfrey, rather than a yellow henbane, as some have thought.

"It hath an upright stalke, not bending any way, thicke, bearded or hairy, and slimy: the leaves are broad and long, greene, drawing somewhat towards a yellow, bearded or hoarrie, but smooth and slimie, having as it were talons, but not either notched or cut in the edges, a great deale bigger downward toward the root than above: while it is young it is leaved, as it were lying upon the ground, but rising to a stalke and growing further, it ceaseth to have such a number of leaves below, and putteth forth branches from half foot to half, and storeth itselfe, by that meanes with leaves, and still riseth higher from the height of four or five foote, unto three or four or five cubits according as is sown in a hot and fat ground, and carefully tilled. The boughs and branches thereof put out at joints, and divide the stalk by distance of halfe a foote: the highest of which branches are bigger than an arme.

"At the tops and ends of his branches and boughs, it putteth foorth flowers almost like those of Nigella, of a whitish and incarnate color, having the fashion of a little bell comming out of a swad or husk, being of the fashion of a small goblet, which husk becometh round, having the fashion of a little apple, or sword's pummell: as soon as the flower is gone and vanished away, it is filled with very small seedes like unto those of yellow henbane, and they are black when they be ripe, or greene, while they are not yet ripe.

"In a hot countree it beareth leaves, flowers, and seeds at the same time, in the ninth or tenth month of the year it putteth foorth young cions at the roote, and reneweth itself by this store and number of cions, and great quantity of sprouts, and yet notwithstanding the roots are little, small, fine thready strings, or if otherwise they grow a little thick, yet remaine they still very short, in respect of the height of the plant. The roots and leaves do yield a glewish and rosinith kind of juice, somewhat yellow, of a rosinlike smell, not unpleasant, and of a sharpe, eager and biting taste, which sheweth that it is by nature hot, whereupon we must gather that it is no kind of yellow henbane as some have thought. Nicotiana craveth a fat ground well stird, and well manured also in this cold countrie (England) that is to say an earth, wherein the manure is so well mingled and incorporated, as that it becometh earthie, that is to say, all turned into earth, and not making any shew any more of dung: which is likewise moist and shadowie, wide and roomy, for in a narrow and straight place it would not grow high, straight, great and well-branched.

"It desireth the South sun before it, and a wall behind it, which may stand in stead of a broad pair of shoulders to keep away the northern wind and to beate backe againe the heat of the sun. It groweth the better if it be oft watered, and maketh itself sport and jolly good cheer with water when the time becometh a little dry. It hateth cold, and therefore to keepe it from dying in winter, it must be either kept in cellars where it may have free benefit of air, or else in some cave made on purpose within the same garden, or else to cover it as with a cloak very well with a double mat, making a penthouse of wicker work from the wall to cover the head thereof with straw laid thereupon: and when the southern sun shineth, to open the door of the covert made for the said herb right upon the said South sun."

The most ludicrous part of "The discourse on Nicotian" will be found in that portion which relates to the making of the plant-bed and transplanting:—

"For to sow it, you must make a hole in the earth with your finger and that as deep as your finger is long, then you must cast into the same hole ten or twelve seeds of the said Nicotiana together, and fill up the hole again: for it is so small, as that if you should put in but four or five seeds the earth would choake it: and if the time be dry, you must water the place easily some five days after: And when the herb is grown out of the earth, inasmuch as every seed will have put up his sprout and stalk, and that the small thready roots are intangled the one within the other, you must with a great knife make a composs within the earth in the places about this plot where they grow and take up the earth and all together, and cast them into a bucket full of water, to the end that the earth may be seperated, and the small and tender impes swim about the water; and so you shall sunder them one after another without breaking of them." * *

THE STALK.

The Tobacco stalk varies with the varieties of the plant. All of the species cultivated in the United States have stalks of a large size—much larger than many varieties grown in the tropics. Those of some species of tobacco are little and easily broken, which to a certain extent is the case with most varieties of the plant when maturing very fast. The stalks of some plants are rough and uneven, while those of others are smooth. Nearly all, including most of those grown in Europe and America, have erect, round, hairy, viscid stalks, and large, fibrous roots; while that of Spanish as well as dwarf tobacco is harder and much smaller. The stalk is composed of a wood-like substance containing a glutinous pith, and is of about the same shade of color as the leaves. As the plant develops in size the stalk hardens, and when fully grown is not easily broken.

TOBACCO STALKS.

The size of the stalk corresponds with that of the leaves, and with such varieties of the plant as Connecticut seed leaf, Virginia, Kentucky, Ohio, St. Domingo, and some others; both will be found to be larger than Spanish, Latakia, and Syrian tobacco, which have a much smaller but harder stalk. It will readily be seen that the stalk must be strong and firm in order to support the large palm-like leaves which on some varieties grow to a length of nearly four feet with a corresponding breadth. The stalk does not "cure down" as fast as the leaves, which is thought now to be necessary in order to prevent sweating, as well as to hasten the curing. Most of the varieties of the plant have an erect, straight stalk, excepting Syrian tobacco, which near the top describes more of a semi-circle, but not to that extent of giving an idea of an entirely crooked plant. The stalk gradually tapers from the base to the summit, and when deprived of its leaves presents a smooth appearance not unlike that of a small tree or shrub deprived of its twigs and leaves.

THE LEAVES.

The Plant bears from eight to twenty leaves according to

DESCRIPTION OF THE LEAVES.

the species of the plant. They have various forms, ovate, lanceolate, and pointed. Leaves of a lanceolate form are the largest, and the shape of those found on most varieties of the American plant. The color of the leaves when growing, as well as after curing and sweating, varies, and is frequently caused by the condition of the soil. The color while growing may be either a light or dark green, which changes to a yellowish cast as the plant matures and ripens. The ground leaves are of a lighter color and ripen earlier than the rest— sometimes turning yellow, and during damp weather rotting and dropping from the stalk. Some varieties of the plant, like Latakia, bear small but thick leaves, which after cutting are very thin and fine in texture; while others, like Connecticut seed leaf and Havana, bear leaves of a medium thickness, which are also fine and silky after curing. But while the color of the plant when growing is either a light or dark green, it rapidly changes during curing, and especially after passing through the sweat, changing to a light or dark cinnamon like Connecticut seed leaf, black like Holland and Perique tobacco, bright yellow of the finest shade of Virginia and Carolina leaf, brown like Sumatra, or dark red like that known by the name of "Boshibaghli," grown in Asia Minor. The leaves are covered with glandular hairs containing a glutinous substance of an unpleasant odor, which characterizes all varieties as well as nearly all parts of the plant.

The leaves of all varieties of tobacco grow the entire length of the stem and clasp the stalk, excepting those of Syrian, which are attached by a long stem. The size of the leaves, as well as the entire plant, is now much larger than when first discovered. One of the early voyagers describes the plant as short and bearing leaves of about the size and shape of the walnut. In many varieties the leaves grow in a semi-circular form while in others they grow almost straight and still others growing erect presenting a singular appearance. The stem or mid-rib running through the leaf is large and fibrous and its numerous smaller veins proportionally larger which on curing become smaller and particularly in

COLOR OF LEAVES.

those kinds best adapted for cigar wrappers. The leaves from the base to the center of the plant are of about equal size but are smaller as they reach the summit, but after

TOBACCO LEAVES.

topping attain about the same size as the others. The color of the leaf after curing may be determined by the color of the leaf while growing—if dark green while maturing in the field, the color will be dark after curing and sweating and the reverse if of a lighter shade of green.

If the soil be dark the color of the leaf will be darker than if grown upon a light loam. Some varieties of the plant have leaves of a smooth glossy appearance while others are rough and the surface uneven—more like a cabbage leaf, a peculiar feature of the tobacco of Syria. The kind of fertilizers applied to the soil also in a measure as well as the soil itself has much to do with the texture or body of the leaf and should be duly considered by all growers of the plant. A light moist loam should be chosen for the tobacco field if a leaf of light color and texture is desired while if a dark leaf is preferred the soil chosen should be a moist heavy loam.

THE FLOWER.

The flowers of the tobacco plant grow in a bunch or cluster on the summit of the plant and are of a pink, yellow, or purple white color according to the variety of the plant. On most varieties the color of the flowers is pink excepting Syrian or Latakia which bears yellow flowers while those of

Shiraz or Persian and Guatemala are white while those of the Japan tobacco, are purple. The segments of the corolla are pointed but on some varieties unequal, particularly that of Shiraz tobacco. The flowers impart a pleasant odor doubtless to all lovers of the weed but to all others a compound of villainous smells among which and above all the rest may be recognized an odor suggestive of the leaves of the plant.

When in full blossom a tobacco field forms a pleasant

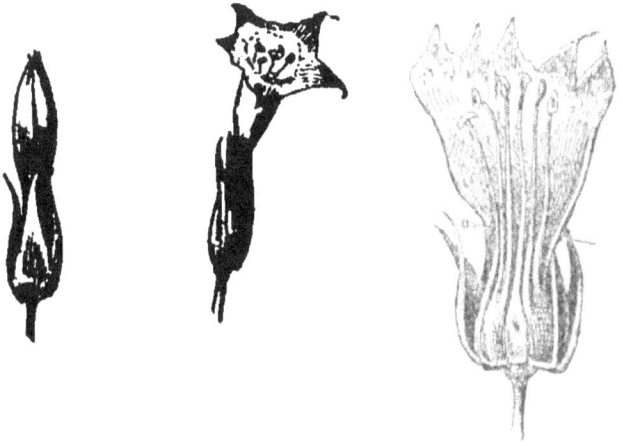

BUD AND FLOWERS.

feature of a landscape which is greatly heightened if the plants are large and of equal size. The pink flowers are the largest while those of a yellow color are the smallest. The plant comes into blossom a few weeks before fully ripe when with a portion of the stalk they are broken off to hasten the ripening and maturing of the leaves. After the buds appear they blossom in a few days and remain in full bloom two or three weeks, when they perish like the blossoms of other plants and flowers. The flowers of Havana tobacco are of a lighter pink than those of Connecticut tobacco but are not as large—a trifle larger however than those of Latakia tobacco. Those varieties of the tobacco plant bearing pink flowers are the finest flavored and are used chiefly for the manufacture of cigars while those bearing yellow flowers are better adapted for cutting purposes and the pipe.

The American varieties of tobacco bear a larger number of

flowers than European tobaccos or those of Africa or Asia. The color of the flowers remain the same whether cultivated in one country or another while the leaves may grow larger or smaller according to the system of cultivation adopted. Those varieties of the plant with heart-shaped leaves have paniculated flowers with unequal cups. The flower stems on the American varieties are much longer than those of European tobaccos and also larger. The season has much to do with the size of the flowers; as if very dry they are usually smaller and not as numerous as if grown under more favorable circumstances.

THE CAPSULE.

As soon as the flowers drop from the fruit bud the capsules grow very rapidly until they have attained full size—which occurs only in those plants which have been left for seed and remain untopped. When topped they are not usually full grown—as some growers top the plants when just coming into blossom, while others prefer to top the plants when in full bloom and others still when the blossoms begin to fall. The fruit is described by Wheeler " as a capsule of a nearly oval figure. There is a line on each side of it, and it contains two cells, and opens at the top. The receptacles one of a half-oval figure, punctuated and affixed to the separating body. The seeds are numerous, kidney-shaped, and rugose."

Most growers of the plant would describe the fruit bud as follows: In form resembling an acorn though more pointed at the top; in some species, of a dark brown in others of a light brown color, containing two cells filled with seeds similar in shape to the fruit bud, but not rugose as described by some botanists. Some writers state that each cell contains about one thousand seeds. The fruit buds of Connecticut, Virginia, Kentucky and Ohio Tobacco as well as of most of the varieties grown within the limits of the United States are much larger than those of Havana, Yara, Syrian, and numerous other species of the plant, while the color of these last named varieties is a lighter shade of brown. The color of the seed also varies according to the varieties of the plant. The seeds of some species are of a dark brown while others are of a lighter shade. The seeds, however, are so small that the variety to which they belong cannot be determined except by planting or sowing them.

The plants selected for seed are usually left growing until late in the season, and at night should be protected from the cold and frost by a light covering of some kind—this may not be absolutely necessary, as most growers of tobacco have often noticed young plants growing around the base or roots of the seed stalk—the seeds of which germinated although

CAPSULES. (FRUIT BUD.)

remaining in the ground during the winter. Strong, healthy plants generally produce large, well filled capsules the only ones to be selected by the grower if large, fine plants are desired. Many growers of tobacco have doubtless examined the capsules of some species of the plant and frequently observed that the capsules or fruit buds are often scarcely more than half-filled while others contain but a few seeds. The largest and finest capsules on the plant mature first, while the smaller ones grow much slower and are frequently several weeks changing from green to brown. Many of the capsules do not contain any seed at all.

THE SUCKER.

The offshoots or suckers as they are termed, make their appearance at the junction of the leaves and stalk, about the roots of the plant, the result of that vigorous growth caused by topping. The suckers can hardly be seen until after the plant has been topped, when they come forward rapidly and in a short time develop into strong, vigorous shoots. Tatham describing the sucker says:

"The sucker is a superfluous sprout which is wont to make its appearance and shoot forth from the stem or stalk, near to the junction of the leaves with the stems, and about the root of the plant, and if allowed to grow, injuring the marketable quality of the tobacco by compelling a division of its nutriment during the act of maturation. The planter is therefore careful to destroy these intruders with the thumb nail, as in the act of topping. This superfluity of vegetation, like that of the top, has been often the subject of legislative care; and the policy of supporting the good name of the Virginia produce has dictated the wisdom of penal laws to maintain her good faith against imposition upon strangers who trade with her."

The ripening of the suckers not only proves injurious to the quality of the leaf but retards their size and maturity and if allowed to continue, prevents them from attaining their largest possible growth.

On large, strong, growing plants the growth of suckers is

SUCKERS.

very rank after attaining a length of from six to ten inches, and when fully grown bearing flowers like the parent stalk. After growing for a length of time they become tough and attached so firmly to the stem of the leaf and stalk that they are broken off with difficulty, frequently detaching the leaf with them. The growth of the suckers, however, determines the quality as well as the maturity of the plants.

Weak, spindling plants rarely produce large, vigorous shoots, the leaves of such suckers are generally small and of a yellowish color. When the plants are fully ripe and ready to harvest the suckers will be found to be growing vigorously around the root of the plant. This is doubtless the best evidence of its maturity, more reliable by far than any other as it denotes the ripening of the entire plant. Suckering the plants hastens the ripening of the leaves, and gives a lighter shade of color, no matter on what soil the plants are grown. Having treated at some length of the various parts of the tobacco plant—stalk, leaves, flowers, capsules and suckers we come now to its nicotine properties. The tobacco plant, as is well known, produces a virulent poison known as Nicotine. This property, however, as well as others as violent is found in many articles of food, including the potato together with its stalk and leaves; the effects of which may be experienced by chewing a small quantity of the latter. The New Edinburgh Encyclopedia says:

"The peculiar effect produced by using tobacco bears some resemblance to intoxication and is excited by an essential oil which in its pure state is so powerful as to destroy life even in very minute quantity."

Chemistry has taught us that nicotine is only one among many principles which are contained in the plant. It is supposed by many but not substantiated by chemical research that nicotine is not the flavoring agent which gives tobacco its essential and peculiar varieties of odor. Such are most probably given by the essential oils, which vary in amount in different species of the plant.

An English writer says:

"Nicotine is disagreeable to the habitual smoker, as is proved by the increased demand for clean pipes or which by some mechanical contrivance get rid of the nicotine."

The late Dr. Blotin tested by numerous experiments the effects of nicotine on the various parts of the organization of man. While the physiological effects of nicotine may be interesting to the medical practitioner, they will hardly interest the general reader unless it can be shown that the effects of nicotine and tobacco should be proved to be indentical.

We are loth to leave this subject, however, as it is so intimately connected with the history of the plant, without treating somewhat of its medicinal properties which to many are of more interest than its social qualities. The Indians not only used the plant socially, religiously, but medicinally. Their Medicine men prescribed its use in various ways for most diseases common among them. The use thus made of the plant attracted the attention of the Spanish and English, far more than its use either as a means of enjoyment or as a religious act. When introduced to the Old World, its claims as a remedy for most diseases gave it its popularity and served to increase its use. It was styled "*Sana sancta Indorum—*" "*Herbe propre à tous maux,*" and physicians claimed that it was "the most sovereign and precious weed that ever the earth tendered to the use of man." As early as 1610, three years after the London and Plymouth Companies settled in Virginia, and some years before it began to be cultivated by them as an article of export, it had attracted the attention of English physicians, who seemed to take as much delight in writing of the sanitary uses of the herb as they did in smoking the balmy leaves of the plant.

Dr. Edmund Gardiner, " Practitioner of Physicke," issued in 1610 a volume entitled, " The Triall of Tobacco," setting forth its curative powers. Speaking of its use he says:

"Tobacco is not violent, and therefore may in my judgement bee safely put in practise. Thus then you plainly see that all medicines, and especially tobacco, being rightly and rationally used, is a noble medicine and contrariwise not in his due time with other circumstances considered, it doth no more than a nobleman's shooe doth in healing the gout in the foot."

Dr. Verner of Bath, in his Treatise concerning the taking the fume of tobacco (1637) says that when " taken moderately and at fixed times with its proper adjunct, which (as they doe suppose) is a cup of sack, they think it be no bad physick." Dr. William Barclay in his work on Tobacco, (1614) declares "that it worketh wonderous cures." He not only defends the herb but the "land where it groweth." At this time the tobacco plant like Indian Corn was very small, possessing but few of the qualities now required to make it merchantable. When first exported to Spain and Portugal from the West Indies and South America, and even by the English from Virginia, the leaf was dark in color and strong and rank in flavor. This, however, seems to have been the standard in regard to some varieties while others are spoken of by some of the early writers upon tobacco as "sweet."

The tobacco (uppowoc) grown by the Indians in America, at the time of its discovery, and more particularly in North America, would compare better with the suckers of the largest varieties of the plant rather than with even the smallest species of the plant now cultivated. At the present time tobacco culture is considered a science in order to secure the colors in demand, and that are fashionable, and also the right texture of leaf now so desirable in all tobaccos designed for wrappers. Could the Indians, who cultivated the plant on the banks of the James, the Amazon and other rivers of America, now look upon the plant growing in rare luxuriance upon the same fields where they first raised it, they could hardly realize them to be the same varieties that they had previously planted.

CHAPTER II.

TOBACCO. ITS DISCOVERY.

NEARLY four hundred years have passed away since the tobacco plant and its use was introduced to the civilized world. It was in the month of November, 1492, that the sailors of Columbus in exploring the island of Cuba first noted the mode of using tobacco. They found the Indians carrying lighted firebrands (as they at first supposed) and puffed the smoke inhaled from their mouths and nostrils.

The Spaniards concluded that this was a method common with them of perfuming themselves; but its frequent use soon taught them that it was the dried leaves of a plant which they burned inhaling and exhaling the smoke. It attracted the attention of the Spaniards no less from its novelty than from the effect produced by the indulgence.

The use of tobacco by the Indians was entirely new to the Spanish discoverers and when in 1503 they landed in various parts of South America they found that both chewing and smoking the herb was a common custom with the natives. But while the Indians and their habits attracted the attention of the Spanish sailors Columbus was more deeply interested in the great continent and the luxuriant tropical growth to be seen on every hand. Columbus himself says of it :—

"Everything invited me to settle here. The beauty of the streams, the clearness of the water, through which I could see the sandy bottom; the multitude of palm-trees of different kinds, the tallest and finest I had ever seen; and an infinite number of other large and flourishing trees; the birds, and the verdure of the plains, are so amazingly beautiful, that this country excelles all others as far as the day surpasses the night in splendor."

Lowe, gives the following account of the discovery of tobacco and its uses:—

"The discovery of this plant is supposed to have been made by Fernando Cortez in Yucatan in the Gulf of Mexico, where he found it used universally, and held in a species of veneration by the simple natives. He made himself acquainted with the uses and supposed virtues of the plant and the manner of cultivating it, and sent plants to Spain, as part of the spoils and treasures of his new-found World."

Oviedo* is the first author who gives a clear account of smoking among the Indians of Hispaniola†. He alludes to it as one of their evil customs and used by them to produce insensibility. Their mode of using it was by inhalation and expelling the smoke through the nostrils by means of a hollow forked cane or hollow reed. Oviedo describes them as "about a span long; and when used the forked ends are inserted in the nostrils, the other end being applied to the burning leaves of the herb, using the herb in this manner stupefied them producing a kind of intoxication."

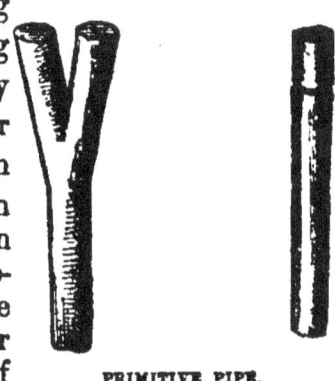

PRIMITIVE PIPE.

Of the early accounts of the plant and its use, Beckman a German writer says:—

"In 1496, Romanus Pane, a Spanish monk, whom Columbus, on his second departure from America, had left in that country, published the first account of tobacco with which he became acquainted in St. Domingo. He gave it the name of Cohoba Cohobba, Gioia. In 1535, the negroes had already habituated themselves to the use of tobacco, and cultivated it in the plantations of their masters. Europeans likewise already smoked it."

An early writer thus alludes to the use of tobacco among the East Indians:—

---

* Historia General de los Indios 1526.
† St. Domingo.

"The East Indians do use to make little balls of the juice of the hearbe tobaco and the ashes of cockle-shells wrought up together, and dryed in the shadow, and in their travaile they place one of the balls between their neather lip and their teeth, sucking the same continually, and letting down the moysture, and it keepeth them both from hunger and thirst for the space of three or four days."

Oviedo says of the implements used by the Indians in smoking:—

"The hollow cane used by them is called tobaco and that that name is not given to the plant or to the stupor caused by its use."

A writer alluding to the same subject says:—

"The name tobacco is supposed to be derived from the Indian tobaccos, given by the Caribs to the pipe in which they smoked the plant."

Others derive it from Tabasco, a province of Mexico; others from the island of Tobago one of the Caribbees; and others from Tobasco in the gulf of Florida.

Tomilson says:—

"The word tobacco appears to have been applied by the caribbees to the pipe in which they smoked the herb while the Spaniards distinguished the herb itself by that name. The more probable derivation of the word is from a place called Tobaco in Yucatan from which the herb was first sent to the New World."

Humboldt says concerning the name:—

"The word Tobacco like maize, savannah, cacique, maguey (agave) and manato, belong to the ancient language of Hayti, or St. Domingo. It did not properly denote the herb, but the tube through which the smoke was inhaled. It seems surprising that a vegetable production so universally spread should have different names among neighboring people. The pete-ma of the Omaguas is, no doubt, the pety of the Guaranos; but the analogy between the Cabre and Algonkin (or Lenni-Lennope) words which denote tobacco may be merely accidental. The following are the synonymes in five languages: Aztec or Mexican, *yetl;* Huron, *oyngona;* Peruvian, *sayri;* Brazil, *piecelt;* Moxo, *sabare.*"

Roman Pane who accompanied Columbus on his second voyage alludes to another method of using the herb. They make a powder of the leaves, which "they take through a cane half a cubit long; one end of this they place in the nose, and the other upon the powder, and so draw it up, which purges them very much."

This is doubtless the first account that we have of snuff-taking; Fairholt says concerning its use:—

"Its effects upon the Indians in both instances seem to have been more violent and peculiar than upon Europeans since."

This may be accounted for from the fact of the imperfect method of curing tobacco adopted by them and all of the natives up to the period of the settlement of Virginia by the English. As nearly all of the early voyagers allude to the plant and especially to its use it would seem probable that it had been cultivated from time immemorial by all the native people of the Orinoco; and at the period of the conquest the habit of smoking was found to be alike spread over both North and South America. The Tamanacs and the Maypures of Guiana wrap maize leaves round their cigars as the Mexicans did at the time of the arrival of Cortez. The Spaniards since have substituted paper for the leaves of maize, in imitation of them.

"The poor Indians of the forests of the Orinoco know as well as did the great nobles at the court of Montezuma, that the smoke of tobacco is an excellent narcotic; and they use it not only to procure their afternoon nap, but also to put themselves in that state of quiescence which they call dreaming with the eyes open or day dreaming."

Tobacco at this period was also rolled up in the leaves of the Palm and smoked. Columbus found the natives of San Salvador smoking after this manner.

NATIVE SMOKING.

Lobel in his History of Plants* gives an engraving

---
* History of Plants, 1576.

of a native smoking one of these rolls or primitive cigars and speaks of their general use by Captains of ships trading to the West Indies.

But not only was snuff taking and the use of tobacco rolls or cigars noted by European voyagers, but the use of the pipe also in some parts of America, seemed to be a common custom especially among the chiefs. Be Bry in his History of Brazil (1590) describes its use and also some interesting particulars concerning the plant. Their method of curing the leaves was to air-dry them and then packing them until wanted for use. In smoking he says:—

"When the leaves are well dried they place in the open part of a pipe of which on burning, the smoke is inhaled into the mouth by the more narrow part of the pipe, and so strongly that it flows out of the mouth and nostrils, and by that means effectually drives out humours."

Fairholt in alluding to the various uses of the herb among the Indians says:—

"We can thus trace to South America, at the period when the New World was first discovered, every mode of using the tobacco plant which the Old World has indulged in ever since."

This statement is not entirely correct—the mode of using tobacco in Norway by plugging the nostrils with small pieces of tobacco seems to have been unknown among the Indians of America as it is now with all other nationalities, excepting the Norwegians.

When Cortez made conquest of Mexico in 1519 smoking seemed to be a common as well as an ancient custom among the natives. Benzoni in his History of the New World[*] describing his travels in America gives a detailed account of the plant and their method of curing and using it. In both North and South America the use of tobacco seemed to be universal among all the tribes and beyond all question the custom of using the herb had its origin among them. The traditions of the Indians all confirm its ancient source; they considered the plant as a gift from the Great Spirit for their

---
[*] From 1541 to 1556.

COMPARATIVE QUALITIES OF TOBACCO.

comfort and enjoyment and one which the Great Spirit also indulged in, consequently with them smoking partook of the character of a moral if not a religious act. The use of tobacco in sufficient quantities to produce intoxication seemed to be a favorite remedy for most diseases among them and was administered by their doctors or medicine-men in large quantities. Benzoni gives an engraving of their mode of inhaling the smoke and says of its use:—

"In La Espanola, when their doctors wanted to cure a sick man, they went to the place where they were to administer the smoke, and when he was thoroughly intoxicated by it, the cure was mostly effected. On returning to his senses he told a thousand stories of his having been at the council of the gods, and other high visions."

It can hardly be supposed that while the custom of using tobacco among the Indians in both North and South America was very general and the mode of use the same, that the plant grown was of the same quality in one part as in another. While the rude culture of the natives would hardly tend to an improvement in quality; the climate being varied would no doubt have much to do with the size and quality of the plant. This would seem the more probable for as soon as its cultivation began in Virginia by the English colonists it had successful rivals in the tobacco of the West Indies and South America. Robertson says:—

"Virginia tobacco was greatly inferior to that raised by the Spaniards in the West Indies and which sold for six times as much as Virginia tobacco." *

But not only has the name tobacco and the implements employed in its use caused much discussion but also the origin of the plant.

Some writers affirm that it came from Asia and that it was first grown in China having been used by the Chinese long before the narcotic properties of opium were known. Tatham in his work on Tobacco says of its origin in substantial agreement with La Bott:—

"It is generally understood that the tobacco plant of

---

* West India tobacco sold for 18 shillings per pound and Virginia for 3 s.

Virginia is a native production of the country; but whether it was found in a state of natural growth there, or a plant cultivated by the Indian natives, is a point of which we are not informed, nor which ever can be farther elucidated than by the corroboration of historical facts and conjectures. I have been thirty years ago, and the greatest part of my time during that period, intimately acquainted with the interior parts of America; and have been much in the unsettled parts of the country, among those kinds of soil which are favorable to the cultivation of tobacco; but I do not recollect one single instance where I have met with tobacco growing wild in the woods, although I have often found a few spontaneous plants about the arable and trodden grounds of deserted habitations. This circumstance, as well as that of its being now, and having been, cultivated by the natives at the period of European discoveries, inclines towards a supposition that this plant is not a native of North America, but may possibly have found its way thither with the earliest migrations from some distant land. This might, indeed, have easily been the case from South America, by way of the Isthmus of Panama; and the foundation of the Choctaw and Chickasaw nations (who we have reasons to consider as descendants from the Tloseolians, and to have migrated to the eastward of the river Mississippi, about the time of the Spanish conquest of Mexico by Cortez), seems to have afforded one fair opportunity for its dissemination."

The first knowledge which the English discoverers had of the plant was in 1565 when they found it growing in Florida, one hundred and seventy-three years after it was first discovered by Columbus on the island of Cuba. Sir John Hawkins says of its use in Florida:—

"The Floridians, when they travel, have a kind of herb dried, which with a cane and an earthen cup in the end, with fire and the dried herbs put together, do suke through the cane the smoke thereof, which smoke satisfieth their hunger, and therewith they live four or five dayes without meat or drinke, and this all the Frenchmen used for this purpose: yet do they holde opinion withall, that it causeth water and steame to void from their stomacks."

This preparation might not have been tobacco as the Indians smoke a kind of bark which they scrape from the killiconick, an aromatic shrub, in form resembling the willow;

they use also a preparation made with this and sumach leaves, or sometimes with the latter mixed with tobacco. Lionel Wafer in his travels upon the Isthmus of Darien in 1699 saw the plant growing and cultivated by the natives. He says:—

"These Indians have tobacco amongst them. It grows as the tobacco in Virginia, but is not so strong, perhaps for want of transplanting and manuring, which the Indians do not well understand, for they only raise it from the seed in their plantations. When it is dried and cured they strip it from the stalks, and laying two or three leaves upon one another, they roll up all together sideways into a long roll, yet leaving a little hollow. Round this they roll other leaves one after another, in the same manner, but close and hard, till the roll be as big as one's wrist, and two or three feet in length. Their way of smoking when they are in company is thus: a boy lights one end of a roll and burns it to a coal, wetting the part next it to keep it from wasting too fast. The end so lighted he puts into his mouth, and blows the smoke through the whole length of the roll into the face of every one of the company or council, though there be two or three hundred of them. Then they, sitting in their usual posture upon forms, make with their hands held together a kind of funnel round their mouths and noses. Into this they receive the smoke as it is blown upon them, snuffing it up greedily and strongly as long as ever they are able to hold their breath, and seeming to bless themselves, as it were, with the refreshment it gives them."

In the year 1534 James Cartier a Frenchman was commissioned to explore the coast of North America, with a view to find a place for a colony. He observed that the natives of Canada used the leaves of an herb which they preserved in pouches made of skins and smoked in stone pipes. It being offensive to the French, they took none of it with them on their return. But writing more particularly concerning the plant he says:—

"In Hochelaga, up the river in Canada there groweth a certain kind of herb whereof in Summer they make a great provision for all the year, making great account of it, and only men use of it, and first they cause it to be dried in the Sune, then wear it about their necks wrapped in a little beast's skine made like a bagge, with a hollow piece of stone or wood like a pipe, then when they please they make powder of it, and then put it in one of the ends of the said Cornet or pipe, and laying a cole of fire upon it, at the other end and suck so long, that they fill their bodides full of smoke, till that it commeth out of their mouth and nostrils, even as out of the Tonnel of a chimney. They say that this doth keepe them warme and in health, they never goo without some of this about them."

Be Bry in his History of Brazil 1590 gives an engraving of a native smoking a pipe and a female offering him a handful of tobacco leaves. The pipe has a modern look and is altogether unlike those found by the English in use among the Indians in Virginia.

OLD ENGRAVING.

An English writer says of the Tobacco using races:—

"From the evidence collected by travellers and archæologists, as to the native arts and relics connected with the use of Tobacco by the Red Indians, it would appear that not one tribe has been found which was unacquainted with the custom,* its use being as well known to the tribes of the North-west and the denizens of the snowy wilds of Canada, as to the races inhabiting Central America and the West India Islands."

Father Francisco Creuxio states that the Jesuit missionaries found the weed extensively used by the Indians of the Seventeenth Century. In 1629 he found the Hurons smoking the dried leaves and stalks of the Tobacco plant or petune. Many tribes of Indians consider that Tobacco is a gift bestowed by the Great Spirit as a means of enjoyment. In consequence of this belief the pipe became sacred, and smoking became a moral if not a religious act, amongst the North American Indians. The Iroquois are of opinion that by burning Tobacco they could send up their prayers to the Great Spirit with the ascending incense, thus maintaining

---

* Arnold in his History of Rhode Island refers to the planting of tobacco by the Indians when the State was first settled. Elliot also says in his History of the same State:—" Tobacco was universal, every man carrying his pipe and bag; and in its cultivation only, did the men condescend to labor; but occasionally all would join, the whole neighborhood, men, women, and children, when some one's field was to be broken up, and they made a loving, sociable, speedy time of it."

communication with the spirit world; and Dr. Daniel Wilson suggests that "the practice of smoking originated in the use of the intoxicating fumes for purposes of divination, and other superstitious rites."

When an Indian goes on an expedition, whether of peace or war, his pipe is his constant companion; it is to him what salt is among Arabs: the pledge of fidelity and the seal of treaties. In the words of a *Review:*

"Tobacco supplies one of the few comforts by which men who live by their hands, solace themselves under incessant hardship."

While the presence, and use of tobacco by the natives of America are among the most interesting features connected with its history, it can hardly be more so than is its early cultivation by the Spaniards, English and Dutch, and afterward by the French. The cultivation of the plant began in the West India Islands and South America early in the Sixteenth Century. In Cuba its culture commenced in 1580, and from this and the other islands large quantities were shipped to Europe. It was also cultivated near Varina in Columbia, while Amazonian tobacco had acquired an enviable reputation as well as Varinian, long before its cultivation began in Virginia by the English. At this period of its culture in America the entire product was sent to Spain and Portugal, and from thence to France and Great Britain and other countries of Europe. The plant and its use attracted at once the attention as well as aroused the cupidity of the Spaniards, who prized it as one of their greatest discoveries.

As soon as Tobacco was introduced into Europe by the Spaniards, and its use became a general custom, its sale increased as extensively as its cultivation. At this period it brought enormous prices, the finest selling at from fifteen to eighteen shillings per pound. Its cultivation by the Spaniards in various portions of the New World proved to them not only its real value as an article of commerce, but also that several varieties of the plant existed; as on removal from one island or province to another it changed in size and quality of leaf. Varinas tobacco at this time was one of the finest tobaccos known,* and large quantities were shipped to Spain and Portugal. The early voyagers little dreamed, however, of the vast proportions to be assumed by the trade in the plant which they had discovered, and which in time proved a source of the greatest profit not only to the European colonies, but to the dealers in the Old World.

Helps, treating on this same subject, says:

"It is interesting to observe the way in which a new product is introduced to the notice of the Old World—a product that was hereafter to become, not only an unfailing source of pleasure to a large section of the whole part of mankind, from the highest to the lowest, but was also to distinguish itself as one of those commodities for revenue, which are the delight of statesmen, the great financial resource of modern nations, and which afford a means of indirect taxation that has perhaps nourished many a war, and prevented many a revolution. The importance, financially and commercially speaking, of this discovery of tobacco—a discovery which in the end proved more productive to the Spanish crown than that of the gold mines of the Indies."

Spain and Portugal in all their colonies fostered and encouraged its cultivation and then at once ranked as the best producers and dealers in tobacco. The varieties grown by them in the West Indies and South America were highly esteemed and commanded much higher prices than that grown by the English and Dutch colonies. In 1620, however, the Dutch merchants were the largest wholesale tobacconists in Europe, and the people of Holland, generally, the greatest consumers of the weed.

The expedition of 1584, under the auspices of Sir Walter Raleigh, which resulted in the discovery of Virginia, also introduced the tobacco plant, among other novelties, to the attention of the English. Hariot,† who sailed with this expedition, says of the plant:

"There is an herb which is sowed apart by itselfe, and is

---

*Trinidad tobacco was then considered the finest.
†A brief and true Report of the New Found Land of Virginia (London, 1588).

called by the inhabitants uppowoc. In the West Indies it hath divers names, according to the severall places and countries where it groweth and is used; the Spaniards generally call it Tobacco. The leaves thereof being dried and brought into powder, they use to take the fume or smoke thereof by sucking it through pipes made of clay into their stomacke and heade, from whence it purgeth superfluous fleame and other grosse humors; openeth all the pores and passages of the body; by which means the use thereof not only preserveth the body from obstructions, but also if any be so that they have not beene of too long continuance, in short time breaketh them; whereby their bodies are notably preserved in health, and know not many grievous diseases wherewithall we in England are oftentimes affected. This uppowoc is of so precious estimation amongest them that they thinke their gods are marvellously delighted therewith; whereupon sometime they make halowed fires, and cast some of the powder therein for a sacrifise. Being in a storme uppon the waters, to pacifie their gods, they cast some up into the aire and into the water: so a weave for fish being newly set up, they cast some therein and into the aire; also after an escape of danger they cast some into the aire likewise; but all done with strange gestures, stamping, sometimes dancing, clapping of hands, holding up of hands, and staring up into the heavens, uttering there withal and chattering strange wordes, and noises.

"We ourselves during the time we were there used to suck it after their manner, as also since our returne, and have found many rare and wonderful experiments of the virtues thereof; of which the relation would require a volume of itselfe; the use of it by so manie of late, men and women, of great calling as else, and some learned phisitions also is sufficient witnes."

The natives also when Drake* landed in Virginia, "brought a little basket made of rushes, and filled with an herbe which they called Tobah;" they "came also the second time to us bringing with them as before had been done, feathers and bags of Tobah for presents, or rather indeed for sacrifices, upon this persuasion that we were gods."

William Strachey† says of tobacco and its cultivation by the Indians:

---

*The World Encompassed. London, 1628.
†"The Historie of Travaile into Virginia Britannica."

"Here is great store of tobacco, which the salvages call apooke: howbeit it is not of the best kynd, it is but poor and weake, and of a byting taste; it grows not fully a yard above ground, bearing a little yellow flower like to henbane; the leaves are short and thick, somewhat round at the upper end; whereas the best tobacco of Trynidado and the Oronoque, is large, sharpe, and growing two or three yardes from the ground, bearing a flower of the breadth of our bell-flower, in England; the salvages here dry the leaves of this apooke over the fier, and sometymes in the sun, and crumble yt into pondre, stalk, leaves, and all, taking the same in pipes of earth, which very ingeniously they can make."

THE CONTRAST.

It would seem then, if the account given by Strachey be correct, that the tobacco cultivated by the Indians of North America was of inferior growth and quality to that grown in many portions of South America, and more particularly in the West India islands. As there are still many varieties of the plant grown in America, so there doubtless was when cultivated by the Indians. While most probably the quality of leaf remained the same from generation to generation, still in some portions of America, owing more to the soil and climate than the mode of cultivating by them, they cured very good tobacco. We can readily see how this might have been, from numerous experiments made with both American and European varieties. Nearly all of the early Spanish, French and English voyagers who landed in America were attracted by the beauty of the country. Ponce De Leon, who sailed from Spain to the Floridas, was charmed by the plants and flowers, and doubtless the first sight of them strengthened his belief in the existence somewhere in this tropical region of the fountain of youth.

The discovery of tobacco proved of the greatest advantage

to the nations who fostered its growth, — and increased the commerce of both England and Spain, doing much to make the latter what it once was, one of the most powerful nations of Europe and possessor of the largest and richest colonies, while it greatly helped the former, already unsurpassed in intelligence and civilization, to reach its present position at the commercial head of the nations of the world.

As Spain, however, has fallen from the high place she once held, her colonial system has also gone down. And while England, thanks to her more liberal policy, still retains a large share of the territory which she possessed at first, Spain, which once held sway over a vast portion of America, has been deprived of nearly all of her colonies, and ere long may lose control of the island on which the discoverer of America first saw the plant.*

It is an historical fact that wherever in the English and Spanish colonies civilization has taken the deepest root, so has also the plant which has become as famous as any of the great tropical products of the earth. The relation existing between the balmy plant and the commerce of the world is of the strongest kind. Fairholt has well said, that "the revenue brought to our present Sovereign Lady from this source alone is greater than that Queen Elizabeth received from the entire customs of the country."

The narrow view of commercial policy held by her successors, the Stuarts, induced them to hamper the colonists of America with restrictions; because they were alarmed lest the ground should be entirely devoted to tobacco. Had not this Indian plant been discovered, the whole history of some portions of America would have been far different. In the West Indies three great products—Coffee, Sugar-Cane, and Tobacco,—have proved sources of the greatest wealth—and wherever introduced, have developed to a great extent the resources of the islands. Thus it may be seen that while the Spaniards by the discovery and colonization

---

* "Spain has doubtless conquered more of the Earth's surface than any other modern nation; and her peculiar national character has also caused her to make the worst use of them. It was always easier for the Moor to conquer than to make a good use of his conquests; and so it has always been with Spain."

of large portions of America strengthened the currency of the world, the English alike, by the cultivation of the plant, gave an impetus to commerce still felt and continued throughout all parts of the globe.

An English writer has truthfully observed that "Tobacco is like Elias' cloud, which was no bigger than a man's hand, that hath suddenly covered the face of the earth; the low countries, Germany, Poland, Arabia, Persia, Turkey, almost all countries, drive a trade of it; and there is no commodity that hath advanced so many from small fortunes to gain great estates in the world. Sailors will be supplied with it for their long voyages. Soldiers cannot (but) want it when they keep guard all night, or upon other hard duties in cold and tempestuous weather. Farmers, ploughmen, and almost all labouring men, plead for it. If we reflect upon our forefathers, and that within the time of less than one hundred years, before the use of tobacco came to be known amongst us, we cannot but wonder how they did to subsist without it; for were the planting or traffick of tobacco now hindered, millions of this nation in all probability must perish for the want of food, their whole livelihood almost depending upon it."

When first discovered in America, and particularly by the English in Virginia, the plant was cultivated only by the females of the tribes, the chiefs and warriors engaging only in the chase or following the warpath. They cultivated a few plants around their wigwams, and cured a few pounds for their own use. The smoke, as it ascended from their pipes and circled around their rude huts and out into the air, seemed typical of the race—the original cultivators and smokers of the plant. But, unlike the great herb which they cherished and gave to civilization, they have gradually grown weak in numbers and faded away, while the great plant has gone on its way, ever assuming more and more sway over the commercial and social world, until it now takes high rank among the leading elements of mercantile and agricultural greatness.

CHAPTER III.

TOBACCO IN AMERICA.

WE do not find in any accounts of the English voyagers made previous to 1584, any mention of the discovery of tobacco, or its use among the Indians. This may appear a little strange, as Captains Amidas and Barlow, who sailed from England under the auspices of Sir Walter Raleigh in 1584, on returning from Virginia, had brought home with them pearls and tobacco among other curiosities. But while we have no account of those who returned from the voyage made in 1602 taking any tobacco with them, it is altogether probable that those who remained took a lively interest in the plant and the Indian mode of use; for we find that in nine years after they landed at Jamestown tobacco had become quite an article of culture and commerce.

Hamo in alluding to the early cultivation of tobacco by the colony, says, that John Rolfe was the pioneer tobacco planter. In his words:

"I may not forget the gentleman worthie of much commendations, which first took the pains to make triall thereof, his name Mr. John Rolfe, Anno Domini 1612, partly for the love he hath a long time borne unto it, and partly to raise commodities to the adventurers, in whose behalfe I intercede and vouchsafe to hold my testimony in beleefe that during the time of his aboade there, which draweth neere sixe years

48 FIRST GENERAL PLANTING.

no man hath laboured to his power there, and worthy incouragement unto England, by his letters than he hath done,

JOHN ROLFE.

witness his marriage with Powhatan's daughter one of rude education, manners barbarous, and cursed generation merely for the good and honor of the plantation."

The first general planting of tobacco by the colony began according to this writer—" at West and Sherley Hundred (seated on the north side of the river, lower than the Bermudas three or four myles) where are twenty-five commanded by capten Maddeson—who are imployed onely in planting and curing tobacco."

This was in 1616, when the colony numbered only three hundred and fifty-one persons. Rolfe, in his relation of the state of Virginia, written and addressed to the King, gives the following description of the condition of the colony in 1616:

"Now that your highness may with the more ease understand in what condition the colony standeth, I have briefly sett downe the manner of all men's several imployments, the number of them, and the several places of their aboad, which places or seates are all our owne ground, not so much by conquest, which the Indians hold a just and lawfull title, but purchased of them freely, and they verie willingly selling it. The places which are now possessed and inhabited are sixe:—Henrico and the lymitts, Bermuda Nether hundred, West and Sherley hundred, James Towne, Kequoughtan, and Dales-Gift. The generall mayne body of the planters are divided into Officers, Laborers, Farmors.

"The officers have the charge and care as well over the farmors as laborers generallie—that they watch and ward for their preservacions; and that both the one and the other's busines may be daily followed to the performance of those imployments, which from the one are reqnired, and the other by covenant are bound unto. These officers are bound to maintayne themselves and families with food and rayment by their owne and their servant's industrie. The laborers are of two sorts. Some employed onely in the generall works, who are fedd and clothed out of the store—others, specially artificers as smiths, carpenters, shoemakers, taylors, tanners, &c., doe worke in their professions for the colony, and maintayne themselves with food ann apparrell, having time lymitted them to till and manure their ground.

"The farmors live at most ease—yet by their good endeavors bring yearlie much plentie to the plantation. They are bound by covenant, both for themselves and servants, to maintaine your Ma'ties right and title in that kingdom, against all foreigne and domestique enemies. To watch and ward in the townes where they are resident. To do thirty-one dayes service for the colony, when they shalbe called thereunto—yet not at all tymes, but when their owne busines can best spare them. To maintayne themselves and families with food and rayment—and every farmor to pay yearlie into the magazine for himself and every man servant, two barrells and a halfe of English measure.

"Thus briefly have I sett downe every man's particular imployment and manner of living; albeit, lest the people—who generallie are bent to covett after gaine, especially having tasted of the sweete of their labors—should spend too much of their tyme and labor in planting tobacco, known to them to be verie vendible in England, and so neglect their tillage of corne, and fall into want thereof, it is provided for

—by the providence and care of Sir Thomas Dale—that no farmor or other, who must maintayne themselves—shall plant any tobacco, unless he shall yearely manure, set and maintayne for himself and every man servant two acres of ground with corne, which doing they may plant as much tobacco as they will, els all their tobacco shalbe forfeite to the colony—by which meanes the magazine shall yearely be sure to receave their rent of corne; to maintayne those who are fedd thereout, being but a few, and manie others, if need be; they themselves will be well stored to keepe their families with overplus, and reape tobacco enough to buy clothes and such other necessaries as are needful for themselves and household. For an easie laborer will keepe and tend two acres of corne, and cure a good store of tobacco—being yet the principall commoditie the colony for the present yieldeth.

"For which as for other commodities, the counncell and company for Virginia have already sent a ship thither, furnished with all manner of clothing, household stuff and such necessaries, to establish a magazine there, which the people shall buy at easie rates for their commodities—they selling them at such prices that the adventurers may be no loosers. This magazine shalbe yearelie supplied to furnish them, if they will endeavor, by their labor, to maintayne it—which wilbe much beneficiall to the planters and adventurers, by interchanging their commodities, and will add much encouragement to them and others to preserve and follow the action with a constant resolution to uphold the same."

The colony at this time was engaged in planting corn and tobacco, "making pitch and tarr, potashes, charcole, salt," and in fishing. Of Jamestown he says:

"At James Towne (seated on the north side of the river, from West and Sherley Hundred lower down about thirty-seven miles) are fifty, under the command of lieutenant Sharpe, in the absence of capten Francis West, Esq., brother to the right ho'ble the L. Lawarre,—whereof thirty-one are farmors; all theis maintayne themselves with food and rayment. Mr. Richard Buck minister there—a verie good preacher."

Rev. Hugh Jones "Chaplain to the Honourable Assembly, and lately Minister of James-Towne and in Virginia," in a work entitled—"The Present State of Virginia," gives the following account of the cultivation of tobacco:

"When a tract of land is seated, they clear it by felling

the trees about a yard from the ground, lest they should shoot again. What wood they have occasion for they carry off, and burn the rest, or let it lie and rot upon the ground. The land between the logs and stumps they hoe up, planting

VIRGINIA TOBACCO FIELD, 1620.

tobacco there in the spring, inclosing it with a slight fence of cleft rails. This will last for tobacco some years, if the land be good; as it is where fine timber, or grape vines grow. Land when hired is forced to bear tobacco by penning their cattle upon it; but cowpen tobacco tastes strong, and that planted in wet marshy land is called nonburning tobacco, which smoaks in the pipe like leather, unless it be of a good age. When land is tired of tobacco, it will bear Indian Corn or English Wheat, or any other European grain or seed with wonderful increase.

"Tobacco and Indian Corne are planted in hills as hops, and secured by worm fences, which are made of rails supporting one another very firmly in a particular manner. Tobacco requires a great deal of skill and trouble in the right management of it. They raise the plants in beds, as we do Cabbage plants; which they transplant and replant upon occasion after a shower of rain, which they call a season. When it is grown up they top it, or nip off the head, succour it, or cut off the ground leaves, weed it, hill it; and when ripe, they cut it down about six or eight leaves on a stalk, which they carry into airy tobacco houses, after it is withered a little in the sun, there it is hung to dry on sticks, as paper at the paper-mills; when it is in proper case, (as they call it) and the air neither too moist, nor too dry, they strike it, or take it down, then cover it up in bulk, or a great heap, where it lies till they have leisure or occasion to strip it (that is pull the leaves from the stalk) or stem it (that is to take out the great fibres) and tie it up in hands, or streight lay it; and so by degrees prize or press it with proper engines into great Hogsheads, containing from about six to eleven hundred pounds; four of which Hogsheads make a tun by dimention, not by weight; then it is ready for sale or shipping.

There are two sorts of tobacco, viz., Oroonoko the stronger, and sweet-scented the milder; the first with a sharper leaf like a Fox's ear, and the other rounder and with finer fibres: But each of these are varied into several sorts, much as Apples and Pears are; and I have been informed by the Indian traders, that the Inland Indians have sorts of tobacco much differing from any planted or used by the Europeans. The Indian Corn is planted in hills and weeded much as tobacco. This grain is of great increase and most general use; for with this is made good bread, cakes, mush, and hommony for the negroes, which with good pork and potatoes (red and white, very nice and different from ours) with other roots and pulse, are their general food."

The cultivation of tobacco increased with the growth of the colony and the increase of price which at this time was sufficient to induce most of the planters to neglect the culture of Corn and Wheat, devoting their time to growing their "darling tobacco." The first thirty years after the colonization of Virginia by the English, the colony made but little progress owing in part to private factions and Indian wars. The horrid massacres by the Indians threatened the extermination of the colony, and for a time the plantations were neglected and even tobacco became more of an article of import than of export, which is substantiated by an early writer of the colony who says:—"A vast quantity of tobacco is consumed in the country in smoking, chewing, and snuff." Frequent complaints were made by the colony of want of strength and danger of imminent famine, owing in

RESTRICTIONS ON TOBACCO-RAISING.

part to the presence of a greater number of adventurers than of actual settlers,—such being the case the resources of the country were in a measure limited.

The demand for tobacco in England increasing each year, together with the high price paid for that from Virginia (3 s. per lb.), stimulated the planters to hazard all their time and labor upon one crop, neglecting the cultivation of the smaller grains, intent only upon curing "a good store of tobacco." The company of adventurers at length found it necessary to check the excessive planting of the weed, and by the consent of the "Generall Assemblie" restraining the plantations to "one hundred plants* ye headd, uppon each of wich plantes there are to bee left butt onely nyne leaves wch portions as neare as could be guessed, was generally conceaved would be agreable with the hundred waight you have allowed."

In 1639 the "Grand Assembly" (summoned the sixth of January) passed a law restricting the growth of the colony to 1,500,000 lbs., and to 1,200,000 in the two years next ensuing. The exporting of the poorer qualities of tobacco by the colony caused much dissatisfaction as will be seen by a letter of the Company dated 11th September, 1621:

"We are assured from our Factor in Holland that except the tobacco that shall next come thence prove to be of more perfection and goodnesse than that was sent home last, there is no hope that it vend att all, for albeit itt passed once yett the wary buyer will not be againe taken, so that we heartily wish that youe would make some provision for the burninge of all base and rotten stuff, and not suffer any but very good to be cured at least sent home, whereby these would certainly be more advanced in the price upon lesse in the quantity; howsoever we hope that no bad nor ill conditioned tobacco shall be by compelling authoritie (abusing its power given for public good to private benefit) putt uppon or Factor, and very earnestly desire that he may have the helpe of justice to constraine men to pay their debts unto him both remaining of the last yeares accompt and what shall this yearse growth deue, and that in Comodities of the same vallew and goodness as shalbe by him contracted for."

---
*Another account is sixty pounds per head.

At this period it appears that tobacco was used as money, and as the measure of price and value. The taxes whether public, county, or parish, were payable in tobacco.

Tatham says, "Even the tavern keepers were compelled to exchange a dinner for a few pounds of tobacco." The law for the regulation of payments in tobacco was passed in the year 1640. From these facts and incidents connected with the culture and commerce of the plant we see how intimately it was connected with both Church and State. Jones well said "the Establishment is indeed tobacco;" the salary of ministers was payable in it according to the wealth of the parish. In most parishes 16000 lbs. was the yearly amount, "and in some 20,000 lbs. of Tobacco; out of which there is a deduction for Cask, prizing, collecting, and about which allowance there are sometimes disputes, as are also differences often about the place, time, and manner of delivering it; but all these things might easily be regulated. Tobacco is more commonly at 20 s. per cent. than at 10; so that certainly it will bring 12 s. 8 d. a hundred, which will make 16000 (the least salary) amount to 100£ per Ann. which it must certainly clear, allowing for all petty charges, out of the lowness of the price stated which is less than the medium between ten and twenty shillings; whereas it might be stated above the medium, since it is oftener at twenty than ten shillings. Besides the payment of the salary, the surplice fees want a better regulation in the payments; for though the allowance be sufficient, yet differences often and illwill arise about these fees, whether they are to be paid in money or tobacco, and when; whereas by a small alteration and addition of a few laws in these and the like respects, the clergy might live more happy, peaceable, and better beloved; and the people would be more easy, and pay never the more dues.

"Some parts of the country make but mean and poor tobacco so that Clergymen don't care to live in such parishes; but there the payment might be made in money, or in the produce of those places, which might be equivalent to the tobacco payments; better for the minister, and as pleasing to the people."

We find further complaints from the London Company of the poor quality of the tobacco "sent home," in a letter addressed to the Governor, bearing date 10th June, 1622:—

"The tobacco sent home by the George for the company

proved very meane and is yett unsold although it hath been offered at 3s. the pound. This we thought fitt to advise you concerning the quantity and the manner how it is raised, in both wich being done contrarie to their directors and extreamly to theire prejudice, the Companie is very ill sattisfied, will write by the next, more largely."

In the year 1620 the difficulties seem first to have been publicly avowed, (though perhaps before felt,) arising from attaching men as permanent settlers to the colony without an adequate supply of women, to furnish the comforts of domestic life; and to overcome the difficulty "a hundred young women" of agreeable persons and respectable characters, were selected in England and sent out, at the expense of the Company, as wives for the settlers. They were very speedily appropriated by the young men of the colony, who paid for the privilege of choice considerable sums as purchase money, which went to replenish the treasury of the Company, from whence the cost of their outfit and passage had been defrayed.

This speculation proved so advantageous to that body, in a pecuniary sense, that it was soon followed up by sending out sixty more, for whom larger prices were paid than for the first consignment; the amount paid on the average for the first one hundred being 120 pounds of tobacco apiece for each, then valued at 3s. per lb., and for the second supply of sixty, the average price paid was 150 lbs. of tobacco, this being the legal currency of the colony, and the standard value by which all contracts, salaries, and prices were paid. In one of the Companies letters dated in London this 12th of August, 1621, we find this account of a portion of the *goods* sent over in the ship Marmaduke:—

"We send you in this ship one widdow and eleven maids for wives for the people in Virginia; there hath been especiall care had in the choise of them for their hath not any one of them beene received but upon good comendations, as by a note herewith sent you may perceive: we pray you all therefore in generall to take them into your care, and most especially we recommend them to you, Mr. Pountes, that at their first landing they may be housed, lodged and provided for of diet till they be marryed for such was the haste of sending them away, as that straightned with time, we had no meanes to putt provisions aboard, which defect shalbe supplied by the magazine shipp; and in case they cannot be presently marryed we desire they may be putt to several householders that have wives till they can be provided of husbands. There are neare fifty more which are shortly to come, we sent by our most honoble Lord William the Earle of Southampton and certain worthy gentlemen who taking into these considerations, that the Plantation can never flourish till families be planted and the respect of wives and children fix the people in the soil; therefore have given this fair beginning for the reimbursing of whose charges, itt is ordered that every man that marries them give 120 lb. waight of best leafe tobacco for each of them, and in case any of them dye that proportion must be advanced to make it upp to those that survive; and this certainly is sett down for that the price sett upon the bages sent last yeare being 20 lb. which was so much money out of purse here, there was returned 66 lb. of tobacco only, and that of the worst and basest, so that fraight and shrinkage reconed together with the baseness of the comoditie there was not one half returned, which injury the company is sensible of as they demand restitution, which accordingly must be had of them that took uppon them the dispose of them the rather that no man may mistake himself, in accomptinge tobacco to be currant 3s. sterling contrary to express orders.

"And though we are desirous that marriadge be free according to the law of nature, yett undervow not to have these maids deterred and married to servants, but only to such freemen or tenants as have means to maintaine them; we pray you therefore to be fathers to them in this business, not enforcing them to marrie against their wills; neither send we them to be servants, but in case of extremitie, for we would have their condition so much better as multitudes may be allured thereby to come unto you; and you may assure such men as marry those women that the first servants sent over by the company shall be consigned to them, it being our intent to preserve families and proper married men before single persons. The tobacco that shall be due uppon the marriadge of these maids we desire Mr. Pountes to receive and returne by the first, as also the little quantities of Pitzarn Rock and Piece of Oare, the copie of whose bill is here returned. To conclude, the company for some weighty reasons too long to relate, have ordered that no man marrying these women expect the proportion of land usually allotted for each head, which to avoid clamor or other trouble hereafter you shall do well to give them notice of."

In another letter written by the company and dated London, September 11th, 1621, they write:—

"By this Shipp and Pinace called the Tyger, we also send as many maids and young women as will make up the number of fifty, with those twelve formerly sent in the Marmaduke, which we hope shalbe received with the same Christian pietie and charitie as they were sent from hence; the providing for them at their first landing and disposing of them in marriage (which is our chief intent), we leave to your care

BUYING WIVES.

and wisdom, to take that order as may most conduce to their good, and satisfaction of the Adventurers, for the charges disbursed in setting them forth, which coming to twelve pounds and upwards, they require one hundred and fiftie of the best leafe tobacco for each of them; and if any of them dye there must be a proportionable addition uppon the rest; this increase of thirty pounds is weight since those sent in the Marmaduke, they have resolved to make, finding the great shrinkage and other losses uppon the tobacco from Virginia will not leave lesse, which tobacco as it shalbe received, we desire may be delivered to Mr. Ed. Blany, who is to keep thereof a particular account. We have used extraordinary care and dilligence in the choice of them, and have received none of whom we have not had good testimony of their honest life and cariadge, which together with their names, we send them inclosed for the satisfaction of such as shall marry them; for whose further encouragement we desire you to give public notice that the next spring we purpose to send over as many youths for apprentices to those that shall now marry any of them and make us due satisfaction.

"This and theire owne good deserts together with your favor and care, will we hope, marry them all unto honest and sufficient men, whose means will reach to present repayment; but if any of them shall unwarily or fondly bestow herself (for the liberty of marriadge we dare not infrindge) uppon such as shall not be able to give present sattisfaction, we desire that at least as soon as ability shalbe, they be compelled to pay the true quantity of tobacco proportioned, and that this debt may have precedence of all other to be recovered.

" For the rest, which we hope will not be many, we desire your best furtherance for providing them fitting services till they may happen uppon good matches, and are here persuaded by many old planters that there will be good maisters now found there, who will readily lay down what charges shall be required, uppon assurance of repayment at their marriadges, which as just and reasonable we desire may be given them. But this and many other things in this business we must refer to your good considerations and fruitful endeavors in opening a work begun here out of pity, and tending so much to the benefitt of the plantation, shall not miscarry for any want of good will or care on your part."

In 1622 a monopoly of the importation of tobacco was granted to the Virginia and Somers Island companies.

" But now at last it hath pleased God for the confirmation no doubt of our hopes and redoubling of our and your courage, to incline His Majestie's Royall heart to grant the sole importation of Tobacco (a thing long and earnestly desired), to the Virginia and Somers Island Companies, and that upon such conditions as the private profit of each man is likely to be much improved and the general state of the plantation strongly secured, while his Majestie's revenue is so closely joyned as together with the colonie it must rise and faile, grow and impair, and that not a small matter neither, but of twenty thousand pounds per annum. (for the offer of so much in certainty hath his majestie been pleased to refuse in favor of the Plantations."

On Friday the 22d of March 1622 the Indians attacked the plantations "and attempted in most places under the color of unsuspected amytie, and by surprise to have cut us all off and to have swept us all away at once throughout the whole lande had itt not pleased God of his abundant mercy to prevent them in many places, for which we can never sufficient magnifie his blessed name."

But notwithstanding this terrible massacre in which nearly four hundred persons were slain the colony increased in wealth and numbers as plantations were laid out and the colonists developed the various resources of the country. From the first planting of tobacco in Virginia by the colony it seemed to meet the royal displeasure of King James the First who falsely and frivolously sought to establish a connection between the balmy plant, and the influences of the Evil One.

In 1622 King James still opposing the cultivation of tobacco sought by every means in his power to discourage its growth and culture. He urged the growing of mulberry trees and the propagation of silk worms, as being of more value than tobacco. In a letter dated 10th June 1622, addressed to the Governor and Council of Virginia by the London Company we find this reproof for neglecting the cultivation of "mulberrie trees":

"His Matie (Majesty) above all things requires from us a proof of silke; sharply reproving the neglect thereof, wherefore we pray you lett that little stock you have be carefully improved, the mulberrie trees preserved and increased, and all other fitt preparations made for, God willing before Christmas you shall receive from us one hundred ounces of Silkworme seed at least, which coming too late from Valentia we have been forced to hatch it here."

In 1623 a letter was prepared for the colony by order of privy council of the king and addressed to Sir Francis Wyatt Knight and Captain General of Virginia and to the rest of the Council of State in which the colony is admonished to pay more attention to "Staple Commodities." That part relating to it reads:

"The carefull and diligent prosecution of Staple Commodities which we promist; we above all things pray you to performe so as we may have speedily the real proof of your cares and endeavors therein, especially in that of Iron, of Vines and Silk the neglect and delay whereof so long is to us here cause of infinit grief and discontent, especially in regard of his Majesties just resentment therein that his Royall grace and love to the Plantation, which after so long a time and long a supply of his Majesties favor hath brought forth no better fruit than Tobacco.

"Yett by the goodness of God inclyning his princely heart, we have received not only from the Lords of his Privy Counsell, but from his Royal mouth such assurance not only of his tender love and care but also of his Royal intentions for the advancement of the Plantation; that we cannot but exceedingly rejoice therein and persuade you with much more comfort and encouragement to go on in the building up of his Royal worke with all sincerity, care and diligence, and that with that perfect love and union amongst yourselves as may really demonstrate that your intentions are all one, the advancement of God's glorie and the service of his Royall Majestie: for the particularities of his Majesties gratious intentions for the future good, you may in part understand them by the courses appointed by the Lords, whereof we here inclosed send the orders.

"And we are further to signifie unto you that the Lords of his Majesties Privy Counsell, having by his Majesties order taken into their considerations the contract made last Sommer by the Company have dissolved the same; and signified that his Majestie out of his gracious and Royall intention and princely favor to the Plantation hath resolved to grant a sole Importion of Tobacco to the two Plantations, with an exception only of 40,000 weight of ye best Spanish Tobacco to be yearly brought in.

"And it hath also pleased his Majesty in favor of the Plantation to reduce ye custom and importing of tobacco to 9d. per pound: And last of all we are to signifie unto you that their Lordships have ordered that all the Tobacco shall be brought in from both Plantations as by their Lordship order whereof we send you a copy, you may perceive."

In 1624 King James prohibited the importation of foreign tobacco as well as the planting of tobacco in England or Ireland. The following is a portion of the proclamation:—

"Whereas our commons, in their last sessions of parliament became humble petitioners to us, that, for many weighty reasons, much concerning the interest of our kingdom, and the trade thereof, we would by our royal power utterly prohibit the use of all foreign tobacco, which is not of the growth of our own dominions: And whereas we have upon all occasions made known our dislike we have ever had of the use of tobacco in general, as tending to the corruption both of the health and manners of our people.

"Nevertheless because we have been often and earnestly importuned by many of our loving subjects, planters, and adventurers in Virginia and the Somer isles; that, as those colonies are yet but in their infancy, and cannot be brought to maturity, unless we be pleased, for a time, to tolerate unto them the planting and vending of their own growth; we have condescended to their desires: and do therefore hereby strictly prohibit the importation of any tobacco from beyond sea, or from Scotland, into England or Ireland other than from our colonies before named; moreover we strictly prohibit the planting of any tobacco either in England or Ireland."

Thus King James by Proclamation and Prohibition set his face sternly against the growth and traffic in the plant, which opposition knew no alteration and continued till his death, which occurred in 1625. James was succeeded by his son Charles I. On ascending the throne Charles manifested the same hostility towards the plant which his father had. He prohibited the importation of all tobacco excepting that grown by the colony, and throughout his reign made no change in the restrictive laws against its growth and sale. He continued its sale, however, as a kingly monopoly, allowing only those to engage in it who paid him for the privilege. The Company had now raised a capital of two hundred thousand pounds, but falling into dispute and disagreeing one with another, Charles thought best to establish a royal government.

Accordingly he dissolved the Company in 1626, "reducing the Country and Government into his own immediate ordering all patents and processes to issue in his own name, reserving to himself a quit-rent of two shillings for every hundred acres of land."

The first act was by proclamation as follows:—

"That whereas, in his royal father's time, the charter of the Virginia Company was by a quo warranto annulled; and whereas his said father was, and he himself also is, of opinion, that the government of that Colony by a company incorporated, consisting of a multitude of persons of various dispositions, amongst whom affairs of the greatest moment are ruled by a majority of votes, was not so proper, for carrying on, prosperously, the affairs of the colony; wherefore, to reduce the government thereof to such a course as might best agree with that form which was held in his royal monarchy; and considering also, that we hold those territories of Virginia and Somer isles, as also that of New England, lately planted, with the limits thereof, to be a part of our royal empire; we ordain that the government of Virginia shall immediately depend on ourself, and not be committed to any company or corporation, to whom it may be proper to trust matters of trade and commerce, but cannot be fit to commit the ordering of state affairs.

"Wherefore our commissioners for those affairs shall proceed as directed, till we establish a council here for that colony; to be subordinate to out council here for that colony. And at our charge we will maintain those public officers and ministers and that strength of men, munition, and fortification, which shall be necessary for the defence of that plantation. And we will also settle and assure the particular rights and interests of every planter and adventurer. Lastly, whereas the tobacco of those plantations (the only present means of their subsisting) cannot be managed for the good of the plantations, unless it be brought into one hand, whereby the foreign tobacco of those plantations may yield a certain and ready price to the owners thereof: to avoid all differences between the planters and adventurers themselves, we resolve to take the same into our own hands, and to give such prices for the same as may give reasonable satisfaction, whereof we will determine at better leisure."

From this time forward the Plantation seemed to prosper, Charles granted lands to all the planters and adventurers who would till them, upon paying the annual sum of two shillings payable to the crown for each hundred acres.

direction, appointing the Governor and Council himself, and
Before the death of King James, however, the cultivation
of tobacco had become so extensive that every other product
seemed of but little value in comparison with it, and the
price realized from its sale being so much greater than that
obtained for "Corne," the latter was neglected and its culture
almost entirely abandoned.

Arthur and Carpenter, in their History of Virginia, give
a graphic and truthful picture of its cultivation during the
reign of King James:—

"The first articles of commerce to the production of which
the early settlers almost exclusively devoted themselves, were
potash, soap, glass and tar. Distance, however, and a want
of the proper facilities to enable them to manufacture cheaply,
rendered the cost of these commodities so great, that exports
of a similar character from Russia and Sweden were still
enabled to maintain their old ascendency in the markets of
Europe. After many fruitless and costly experiments in the
cultivation of the vine, the growing demand for tobacco
enabled the planters to turn their labor into a profitable
channel. As the demand increased the profits became correspondingly great, and every other species of labor was abandoned for the cultivation of tobacco.

"The houses were neglected, the palisades suffered to rot
down, the fields, gardens and public squares, even the very
streets of Jamestown were planted with tobacco. The townspeople, more greedy of gain than mindful of their own
security, scattered abroad into the wilderness, where they
broke up small pieces of rich ground and made their crop
regardless of their proximity to the Indians, in whose good
faith so little reliance could be placed."

During the reign of Charles I. many families of respectable connection joined the colony, and from this time
forward the colony increased in wealth as well as numbers.
King Charles, to use the language of another, had now commenced "as a tobacco merchant and monopolist," and in 1627
issued a proclamation renewing his already strong monopoly
more effectually, by appointing certain officers of London
"to seize all foreign tobacco, not of the growth of Virginia
or 'Bermudas, for his benefit, agreeable to a former commission: also to buy up for his use all the tobacco coming from

our said plantations, and to sell the same again for his benefit."

Again in 1630 King Charles issued another proclamation,

GROWING TOBACCO IN THE STREETS.

and among other restrictions limited the importation of it from the colony. Quickly following this the King issued in 1632 another proclamation regulating the retailing of tobacco. In 1634 he also prohibited the landing of tobacco any where except at the quay near the custom house in London.

In 1636 Charles appoined Sir John Harvey to be continued governor of the Plantation. In 1643 parliament laid a tax for the year 1644, calling it Excise, and also laid a duty of four shillings per pound on foreign, and two shillings per pound on English tobacco. From what has already been written, it will be seen that both King James and his son

Charles I. enacted the most stringent laws against its importation, nearly suppressing the trade, which caused the English farmers to cultivate it for home use; but another law was now added to suppress its growth on English soil.

Fairholt in speaking of the hostility of King James to the plant says:

"When Kings make unnecessary and unjust laws, subjects naturally study how to evade them: it is a mere system of self-defence; and as James nearly suppressed the importation of tobacco the English began to grow it on their own land. But the Scottish Solomon who was on the alert, added another law restraining its cultivation 'to misuse and misemploy the soil of this fruitful Kingdom.' As this enforced the trade with the English colony of Virginia alone, it was soon found that Spanish and Portuguese tobacco might be brought into port on the payment of the old duty of twopence a pound; thus a large trade was carried on with their planters to the injury of the British colonists.

"Its use increased in spite of all legislative laws and enactments and James ended by prohibiting any person from dealing in the article who did not hold his letters patent. By this means the trade was monopolized, the consumers oppressed, importation diminished, and the London Company of Virginia traders ultimately ruined. Those who are fond of excusing the evil acts of one of the worst of English Kings, pretend to see James' care for his subjects' health and wealth in these restrictions, totally regardless of the fact that James cared for neither when the monopoly brought large sums into his own pocket."

In 1632 Charles I. granted to Sir George Calvert (who about this time was made Lord Baltimore) the territory now known as Maryland; soon after receiving the grant he died, when his son took the grant in his own name. The next year he sailed from England with two hundred persons and settled in his new possessions. The colony from the first, prospered far better than the colony of Virginia and soon laid the foundation of a strong and substantial government. Like the Virginians they soon engaged in the cultivation of tobacco which seemed as well adapted to the soil as the other products, corn and English wheat. The Indians were found here as in the Plantation of Virginia planting tobacco as they did Indian corn and cultivating little patches of it near their wigwams choosing the most fertile soil the females of the tribe being the actual cultivators.

From this time forward both colonies developed into

NATIVES GROWING TOBACCO.

strong and flourishing plantations and with each succeeding year increased the cultivation of tobacco which had now become more extensively cultivated than all the other products combined. Its culture however was looked upon with the same disapproval by Charles II. who confirmed the old laws against its sale and cultivation. But notwithstanding the remonstrances of the Stuarts the plant grew in use and favor and could not be uprooted even by a kingly hand. The early cultivators of the plant received a fresh impetus from the importation of a new species of labor in the form of Negro slaves brought from the West India islands. They arrived in the Ship Treasurer "being manned by the best men of the colony who set out on roving in ye Spanish dominions in the West Indies" and after a successful cruise against the Spaniards returned with their spoils including a certain number of Negroes. Rolfe in alluding to the importation of Negroes says:

"About the last of August came in, a Dutch man-of-warre that sold us twenty negars."

Most writers are of the opinion that this was in 1620, one of whom says "in the same year that the Pilgrims landed at Plymouth, slaves landed in Virginia." Another writer says of the introduction of slave labor into the Plantations, "Is there not a probability that the vessel was under control of Argall, if not the ship Treasurer? If twenty negroes came in 1619, as alleged, their increase was very slow, for according to a census of 16th of February, 1624, there were but twenty-two then in the colony, distributed as follows: eleven at Flourdiew Hundred, three at James City, one at James Island, one at the plantation opposite James City, four at Warisquoyok, and two at Elizabeth City."

About the same time that "negars" landed in the colony, commenced the arrival of starving boys and girls picked up out of the streets of London. The "negars" are described as follows by an early writer of the colony. "The negroes live in small cottages called quarters, in about six in a gang, under the direction or an overseer or baliff; who takes care that they tend such land as the owner allots and orders, upon which they raise Hogs and Cattle, plant Indian Corn (or maize) and Tobacco for the use of their Master; out of which the overseer has a dividend (or share) in proportion to the number of hands including himself; this with several privileges in his salary, and is an ample recompense for his pains, and encouragement of his industrious care, as to the labor, health, and provision of the negroes. The negroes are very numerous, some gentlemen having hundreds of them of all sorts, to whom they bring great profit; for the sake of which they are obliged to keep them well, and not overwork, starve, or famish them, besides other inducements to favor them, which is done in a great degree, to such especially that are laborious, careful, and honest; though indeed some Masters, careless of their own interest and reputation, are too cruel and negligent.

"The negroes are not only increased by fresh supplies from Africa and the West India Islands, but also are very prolific among themselves; and they that are born there talk good English, and effect our language, habits, and customs; and tho' they be naturally of a barbarous and cruel temper, yet are they kept under by severe discipline upon occasion, and by good laws are prevented from running away, injuring the English or neglecting their business. Their work (or chimerical hard slavery) is not very laborious; their greatest hardship consisting in that they and their posterity are not at their own liberty or disposal, but are the property of their owners; and when they are free they know not how to provide so well for themselves generally; neither did they live so plentifully nor (many of them) so easily in their own country where they are made slaves to one another, or taken captive by their enemies. Their work is to take care of the stock, and plant Corn, Tobacco, Fruits and which is not harder than thrashing, hedging, or ditching; besides, though they are out in the violent heat, wherein they delight, yet in wet or cold weather there is little occasion for their working in the fields, in which few will let them be abroad, lest by this means they might get sick or die, which would prove a great loss to their owners, a good Negroe being sometimes worth three (nay four) score pounds sterling, if he be a tradesmen; so that upon this (if upon no other account) they are obliged not to overwork them, but to clooth and feed them sufficiently, and take care of their health."

The planters, supplied with greater facilities for the work, now increased the size of their tobacco plantations, "taking up new ground" (clearing the land) and planting a much larger area. The first exportation of the colony's tobacco was brought into competition with that of much finer flavor, which had acquired an established reputation long before the English began the culture of the plant in the New World. The Spanish, Dutch and Portuguese had long monopolized its culture and trade, and brought from St. Domingo, Jamaica, St. Thomas, the Phillippine Islands, West Florida, and various parts of South America, several varieties of tobacco of excellent quality, and which sold at an exorbitant price. On testing the tobacco grown by the London and Plymouth companies it was found to be sweet and mild in flavor, of a light color, and well adapted for smoking. On its first introduction into England it sold for 3s. per pound, but as its culture increased the price lessened, until it was sold at one-half this amount.

The planters, who at first cultivated small patches, now planted large fields of tobacco, and such was the greed for gain that some planters gathered a second crop upon the same field from the suckers left growing upon the parent stalk. Tatham* says in regard to it:—

"It has been customary in former ages to rear an inferior plant from the sucker which projects from the root after the cutting of an early plant; and thus a second crop has often been obtained from the same field by one and the same course of culture; and although this scion is of a sufficient quality for smoking, and might become preferred in the weaker kinds of snuff, it has been (I think very properly) thought eligible to prefer a prohibitory law, to a risk of imposition by means of similitude. The practice of cultivating suckers is on these accounts not only discountenanced as fraudulent, but the constables are strictly enjoyned *ex officio* to make diligent search, and to employ the *posse commitatus* in destroying such crops; a law indeed for which, to the credit

DESTROYING SUCKERS.

of the Virginians, there is seldom occasion; yet some few instances have occurred, within my day, where the constables have very honorably carried it into execution in a

---

*Essay on Tobacco, London, 1800.

manner truly exemplary, and productive of public good."

Fairholt says of the same subject:—

"It was sometimes the custom with planters to reset the suckers, and thus grow a double crop on one field, such conduct was disallowed; for the reason that the crop was inferior, and the more honest grower, who conscientiously cleared his plants, and gave them abundance of room to grow, was dishonestly competed with; and the first rate character of the Virginian crop prejudiced by the action."

Fairholt makes a mistake in speaking of the planter as re-setting the suckers, and his statement shows him to be entirely unacquainted with the habits of the plant. As soon as the plants are harvested, the stump of the plant remaining in the ground puts forth one or more vigorous suckers or shoots, which often in a good season grow almost as high as the parent stalk. In some tobacco-growing sections one or two crops of suckers are gathered besides the first crop.

The Creole planters in Louisiana are said to grow three crops in this manner, the first or parent crop and two growths of suckers. The quality of leaf, however, is greatly inferior, as it is small and thin and lacking in all the qualities necessary for a fine leaf. The planters now adopted new methods of culture, and cultivated several species of the plant known as Oronoko and little Frederick, although they did not fertilize the fields, even when the soil became impoverished, but simply took new fields for its culture.

Hugh Jones says of the kinds of tobacco grown in Virginia:—

"The land between the James and York rivers seemes nicely adapted for sweet scented tobacco; for 'tis observed that the goodness decreaseth the farther you go to the northward of the one, and the southward of the other; but this may be (I believe) attributed in some measure to the seed and management, as well as to the land and latitude: For on York river in a small tract of land called Diggens neck, which is poorer than a great deal of other land in the same latitude, by a particular seed and management, is made the famous crop known by the name of E Dees, remarkable for its mild taste and fine smell." He speaks of the planters and their plantations as follows:—"Neither the interests nor inclinations of the Virginians induces them to cohabit in towns: so that they are not forward in contributing their assistance towards the making of particular places, every plantation affording the owner the provision of a little market; wherefore they most commonly build upon some convenient spot or neck of land in their own plantation, though towns are laid out and established in each county.

"The whole country is a perfect forest, except where the woods are cleared for plantations, and old fields, and where have been formerly Indian towns, and poisoned fields and meadows, where the timber has been burnt down in fire hunting and otherwise; and about the creeks and rivers are large rank morasses or marshes, and up the country are poor savannahs. The gentlemen's seats are of late built for the most part of good brick, and many of timber very handsome, commodious, and capacious; and likewise the common planters live in pretty timber houses, neater than the farm houses are generally in England: With timber also are built houses for the overseers and out-houses; among which is the kitchen apart from the dwelling house, because of the smell of hot victuals, offensive in hot weather.

"The habits, life, customs, computations of the Virginians, are much the same as about London, which they esteem their home; and for the most part have contemptible notions of England, and wrong sentiments of Bristol, and the other out-posts, which they entertain from seeing and hearing the common dealers, sailors, and servants that come from those towns, and the country places in England and Scotland, whose language and manners are strange to them; for the planters and even the native negroes generally talk good English without idiom and tone, and can discourse handsomely upon most common subjects: and conversing with persons belonging to trade and navigation from London, for the most part they are much civilized, and wear the best of clothes according to their station; nay, sometimes too good for their circumstances, being for the generality, comely handsome persons of good features and fine complexions (if they take care) of good manners and address.

"They are not very easily persuaded to the improvement of useful inventions (except a few, such as sawing mills) neither are they great encouragers of manufactures, because of the trouble and certain expense in attempts of this kind, with uncertain prospect of gain; whereas by their staple commodity, tobacco, they are certain to get a plentiful provision; nay, often very great estates. Upon this account they think it folly to take off their hands (or negroes) and employ their care and time about anything that may make them lessen their crop of tobacco. So that though they are apt to learn, yet they are fond of and will follow their own ways, humors and notions, being not easily brought to new projects and schemes; so that I question if they would have been improved upon by the Mississippi or South sea, or any other such monstrous bubbles. The common planters leading easy lives without much labor, or any manly exercise, except horse-racing, nor diversion, except cock-fighting, in which some greatly delight.

"This easy way of living, and the heat of the summer. makes some very lazy, who are then said to be climate-struck They are such lovers of riding, that almost every ordinary person keeps a horse; and I have known some spend the morning in ranging several miles in the woods to find and catch their horses to ride only two or three miles to the Church, to the Court-House or to a Horse-Race, where they generally appoint to meet upon business; and are more certain of finding those that they want to speak or deal with, than at their home. No people can entertain their friends with better cheer and welcome; and stranger and traveler is here treated in the most free, plentiful, and hospitable manner; so that a few Inns or Ordinaries on the road are sufficient."

This is no doubt a correct picture of the early planters of Virginia. Many of them became the owners of large plantations and all those who were successful growers of tobacco became wealthy in proportion to the quality of leaf produced.

The merchants, factors or store-keepers bought up the tobacco of the planters paying in goods or "current Spanish money, or with sterling bills payable in Great Britain." At first the cultivation of tobacco by the colony was confined to Jamestown and the immediate vicinity, but as the colony increased and the country became more densely populated, plantations were laid out in the various counties and a large quantity was produced some ways from the great center Jamestown; accordingly various methods were adopted to get the tobacco to market, some of which was sent by boats or canoes down the rivers, while some was conveyed in carts and wagons while another method was by rolling in hoops.

Tatham in his interesting work on tobacco, gives the following description of the method:

"I believe rolling tobacco the distance of many hundred miles, is a mode of conveyance peculiar to Virginia; and for which the early population of that country deserve a very handsome credit. Necessity (that very prolific mother of invention), first suggested the idea of rolling by hand; time and experience have led to the introduction of horses, and have ripened human skill, in this kind of carriage, to a degree of perfection which merits the adoption of the mother country, but which will be better explained under the next head of this subject.

"The hogsheads, which are designed to be rolled in common hoops, are made closer in the joints than if they were intended for the wagon; and are plentifully hooped with strong hickory hoops (which is the toughest kind of wood), with the bark upon them, which remains for some distance a

CARRYING TOBACCO TO MARKET.

protection against the stones. Two hickory saplings are affixed to the hogshead, for shafts by boring an auger-hole through them to receive the gudgeons or pivots, in the manner of a field rolling-stone; and these receive pins of wood, with square tapered points, which are admitted through square mortises made central in the heading, and driven a considerable depth into the solid tobacco. Upon the hind part of these shafts, between the horses and the hogshead, a few light planks are nailed, and a kind of little cart body is constructed of a sufficient size to contain a bag or two of provender and provision, together with an axe, and such other tools as may be needed upon the road, in case of accident. In this manner they set out to the inspection in companies, very often joining society with the wagons, and always pursuing the same method of encamping."

The methods of making the plant bed, cultivating and harvesting, by the early planters may be interesting to all growers of the plant and are here described as showing the progress made in cutting tobacco from that time until now.

"In spring red seed, in preference to the white, is put into a clean pot; milk or stale beer is poured upon it, and it is left for two or three days in this state; it is then mixed with a quantity of fine fat earth, and set aside in a hot chamber, till the seeds begin to put out shoots. They are then sown in a hot-bed. When the young plants have grown to a finger's length, they are taken up between the fifteenth and twenty-second of May, and planted in ground that has been previously well manured with the dung of doves or swine. They are placed at square distances of one and a half-foot from one another. In dry weather, they are now to be watered with lukewarm water softly showered upon them, between sunset and twilight. When these plants are full two feet high, the top of the stems are broken off, to make the leaves grow thicker and broader. Here and there are left a few plants without having their tops broken off, in order that they may afford seeds for another year. Throughout the summer the other plants are from time to time, pruned at the top, and the whole field is carefully weeded to make the growth of the leaf so much the more vigorous.

"In the month of September, from the sixteenth day, and between the hours of ten in the morning and four in the afternoon, the best leaves are to be taken off. It is more advantageous to pluck the leaves when they are dry than when they are moist. When plucked they are to be immediately brought home, and hung upon cords within the house to dry, in as full exposure as is possible to the influence of the sun and air; but so as to receive no rain. In this exposure they remain till the months of March and April following; when they are to be put up in bundles, and conveyed to the store-house, in which they may be kept, that they may be there till more perfectly dried by a moderate heat. Within eight days they must be removed to a different place, where they are to be sparingly sprinkled with salt water, and left till the leaves shall be no longer warm to the feeling of the hand. A barrel of water with six handfuls of salt are the proportions. After all this the tobacco leaves may be laid aside for commercial exportation. They will remain fresh for three years."

In Maryland they formerly prepared the land for a plant-bed by burning upon it a great quantity of brush-wood,

ENRICHING PLANT-BED.

afterwards raking the surface fine; the seed was then sowed broadcast. The young plants were kept free from weeds, and were transplanted when about two inches high.

The cultivation of tobacco gradually spread from one State to another. From Virginia it was introduced into North Carolina and Maryland and finally Kentucky which is now the largest producing tobacco State in the Union. The demand for Virginia tobacco continued to increase and long before the Revolutionary war, Virginia exported annually thousands of hogsheads of leaf tobacco. Half a century ago the plant began to be cultivated in Ohio and from the first grew remarkably well, producing a leaf adapted for both cutting and cigar purposes.

Tobacco was planted in New Netherland (New York) by

the early Dutch settlers and in 1638 "had become a staple production." In 1639 "from Virginia numbers of persons whose terms of service had expired, were attracted to Manhattan, where they introduced improved modes of cultivating tobacco." Van Twiller was himself a grower of the plant and had his tobacco farm at Greenwich. Soon after its cultivation began it was subjected to Excise; and regulations were published to check the abuses which injured "the high name" it had gained in foreign countries. *

Wailes says of the early cultivation of tobacco in Mississippi:

"When the country came under the dominion of Spain, a market was opened in New Orleans; a trade in tobacco was established, and a fixed and remunerating price was paid for it, delivered at the king's warehouses. Tobacco thus became the first marketable staple production of Missisippi." †

An English writer has the following account of the culture of tobacco in Louisiana by the French:

"Tobacco is another plant indigenous to this part of America; the French colonists cultivated it with such success that had they received any encouragement from their government they might soon have rivalled Virginia and Maryland; but instead of this they were taxed heavily for cultivating it, by duties laid on the trade; what they produced was of so excellent a quality, as to sell some at five shillings a pound. There is one advantage in this culture here which ought not to be forgotten; in Louisiana the French planters after the tobacco is cut, weeded and cleaned the ground on which it grew the roots, push forth fresh shoots, which are managed in the same manner as the first crop. By this means a second crop is made on the same ground, and sometimes a third. These seconds indeed, as they are called, do not usually grow so high as the first plant, but notwithstanding they make very good tobacco."

During the reign of the Stuarts, the plant was first cultivated in New England but only in small quantities ‡ and

---

* Jacob van Churler and David Provoost were appointed inspectors of the new staple tobacco. "In 1652 the commonalty at Manhattan was informed that, to show their good intentions, the Amsterdam directors had determined to take off the export duty of tobacco."

† In 1783 Mr. Wm. Dunbar writes: "The soil of Natchez is particularly favorable for tobacco and there are overseers there, who will almost engage to produce you between two and three hogsheads to the hand besides provisions."

‡ "Every farmer plants a quantity of tobacco near his house in proportion to the size of his family. It is likewise very necessary that they should plant tobacco, because it is so universally smoked by the common people."—*Kalm's travels in North America,* 1772.

used solely for smoking. About 1835 the plant received more attention from the farmers living in the Connecticut valley containing some of the finest tobacco land in the country. They found by repeated trials that the soil was well adapted to the production of a finer leaf tobacco than any they had ever seen. At this time Kentucky and Havana tobacco were used in the manufacture of cigars, but on testing American tobacco or as it is now known " Connecticut seed leaf" it was found to make the finest wrappers yet produced, and consequently the best looking cigars. From that time its reputation has kept pace with its cultivation, until it now enjoys a world wide popularity. As a wrapping tobacco it towers far above the seed products of other states and can never have a successful competitor in the other varieties now cultivated in the Middle and Western States. Doubtless America furnishes the finest varieties of the plant now cultivated, suited for all kinds of manufacturing, and adapted to all the various forms in which it is used.

The great diversity of soil and climate renders this probable while actual experiments and improved methods of culture have demonstrated it to a certainty. Thousands of hogsheads, cases, and bales are annually shipped to all parts of the world and the demand for American tobacco is greater than for the varieties grown in the Old World. More than two hundred and fifty years have passed since the London and Plymouth Companies began its cultivation in the Old Dominion, and on the same soil where the red man grew his "uppowac." Virginia leaf still continues to flourish, and to-day it is the great agricultural product of the State.

From a small beginning, like the plant itself it has developed into a great and increasing industry and its culture become a source of wealth unprecedented in agricultural history. Could the sapient James I. and his successors the Stuarts, now look upon this cherished production of the world, they would discover a commercial prosperity connected with those nations which have fostered and encouraged its growth far in advance of those who have frowned upon the plant and prohibited or hindered its cultivation. Saint Pierre alluding to the beneficence of nature and of the folly and cruelty of man as contrasted says:

"When the princes of Europe went Gospel in hand, to

SHIPPING TOBACCO.

lay waste Asia, they brought back the plague, the leprosy and the small-pox, but nature showed to a Dervish the coffee tree in the mountains of Yemen, and at the moment when nature brought curses on us through the Crusaders, it brought delights to us through the cup of a Mohammedan Monk. The descendants of those princes took possession of America, and transmitted to us by this conquest, an inexhaustible succession of wars and maladies. While they were exterminating the inhabitants of America with cannon, a Carib invited sailors to smoke his Calumet as a signal of peace. The perfume of the tobacco vanquished their torments and their troubles, and the use of tobacco was spread all over the earth. While the afflictions of the two worlds came from artillery, which kings call their last resort, the consolations of civilized nations flowed from the pipe of a savage."

It seems hardly possible to draw a more graphic picture of the blessings diffused by the balmy plant, than that just given. Its peculiar charms and soothing influence are well calculated to inspire in the breast of man, feelings of peace and happiness, rather than elements of discord and strife. The pipe of a king burns not more freely the shreds of the plant, than it does the last remnant of hostile feelings and the recollections of bitter wrongs; while the snuff-box of the diplomat contains the precious dust that has soothed the fierce hatred of rival houses and cemented the divided factions of a tottering throne.

CHAPTER IV.

TOBACCO IN EUROPE.

THE discovery of the tobacco plant in America by European voyagers aroused their cupidity no less than their curiosity. They saw in its use by the Indians a custom which, if engrafted upon the civilization of the Old World, would prove a source of revenue commensurate with their wildest visions of power and wealth. This was particularly the case with the Spanish and Portuguese conquerors, whose thirst for gold was gratified by its discovery. The finding by the Spaniards of gold, silver, and the balmy plant, and by the Portuguese of valuable and glittering gems, opened up to Spain and Portugal three great sources of wealth and power. But while the Spaniards were the first discoverers of the plant there seems to be conflicting opinions as to which nation first began its culture, and whether the plant was cultivated first in the Old World or in the New. Humboldt says:—

"It was neither from Virginia nor from South America, but from the Mexican province of Yucatan that Europe received the first tobacco seeds about the year 1559.* The Spaniards became acquainted with tobacco in the West India Islands at the end of the 15th Century, and the cultivation of Tobacco preceded the cultivation of the potato in Europe more than one hundred and twenty years. When Sir Walter Raleigh brought tobacco from Virginia to England in 1586,

---

*Mussey in his Essay on Tobacco records " That Cortez sent a specimen of the plant to the king of Spain in 1519. Yucatan was discovered by Hernandez Cordova in 1517, and in 1519 was first settled.

whole fields of it were already cultivated in Portugal.* It was also previously known in France."

Another author says of its introduction into Europe:—

"The seeds of the tobacco plant were first brought to Europe by Gonzalo Hernandez de Oviedo, who introduced it into Spain, where it was first cultivated as an ornamental plant, till Monardes† extolled it as possessed of medicinal virtues."‡

Murray says of the first cultivation of tobacco and potatoes in the Old World:—

"Amidst the numerous remarkable productions ushered into the Old Continent from the New World, there are two which stand pre-eminently conspicuous from their general adoption. Unlike in their nature, both have been received as extensive blessings—the one by its nutritive powers tends to support, the other by its narcotic virtues to soothe and comfort the human frame—the potato and tobacco; but very different was the favor with which these plants were viewed. The one long rejected, by the slow operation of time, and, perhaps, of necessity, was at length cherished, and has become the support of millions, but nearly one hundred and twenty years passed away before even a trial of its merits was attempted; whereas, the tobacco from Yucatan, in less than seventy years after the discovery, appears to have been extensively cultivated in Portugal, and is, perhaps, the most generally adopted superfluous vegetable product known; for sugar and opium are not in such common use. The potato by the starch satisfies the hunger; the tobacco by its morphia calms its turbulence of the mind. The former becomes a necessity required, the latter a gratification sought for."

It would appear then that the year 1559 was about the period of the introduction of tobacco into Europe. Phillip II. of Spain sent Oviedo to visit Mexico and note its productions and resources; returning he presented "His Most Catholic Majesty" with the seeds of the plant. In the following year it was introduced into France and Italy. It was first brought to France by Jean Nicot of Nismes in Languedoc, who was sent as ambassador to Sebastian, King of

---

*Spain began its culture in Mexico on the coast of Caraccas at the islands of St. Domingo and Trinidad, and particularly in Louisiana.
†Pourchat declares that the Portuguese brought it into Europe from Tobago, an island in the West Indies; but this is hardly probable, as the island was never under the Portuguese dominion.
‡Monardes wrote upon it only from the small account he had of it from the Brazilians."

Portugal, and who obtained while at Lisbon some tobacco seed from a Dutch merchant who had brought it from Florida.* Nicot returned to France in 1561, and presented the Queen, Catherine de Medicis, with a few leaves of the plant.†

As the history of Nicot is so intimately connected with that of the plant, a short sketch of this original importer will doubtless be interesting to all lovers of the weed:—

"John Nicot, Sieur de Villemain, was born at Nismes in 1530, and died at Paris in 1600. He was the son of a notary at Nismes, and started in life with a good education, but with no fortune. Finding that his native town offered no suitable or sufficient field for his energies, he went to Paris and strove hard to extend his studies as a scholar and his connections as an adventurer. He made the acquaintance of some courtiers, who felt or affected an interest in learning and in learned men. His manners were insinuating; his character was pliable. When presented at court he succeeded in gaining the esteem and confidence of Henry II., the husband of Catherine de Medicis. Francis II., the son of Henry II., and the first husband of Mary Stuart, continued to Nicot the favor of which Henry II. had deemed him worthy, and sent him in 1560 as ambassador to Sebastian, King of Portugal. He was successful in his mission. But it was neither his talents as a diplomatist, nor his remarkable mind, nor his solid erudition, which made Nicot immortal. It was by popularizing tobacco in France that he gained a lasting fame.

"It is said that it was at Lisbon that Nicot became acquainted with the extraordinary properties of tobacco. But it is likewise stated with quite as much confidence, that a Flemish merchant, who had just returned from America, offered Nicot at Bordeaux, where they met, some seeds of the tobacco, telling him of their value. The seeds Nicot sent to Catherine de Medicis, and on arriving in Paris he gave her some leaves of tobacco. Hence, when tobacco began to creep into use in France it was called Queen's Herb or Medicean Herb.‡ The cultivation of tobacco, except as a fancy plant, did not begin in France till 1626; and John

---

*Parkinson in his Herball [London, 1640] says:—"It is thought by some that John Nicot, this Frenchman, being agent in Portugall for the French King, sent this sort of tobacco [Brazil] and not any other to the French Queene, and is called therefore herba Regina, and from Nicotiana, which is probably because the Portugalls and not the Spaniards were masters of Brazile at that time."

†"Sir John Nicot sent some seeds of it into France, to King Francis II., the Queen Mother, and Lord Jarnac, Governor of Rochel, and several others of the French Lords."

‡The Abbe Jacques Gohory, the author of the first book written on tobacco, proposed to call it Catherinaine or Medicee, to record the name of Medicis and the medicinal virtues of the plant; but the name of Nicot superseded these, and botanists have perpetuated it in the genus *Nicotiana*."—*Le Maout and Decaisne.*

Nicot could have had no presentiment of the agricultural, commercial, financial and social importance which tobacco was ultimately to assume. Nicot published two works. The first was an edition of the History of France or of the Franks, in Latin, written by a Monk called Aimonious, who lived in the tenth century. The second was a 'Treasury of the French Language, Ancient and Modern.'"

Stevens and Liebault in the "Country Farm"* give the following account of its early introduction into France and the wonderful cures produced by its use:

"Nicotiana though it have (has) beene but a while knowne in France yet it holdeth the first and principall place amongst Physicke herbs, by reason of his singular and almost diuine (divine) vertues, such as you shall heare of hereafter, whereof (because none either of the old or new writers that have written of the nature of plants, have said anything), I am willing to lay open the whole history, as I have come by it through a deere friend of mine, the first author, inventor, and bringer of this herb into France: as also of many both Spaniards, Portugals, and others which have travelled into Florida, a country of the Indians, from whence this herbe came, to put the same in writing to relieve such griefe and travell, as have heard of this herbe, but neither know it nor the properties thereof. This herbe is called Nicotiana of the name of an ambassador which brought the first knowledge of it into this realme, in like manner as many plants do as yet retaine the names of certaine Greekes and Romans, who being strangers in divers countries, for their common-wealth's service, have from thence indowed their own countree with many plants, whereof there was no knowledge before. Some call it the herbe of Queen mother, because the said ambassador Lord Nicot did first send the same unto the Queen mother, † (as you shall understand by and by) and for being afterwards by her given to divers others to plant and make to grow in this country. Others call it by the name of the herbe of the great Prior, because the said Lord a while after sailing into these western seas, and happening to lodge neere unto the said Lord ambassador of Lisbone, gathered divers plants thereof out of his garden, and set them to increase here in France, and there in greater quantitie, and with

---

* London 1606.
† George Buchanan, the Scotch Philosopher and poet tutor of James I., had a strong aversion to Catherine of Medicis, and in one of his Latin epigrams, alludes to the herb being called *Medicie*, advising all who valued their health to shun it, not so much from its being naturally hurtful, but that it needs must become poisonous if called by so hateful a name.

more care than any other besides him, he did so highly esteeme thereof for the exceeding good qualities sake.

"The Spaniards call it Tobaco, it were better to call it Nicotiana, after the name of the Lord who first sent it into France, to the end that we may give him the honor which he hath deserved of us, for having furnished our land with so rare and singular an herbe: and thus much for the name, now listen unto the whole historie: Master John Nicot, one of the king's counsell, being ambassador for his Maiestie (Majesty) in the realme of Portiugall, in the yeere of our Lord God, 1559. 60. and 61. went on a day to see the monuments and northie places of the said king of Portiugall: at which time a gentleman keeper of the said monuments presented him with this herbe as a strange plant brought from Florida. The nobleman Sir Nicot having procured it to growe in his garden, where it had put forth and multiplied very greatly, was aduertifed (notified) on a daie by one of his pages, that a yoong boie kinsman of the said page, had laide (for triall sake) the said herbe, pressed, the substance and juice and altogether, upon an ulcer which he had upon his cheeke, neere unto his nose, next neighbor to a *Noli me tangere*, (a cancer) as having already seazed upon the cartilages, and that by the use thereof it was become marvellous well: upon this occasion the nobleman Nicot called the boie to him, and making him to continue the applying of this herbe for eight or ten days, the *Noli me tangere* became thoroughly kild: nowe they had sent oftetimes unto one of the king's most famous phisitions, the said boie during the time of this worke and operation to make and see the proceeding and working of the said Nicotiana, and having in charge to do the same until the end of ten days, the said phisition then beholding him, assured him that the *Noli me tangere* was dead, as indeed the boie never felt anything of it at any time afterward.

"Some certain time after, one of the cooks of the said ambassador having almost all his thombe (thumb) cut off from his hand, with a great kitchin knife, the steward running unto the said Nicotiana, made to him use of it five or six dressings, by the ende of which the wounde was healed. From this time forward this herbe began to become famous in Lisbon, where the king of Portiugal's court was at that time, and the vertues thereof much spoken of, and the common people began to call it the ambassador's herbe. Now upon this occasion there came certain days after, a gentleman from the fields being father unto one of the pages of the said Lord ambassador, who was troubled with an ulcer in his legge of two years continuance, and craved of the said Lord some of his herbe, and using it in manner afore mentioned, he was healed by the end of ten or twelve daies. After this yet the herbe grewe still in greater reputation, inasmuch as that many hasted out of all corners to get some of this herbe. And among the rest, there was one woman which had a great ring worme, covering all her face like a mask, and having taken deepe roote, to whom the said Lord caused this Petum to be given, and withall the manner of using it to be told her, and at the end of eight or ten daies, this woman being thoroughly cured, came to shewe herself unto the said Lord, and how that she was cured. There came likewise a captain bringing with him his son diseased with the king's evill, unto the said Lord Ambassador, for to send him into France, upon whom there was some triall made of the said herbe, whereupon within four daies he began to show great signs and tokens of healing, and in the end was thoroughly cured of his king's evil."

Italy received the first plant from Santa Croce,* who, like Nicot, obtained the seed in Lisbon. In 1575 first appeared a figure of the plant in Andre Theret's "Cosmographie," which was but an imperfect representation of the plant. It was supposed by many on its discovery to grow like the engraving given—in form resembling a tree or shrub rather than an herb. Tobacco was first brought to England by Sir John Hawkins, who obtained the plant in Florida in 1565, and afterwards by Sir Francis Drake.† The first planters of it in England were said to be Captain Grenfield and Sir Francis Drake. One account of its introduction into England is as follows:

"The plant was first used by Sir Walter Raleigh and others, who had acquired a taste for it in Virginia.‡ Among

---

*The Pied Bull Inn, at Islington, was the first house in England where tobacco was smoked, while Moll Cut-Purse, a noted pickpocket who flourished in the time of Charles II., is said to have been the first Englishwoman who smoked tobacco.

†"It was introduced, about 1520, into Portugal and Spain by Doctor Hernandez of Toledo; into Italy by Thornabon and the Cardinal de Sainte-Croix, into England by Captain Drake and into France by Andre Theret, a gray friar."—Le Maout and Decaisne's General System of Botany (Paris 1868).

‡Short says of its introduction into England: "Sir Walter Raleigh's Marriners, under Mr. Ralph Lane, his Agent in Virginia first brought this Commodity into England Anno 1584; and that famous Proprietor of this Plantation foresaw good reasons to introduce the use of it, however King James might afterwards, through his own personal Distaste both of it and him, wrote his Counterblast against it; a work surely consistent with the Pen of no Prince, but one of his Politicks."

the natives the usual mode employed in smoking the plant was by means of hollow canes, and pipes made of wood and decorated with copper and green stones. To deprive it of its acidity, some of the natives were wont to pass the smoke through bulbs containing water, in which aromatic and medicinal herbs had been infused."

Neander ascribes this invention to the Persians; but Magnenus rather attributes it to the Dutch and English, to the latter of whom attaches the credit of having invented the clay pipes of modern times. Some writers have concluded that the plant served as a narcotic in some parts of Asia. Liebaut thinks it was known in Europe* many years before the discovery of the New World, and asserts that the plant had been found in the Ardennes. Magnenus, however, claims its origin as transatlantic and affirms as his belief that the winds had doubtless carried the seeds from one continent to the other. Pallos says that among the Chinese, and among the Mongol tribes who had the most intercourse with them, the custom of smoking is so general, so frequent, and has become so indispensable a luxury; the tobacco purse affixed to their belt so necessary an article of dress; the form of the pipes, from which the Dutch seem to have taken the

OLD ENGRAVING OF TOBACCO.

---

* James the First also inclines to this belief, declaring tobacco to be "a common herb which (though under divers names) grows almost everywhere."

model of theirs, so original; and, finally, the preparation of the leaves so peculiar, that they could not possibly derive all this from America by way of Europe, especially as India, where the practice of smoking is not so general, intervenes between Persia and China. Meyen also states that the consumption of tobacco in the Chinese empire is of immense extent, and the practice seems to be of great antiquity, "for on very old sculptures I have observed the very same tobacco pipes which are still used." Besides, we now know that the plant which furnishes the Chinese tobacco is even said to grow wild in the East Indies.

"Tobacco," says Loudon, "was introduced into the county of Cork, with the potatoe, by Sir Walter Raleigh." A quaint writer of this period says of the plant: "Tobacco, that excellent plant, the use whereof (as of fifth element) the world cannot want, is that little shop of Nature, wherein her whole workmanship is abridged; where you may see earth kindled into fire, the fire breathe out an exhalation, which entering in at the mouth walks through the regions of a man's brain, drives out all ill vapors but itself, draws down all bad humors by the mouth, which in time might breed a scab over the whole body, if already they have not; a plant of singular use; for, on the one side Nature being an enemy to vacuity and emptiness and on the other, there being so many empty brains in the world as there are, how shall Nature's course be continued? How shall those empty brains be filled but with air, Nature's immediate instrument to that purpose? If with air, what so proper as your fume; what fume so healthful as your perfume, what perfume so sovereign as tobacco. Besides the excellent edge it gives a man's wit, as they but judge that have been present at a feast of tobacco, where commonly all good wits are consoled; what variety of discourse it begets, what sparks of wit it yields?" *

The name of Sir Walter is intimately connected with the history of tobacco, and is associated with many of the brilliant exploits and explorations during the reign of the illustrious Elizabeth.† His name has come down to us as

---

*A writer in the "New England Magazine" says in a different strain: "This is the enemy that men put in their mouths, to steal away their health. This has filled the camp, the court, the grove. It is found in the pulpit, the senate, the bar and the boudoir."

†Thorpe, in his "History and Mystery of Tobacco," relates the following anecdote: "Tradition says, that in the time of Queen Elizabeth Sir Walter Raleigh used to sit at his door with Sir Hugh Middleton and smoke."

being that of the first smoker of tobacco in England,* and many amusing anecdotes are told of him and the new custom which he introduced and sanctioned. Dixon has given us the following vivid picture of the great Elizabethan navigator:

"In a pleasant room of Durham House, in the Strand,—a room overhanging a lovely garden, with the river, the old bridge, the towers of Lambeth Palace, and the flags of Paris Garden and the Globe in view,—three men may have often met and smoked a pipe in the days of Good Queen Bess, who are dear to all readers of English blood; because, in the first place, they were the highest types of our race in genius and in daring; in the second place because the work of their hands has shaped the whole after-life of their countrymen in every sphere of enterprise and thought. That splendid Durham House, in which the nine-days queen had been married to Guilford Dudley, and which had afterwards been the town-house of Elizabeth, belonged to Sir Walter Raleigh, by whom it was held on leave from the queen. Raleigh, a friend of William Shakespeare and the players, was also a friend of Francis Bacon and the philosophers. Raleigh is said to have founded the Mermaid Club; and it is certain that he numbered friends among the poets and players. The proofs of his having known Shakespeare, though indirect, are strong. Of his long intercourse with Bacon every one is aware. Thus it requires no effort of the fancy to picture these three men as lounging in a window of Durham House, puffing the new Indian weed from silver bowls, discussing the highest themes in poetry and science, while gazing on the flower-beds and the river, the darting barges of dames and cavalier, and the distant pavilions of Paris Garden and the Globe."

Its use by so distinguished a person as Raleigh was equivalent to its general introduction.† Aubrey says:

"He was the first that brought tobacco into England, and into fashion. In our part—Malmsbury Hundred—it came

---

*Dr. Thomas Short, in his work "Discourses on Tea, Tobacco, Punch, &c.," (London 1750,) says of the original smoker: "Sir Walter was the first that brought the Custom of smoking it into Britain, upon his return from America; for he saw the natives of Florida, Brazil and other places of the Indies, smoak it thus, they hung about their Necks little Pipes or Horns made of the Leaves of the Date Tree, or of Reeds or Rushes; and at the ends of them they put several dry Tobacco Leaves twisted and broken, and set the ends of them on fire, and sucked in as much of the smoak as they could."

†So common was the indulgence that in 1600, only seventeen years after Sir Francis Drake returned from America, and set the example of using tobacco, the French Embassador writes in his dispatches to Paris, that the peers, while engaged in the trials of Essex and Southampton, deliberated upon their verdicts with pipes in their mouths!

first into fashion by Sir Walter Long. They had first silver pipes. The ordinary sort made use of a walnut shell and a

SIR WALTER RALEIGH.

strawe. I have heard my grandfather Lyte say that one pipe was handed from man to man round the table. Sir Walter Raleigh standing in a stand at Sir Ro. Poyntz parke at Acton tooke a pipe of tobacco, which made the ladies quitte it till he had donne."

A writer has truthfully said in regard to associating the name and use of the plant with the primitive users of it.

"The ambitious sought fame by associating themselves with the introduction of the plant and its cultivation; hence we find it named after cardinals, legates, and embassadors, while in compliment to Catherine, wife of Henry the Second, it was called the Queen's herb."

Kings now rushed into the tobacco trade. Those of Spain took the lead, and became the largest manufacturers of snuff

---

*Savary says that tobacco has been known among the Persians for upwards of 400 years, and supposes that they received it from Egypt, and not from the East Indies.

and cigars in Christendom, and the royal workshops of Seville are still the most extensive in Europe. Other monarchs monopolized the business in their dominions, and all began to reap enormous profits from it, as most do at this day. In the year 1615 tobacco was first planted in Holland; and in Switzerland in 1686. As soon as its cultivation became general in Spain and Portugal the tobacco trade was "farmed out," bringing an enormous revenue to those kingdoms. About the beginning of the Seventeenth Century the Portuguese introduced into Hindostan and Persia* two things, pine-apples and tobacco. To the pine-apples no objection seems to have been made; but to the tobacco the most strenuous resistance was offered by the sovereigns of the two countries. Spite, however, of punishments and prohibitions the use of tobacco spread with the rapidity of lightning.

In England, tobacco taking soon became a favorite custom not only with the loiterers about taverns and other public places, but among the courtiers of Elizabeth. Smoking was called drinking tobacco, as the fashionable method was to "put it through the nose" or exhale it through the nostrils. At this period tobacco seemed to have nearly the same effect as it did upon the Indian, producing a sort of intoxication; thus in "The Perfuming of Tobacco" (1611) it is said: "The smoke of tobacco drunke or drawen by a pipe, filleth the membranes of the braine, and astonisheth and filleth many persons with such joy and pleasure, and sweet losse of senses, that they can by no means be without it."

The term "drinking tobacco" was not confined to England, but was used in Holland, France, Spain and Portugal, as the same method of blowing the smoke through the nostrils, seemed to be everywhere in vogue.

The use of tobacco increased very rapidly soon after its importation from Virginia. The Spaniards and Portuguese had hitherto monopolized the trade, so that it brought enormous prices, some kinds selling for its weight in silver. As soon as its culture commenced in Virginia the demand for West India tobacco lessened and Virginia leaf soon came into favor, owing not more to the lowering of price than to the quality of the leaf.* This was about 1620, which some writers have called the golden age of tobacco. It had now become a prime favorite and was used by nearly all classes. Poets and dramatists sung its praises, while others wrote of its wonderful medicinal qualities.† Fops and knaves alike indulged in its use.

"About the latter end of the sixteenth century, tobacco was in great vogue in London, with wits and 'gallants,' as the dandies of that age were called. To wear a pair of velvet breeches, with panes or slashes of silk, an enormous starched ruff, a gilt handled sword, and a Spanish dagger; to play at cards or dice in the chambers of the groom-porter, and smoke tobacco in the tilt-yard or at the play-house, were then the grand characteristics of a man of fashion. Tobacconists' shops were then common; and as the article, which appears to have been sold at a high price, was indispensable to the gay 'man about town,' he generally endeavored to keep his credit good with his tobacco-merchant. Poets and pamphleteers laughed at the custom, though generally they seem to have no particular aversion to an occasional treat to a sober pipe and a poute of sack. Your men of war, who had served in the Low Countries, and who taught young gallants the noble art of fencing, were particularly fond of tobacco; and your gentlemen adventurers, who had served in a buccaneering expedition against the Spaniards, were no less partial to it. Sailors—from the captain to the ship-boy—all affected to smoke, as if the practice was necessary to their character; and to 'take tobacco' and wear a silver whistle, like a modern boatswain's mate, was the pride of a man-of-war's man.

"Ben Jonson, of all our early dramatic writers, most frequently alludes to the practice of smoking. In his play of 'Every Man in his Humour,' first acted in 1598, Captain Bobadil thus extols in his own peculiar vein the virtues of tobacco; while Cob, the water carrier, with about equal truth, relates some startling instances of its pernicious effects.

---

*Neander, in his work on "Tobacologia" (London, 1622), mentions eighteen varieties of tobacco, or at least localities from where it was shipped to London, among which are the following: Varinas (considered the best), Brazil, Maracay, Orinoco, Margarita, Caracas, Cumana, Amazon, Virginia, Phillipines, St. Lucia, Trinidad, and St. Domingo.

†"The first author (says an English writer) who wrote of this Plant was Charles Stephanus, in 1564. This was a mean, short, inaccurate Draught, till Dr. John Liebault wrote a whole Discourse of it next year, and put it into his second Book of Husbandry, which was every year reprinted with additions and alterations, for twenty years after. He had a large Correspondence, a good Intelligence, and wrote the best of the age, and gathered the greatest stock of experience about this new Plant."

"'*Bobadil.* Body o' me, here's the remainder of seven pound since yesterday was seven-night! 'Tis your right Trinadado! Did you never take any, Master Stephen?

"'*Stephen.* No, truly, Sir; but I'll learn to take it since you commend it so.

"'*Bobadil.* Sir, believe me upon my relation,—for what I tell you the world shall not reprove. I have been in the Indies where this herb grows, where neither myself, nor a dozen gentlemen more of my knowledge, have received the taste of any other nutriment in the world, for the space of one and twenty weeks, but the fume of this simple only. Therefore, it cannot be but 'tis most divine. Further, take it, in the true kind, so, it makes an antidote, that had you taken the most deadly poisonous plant in all Italy, it should expel it and clarify you with as much ease as I speak. And for your greenwound, your balsamum, and your St. John's-wort, are all mere gulleries and trash to it, especially your Trinidado: your Nicotian is good too. I could say what I know of it for the expulsion of rheums, raw humours, crudities, obstructions, with a thousand of this kind, but I profess myself no quack-salver: only thus much, by Hercules; I do hold it, and will affirm it before any prince in Europe, to be the most sovereign and precious weed that ever the earth tendered to the use of man.'

Cob. "'By gad's me, I mar'l what pleasure or felicity they have in taking this roguish tobacco! It's good for nothing but to choke a man and fill him full of smoke and embers. There were four died out of one house last week with taking of it, and two more the bell went for yesternight; one of them, they say, will ne'er 'scape it: he voided a bushel of soot yesterday, upward and downward. By the stocks! an' there were no wiser men than I, I'd have it present whipping, man or woman that should but deal with a tobacco-pipe; why, it will stifle them all in the end, as many as use it; it's little better than rats-bane or rosaker.'"*

From the first announcement that English navigators had discovered tobacco in Virginia, until the London and Plymouth companies sailed for the New World, the deepest interest was taken in the voyagers. Drayton, the poet, wrote of "The Virginian Voyage," while Chapman and other dramatists wrote plays in which allusions were made to Virginia. In the "Mask of Flowers," performed at White Hall

---

* A preparation of arsenic.

upon Twelfth Night, 1613–14, one of the characters challenges another, and asserts that wine is more worthy than tobacco. The costumes were exceedingly grotesque and suggestive of the New rather than of the Old World. Kawosha one of the principal characters rode in, wearing on his head a cap of red-cloth of gold, from his ears were pendants, a glass chain was about his neck, his body and legs were covered with olive-colored stuff, in his hands were a bow and arrows, and the bases of tobacco-colored stuff cut like tobacco leaves. The play abounds with allusions to the "Indian weed."

> "*Silenus.*— Kawosha comes in majestie,
>   Was never such a God as he;
>   He's come from a far countrie
>   To make our nose a chimney.
>
> *Kawosha.*—The wine takes the contrary way
>   To get into the hood;
>   But good tobacco makes no stay
>   But seizeth where it should.
>   More incense hath burned at
>   Great Kawoshae's foote
>   Than to Silen and Bacchus, both,
>   And take in Jove to boote.
>
> *Silenus.*—The worthies they were nine tis true,
>   And lately Arthur's knights I knew;
>   But now are come up Worthies new,
>   The roaring boys Kawoshae's crew.
>
> *Kawosha.*—Silenus toppes the barrel, but
>   Tobacco toppes the braine
>   And makes the vapors fire and soote,
>   That mon revise againe.
>   Nothing but fumigation
>   Doth charm away ill sprites,
>   Kawosha and his nation
>   Found out these holy rites."

The writers of this period abound in allusions to tobacco and its use. The poets and dramatists found in it a fertile field for the display of their satire, and from 1600 to 1650 stage plays introduced many characters as either tobacco drinkers or sellers. It had now become so great a custom and had increased so fast after the importation of Virginia tobacco that it afforded them no insignificant theme for the display of their genius.* The plays of Jonson, Decker, Rowland, Heywood, Middleton, Fields, Fletcher, Hutton, Lodge, Sharpham, Marston, Lilly (court poet to Elizabeth), the Duke of Newcastle and others are full of allusions to the plant and those who indulged in its use. Shakespeare,† however, does not once allude to its use, and his silence on this then curious custom has provoked much conjecture and inquiry. Some affirm that he wrote to please royalty, but if so why did he not condemn the custom to appease the wrath of a sapient king. Others say he kept silence because he was the friend of Raleigh, and though he would have gladly held up the great smoker and his favorite indulgence, feared to add to the popularity of the custom by displeasing his royal master. Another class affirm that as the stories of his plays are all antecedent to his own time, therefore he never mentions either the drinking of tobacco, or the tumultuous scenes of the ordinary which belonged to it, and which are so constantly met with in his contemporary dramatists. Says one:

"How is it that our great dramatist never once makes even the slightest allusion to smoking? Who can suggest a reason? Our great poet knew the human heart too well, and kept too steadily in view, the universal nature of man to be afraid of painting the external trapping and ephemeral customs of his own time. Does he not delight to moralize on false hair, masks, rapiers, pomanders, perfumes, dice, bowls, fardingales, etc? Did he not sketch for us, with enjoyment and with satire, too, the fantastic fops, the pompous stewards, the mischievous pages, the quarrelsome revellers, the testy gaolers, the rhapsodizing lovers, the sly cheats, and the ruffling courtiers that filled the streets of Elizabethan London, persons who could have been found nowhere else

---

* "Never did nature produce a Plant that in a short Time became so universally used, for it was but a short while known in Europe, till it was taken almost everywhere, either chewed, smoked, or snuffed. A pipe of tobacco is now the general and most frequent companion of, Mug. Bottle, or Punch-bowl."—*T. Short.*

† Gifford has also remarked that Shakespeare is the only one of the dramatic writers of the age of James who does not condescend to notice tobacco; all the others abound in allusions to it. In Jonson we find tobacco in every place—in Cob the waterman's house, and in the Apollo Club-room, on the stage, and at the ordinary. The world of London was then divided into two classes—the tobacco-lovers and the tobacco-haters.

nor in any other age? No one can dispute that he drew the life that he saw moving around him. He sketched these creatures because they were before his eyes and were his enemies or his associates; they live still because their creator's genius was Promethean, and endowed them with immortality. Bardolph, Moth, Slender, Abhorson, Don Armado, Mercutio, etc., are portraits, as everyone knows and feels who is conversant with the manners of the Elizabethan times as handed down in old plays.

"If Shakespeare's contemporaries were silent about the then new fashion of smoking, we should not so much wonder at Shakespeare's taciturnity. But Decker's and Ben Jonson's works abound in allusions to tobacco, its uses and abuses. The humorist and satirist lost no opportunity of deriding the new fashion and its followers. The tobacco merchant was an important person in London of James the First's time—with his Winchester pipes, his maple cutting-blocks, his juniper-wood charcoal fires, and his silver tongs with which to hand the hot charcoal to his customers, although he was shrewdly suspected of adulterating the precious weed with sack lees and oil. It was his custom to wash the tobacco in muscadel and grains, and to keep it moist by wrapping it in greased leather and oiled rags, or by burying it in gravel. The Elizabethan pipes were so small that now when they are dug up in Ireland the poor call them 'fairy pipes' from their tininess. These pipes became known by the nickname of 'the woodcock's heads.' The apothecaries, who sold the best tobacco, became masters of the art, and received pupils, whom they taught to exhale the smoke in little globes, rings, or the 'Euripus.' 'The slights' these tricks were called. Ben Jonson facetiously makes these professors boast of being able to take three whiffs, then to take horse, and evolve the smoke—one whiff on Hounslow, a second at Staines, and a third at Bagshot.

"The ordinary gallant, like Mercutio, would smoke while the dinner was serving up. Those who were rich and foolish carried with them smoking apparatus of gold or silver—tobacco-box, snuff-ladle, tongs to take up charcoal, and priming irons. There seems, from Decker's 'Gull's Horn-Book,' to have been smoking clubs, or tobacco ordinaries as they were called, where the entire talk was of the best shops for buying Trinidado, the Nicotine, the Cane, and the Pudding, whose pipe had the best bore, which would turn blackest, and which would break in the browning. At the theatres, the rakes and spendthrifts who crowded the stage of Shakespeare's time sat on low stools smoking; they sat with their three sorts of tobacco beside them, and handed each other lights on the points of their swords, sending out their pages for more Trinidado if they required it. Many gallants 'took' their tobacco in the lords room over the stage, and went out to (Saint) Paul's to spit there privately. Shabby sponges and lying adventurers, like Bobadil, bragged of the number of packets of 'the most divine tobacco' they had smoked in a week, and told enormous lies of living for weeks in the Indies on the fumes alone. They affirmed it was an antidote to all poison; that it expelled rheums, sour humours, and obstructions of all kinds. Some doctors were of opinion that it would heal gout* and the ague, neutralise the effects of drunkenness, and remove weariness and hunger. The poor on the other hand, not disinclined to be envious and detracting when judging rich men's actions, laughed at men who made chimneys of their throats, or who sealed up their noses with snuff.

"Ben Jonson makes that dry, shrewd, water carrier of his, Cob, rail at the 'roguish tobacco:' he would leave the stocks for worse men, and make it present whipping for either man or woman who dealt with a tobacco-pipe. But King James, in his inane 'Counterblast,' is more violent than even Cob. He argues that to use this unsavory smoke is to be guilty of a worse sin than that of drunkenness, and asks how men, who cannot go a day's journey without sending for hot coals to kindle their tobacco, can be expected to endure the privations of war. Smoking, the angry and fuming king protests, had made our manners as rude as those of the fish-wives of Dieppe. Smokers, tossing pipes and puffing smoke over the dinner-table, forgot all cleanliness and modesty. Men now, he says, cannot welcome a friend but straight they must be in hand with tobacco. He that refused a pipe in company was accounted peevish and unsociable. 'Yea,' says the royal coxcomb and pedant, 'the mistress cannot in a more mannerly kind entertain her servant than by giving him out of her fair hand a pipe of tobacco.' The royal reformer (not the most virtuous or cleanly of men) closes his denunciation with this tremendous broadside of invective:

'Have you not reason, then' he says, 'to be shamed and to forbear this filthy novelty, so basely grounded, so foolishly received, and so grossly mistaken in the right use thereof?

---

* "Some hold it for a singular remedie against the gowte (gout), to chaw every morning the leaves of Petum (tobacco), because it voideth great quantitie of flegme out at the mouth, hindering the same from falling upon the joints, which is the very cause of the gowte." *Dr. Richard Surflet* (1605).

CURATIVE QUALITIES.

To your abuse thereof sinning against God, harming yourself both in persons and goods, and taking also thereby the notes and marks of vanity upon you by the custom thereof, making yourselves to be wondered at by all, foreign civil nations and by all strangers that come among you, and be scorned, and contemned; a custom both fulsome to the eye, hateful to the nose, harmful to the brain, dangerous to the lungs, and in the black stinking fume thereof nearest resembling the horrible Stigian smelle of the pit that is bottomless."

The supposed curative virtues of the tobacco plant had much to do with its use in Europe while the singular mode of exhaling through the nostrils added to its charms, and

EXHALING THROUGH THE NOSE.

doubtless led to far greater indulgence. Spenser in his Fairy Queen makes one of the characters include it with other herbs celebrated for medicinal qualities.

> "Into the woods thence-forth in haste she went,
>   To seek for herbes that mote him remedy;
>   For she of herbes had great intendiment,
>   Taught of the Nymph which from her infancy,
>   Had nursed her in true nobility:
>   There whether it divine Tobacco were,
>   Or Panachæ, or Polygony,
>   She found and brought it to her patient deare,
>   Who all this while lay bleeding out his heart-blood neare."

Lilly also a little later, in his play of The Woman in the Moone (1597(, speaks of it (through one of the characters) as being a medicinal herb—

> "Gather me balme and cooling violets
> And of our holy herbe nicotian,
> And bring withall pure honey from the hive
> To heale the wound of my unhappy hand."

Barclay, in his tract on "The Vertues of Tobacco," recommends its use as a medicine. The following is one of the modes of use:

"Take of leafe Tobacco as much as, being folded together, may make a round ball of such bignesse that it may fill the patient's mouth, and inclyne his face downwards toward the ground, keeping the mouth open, not mouthing any whit with his tongue, except now and then to waken the medicament, there shall flow such a flood of water from his brain and his stomacke, and from all the parts of his body that it shall be a wonder. This must he do fasting in the morning, and if it be for preservation, and the body be very cacochyme, or full of evil humors, he must take it once a week, otherwise once a month. He gives the plant the name of 'Nepenthes,' and says of it, that 'it is worthy of a more loftie name.'" He writes the following verse addressed to:

"THE ABUSERS OF TOBACCO."

> "Why do you thus abuse this heavenly plant,
> The hope of health, the fuel of our life?
> Why do you waste it without fear of want,
> Since fine and true tobacco is not ryfe?
> Old Enclio won't foul water for to spair,
> And stop the bellows not to waste the air."

He also alludes to the quality of tobacco and says: "The finest Tobacco is that which pearceth quickly the odorat with a sharp aromaticke smell, and tickleth the tongue with acrimonie, not unpleasant to the taste, from whence that which draweth most water is most veituous, whether the substance of it be chewed in the mouth, or the smoke of it reccived."

He speaks of the countries in which the plant grows, and prefers the tobacco grown in the New World as being superior to that grown in the Old. In his opinion, "only that which is fostered in the Indies, and brought home by Mariners and Traffiquers, is to be used." But not alone were Poets and Dramatists inspired to sing in praise or dispraise of tobacco, Physicians and others helped to swell in broadsides, pamphlets and chap-books, the loudest praises or the most bitter denunciation of the weed. Taylor, the water poet, who lost his occupation as bargeman when the coach came into use, thought that the devil brought tobacco into England in a coach. One of the first tracts wholly devoted to tobacco is entitled Nash's "Lenten Stuffe." The work is dedicated to Humphrey King, a tobacconist, and is full of curious sayings in regard to the plant. Another work, entitled "Metamorphosis of Tobacco," and supposed to have been written by Beaumont, made its appearance about this time. Samuel Rowlands, the dramatist, wrote two works on tobacco; the first is entitled "Look to it, for I'll Stabbe Ye," written in 1604; the other volume is a small quarto, bearing this singular title: "A whole crew of Kind Gossips, all met to be Merry." This is a satire on the time and manners of the period, and is written in a coarse style worthy of the author. In 1605 there appeared a little volume bearing for its title, "Laugh and Lie Down, or the World's Folly." This work describes the fops and men of fashion of its time, and shows how popular the custom of tobacco taking had become. In 1609, in "The Gull's Horne Book," a gallant is described as follows:

"Before the meate comes smoaking to the board our Gallant must draw out his tobacco box, the ladle for the cold snuff into his nostrils, the tongs and the priming iron. All this artillery may be of gold or silver, if he can reach to the price of it; it will be a reasonable, useful pawn at all times when the current of his money falles out to rune low. And here you must observe to know in what state tobacco is in town, better than the merchants, and to discourse of the potecaries where it is to be sold as readily as the potecary himself."

One of the severest tirades against tobacco appeared in 1612, "The Curtain Drawer of the World." In speaking of the users of the weed, and especially noblemen, he says:

"Then noblemen's chimneys used to smoke, and not their noses; Englishmen without were not Blackamoores within, for then Tobacco was an Indian, unpickt and unpiped,—now made the common ivy-bush of luxury, the curtaine of dishonesty, the proclaimer of vanity, the drunken colourer of Drabby solacy."

In the "Soule's Solace, or Thirty-and-One Spiritual Emblems," by Thomas Jenner, occurs the following verses:

> "The Indian weed, withered quite,
> Greene at noone, cut down at night,
> Shows thy decay; all flesh is hay;
> Thus thinke, then drinke Tobacco.
>
> The Pipe that is so lily-white,
> Show thee to be a mortal wight,
> And even such, gone with a touch,
> Thus thinke, then drinke Tobacco.
>
> And when the smoake ascends on high,
> Thinke thou beholdst the vanity
> Of worldly stuffe, gone with a puffe,
> Thus thinke, then drinke Tobacco.
>
> And when the Pipe grows foul within,
> Thinke on thy soul defiled with sin,
> And then the fire it doth require;
> Thus thinke, then drinke Tobacco.
>
> The ashes that are left behind,
> May serve to put thee still in mind,
> That unto dust return thou must;
> Thus thinke, then drinke Tobacco."

Buttes, in a little volume entitled "Dyets Dry Dinner," (1599) says that "Tobacco was translated out of India in the seede or roote; native or sative in our own fruitfullest soils. It cureth any griefe, dolour, imposture, or obstruction proceeding of colde or winde, especially in the head or breast. The fume taken in a pipe is good against Rumes, ache in the head, stomacke, lungs, breast; also in want of meate, drinke, sleepe, or rest."

The introduction of tobacco from the colony of Virginia was followed soon after by a reduction of price that led to more frequent use among the poorer classes, such as grooms and hangers on at taverns and ale-houses, who are alluded to in Rich's "Honestie of this Age:

"There is not so base a groome that comes into an alehouse to call for his pott, but he must have his pipe of tobacco; for it is a commodity that is nowe as vendible in every tavern, wine and ale-house, as eyther wine, ale or beare;

OLD LONDON ALE-HOUSE.

and for apothecaerie's shops, grocer's shops, chandler's shops, they are never without company, that from morning till night, are still taking of tobacco. What a number are there besides, that doe keepe houses, set open shoppes, that have no other trade to live by, but by selling of tobacco. I have heard it told, that now very lately there hath been a catalogue of all those new erected houses that have sett up that trade of selling tobacco in London, and neare about London; and if a man may believe what is confidently reported, there are found to be upwards of seven thousand of houses that doth live by that trade.

"If it be true that there be seven thousand shops in and about London, that doth vend tobacco, as it is credibly reported that there be over and above that number, it may well be supposed to be but an ill customed shop, that taketh not five shillings a day, one day with another throughout the whole year; or, if one doth take lesse, two other may take more; but let us make our account, but after two shillings sixpence a day, for he that taketh lesse than that would be ill able to pay his rent, or to keepe open his shop windows; neither would tobacco houses make such a muster as they do, and that almost in every lane, and in every by-corner round about London."

"A Tobacco seller is described after this manner by

Blount in a volume "Micro-Cosmographie; Or A Piece of of the World discovered; in Essays and Characters" (1628).

"A tobacco seller is the only man that finds good in it which others brag of, but doe not, for it is meate, drinke, and clothes to him. No man opens his ware with greater seriousness, or challenges your judgment more in the operation. His Shop is the Randenvous of spitting, where men dialogue with their noses, and their conversation is smoke. It is the place only where Spain is commended, and preferred before England itself.

"He should be well experienced in the World; for he has daily tryall as men's nostrils, and none is better acquainted with humour. His is the piecing commonly of some other trade, which is bawd to his Tobacco, and that to his wife, which is the flame that follows the smoke."

Early in the Seventeenth Century began the persecution by royal haters of the plant, others, however, had denounced the weed and its use and users, but venting nothing more than a tirade of words against it, had but little effect in breaking up the trade or the custom.* James I. sent forth his famous "Counterblast" and in the strongest manner condemned its use. A portion of it reads thus:

"Surely smoke becomes a kitchen fane better than a dining chamber: and yet it makes a kitchen oftentimes in the inward parts of men, soyling and injecting with an unctuous oyly kind of roote as hath been found in some great tobacco takers, that after death were opened. A custom loathsome to the eye, harmful to the braine, dangerous to the lungs, and the black stinking fume thereof, nearest resembling the horrible Stygian smoke of the pit that is bottomless." †

Quaint old Burton in his "Anatomy of Melancholy," recognizes the virtues of the plant while he anathematizes its abuse. He says:—

"Tobacco, divine, rare, superexcellent tobacco, which goes far beyond all their panacetas, potable gold, and philosophers' stones, a soveraign remedy to all diseases. A good vomit, I

---

* Elizabeth during her reign, published an edict against its use, assigning as a reason, that her subjects, by employing the same luxuries as barbarians, were likely to degenerate into barbarism.
"From the first introduction of the weed, the votaries of the pipe have enjoyed all the blessings of persecution. Kings have punished, priests have anathematized, satirists satirized and women scolded; but still the weed, with its divers shapes and different names, reigns supreme among narcotics in every region of the globe."—*Emerson's Magazine.*
† Another writer in the same censorious manner says of the use of tobacco. "Smoking is the jovial repast of Cannibals or Man-eaters, and the grand entertainment of idolatrous Pagan Festivals. Masters will not permit the use of it to their servants or slaves and such as use it can hardly find masters or buyers."

confesse, a vertuous herb, if it be well qualified, opportunely taken, and medicinally used; but, as it is commonly abused by most men, which take it as tinkers do ale, 'tis a plague, a mischief, a violent purger of goods, lands, health, hellish, divelish and damned tobacco, the ruine and overthrow of body and soul."

The duty on importation had been only twopence per pound, a moderate sum in view of the prices realized by the sale of it.

The King now increased it to the enormous sum of two shilling and ten pence. James termed the custom of using tobacco an "evil vanitie" impairing "the health of a great number of people their bodies weakened and made unfit for labor, and the estates of many mean persons so decayed and consumed, as they are thereby driven to unthriftie shifts only to maintain their gluttonous exercise thereof." * Brodigan says of the "Counterblast:"

"However absurd his reasoning may appear, it unfortunately happened that he possessed the power to reduce his aversion to practice, and he may be considered as the author of that unwarrantable persecution of the tobacco plant, which under varying circumstances, has been injudiciously continued to the present time."

Other royal haters of the plant issued the most strenous laws† and affixed penalties of the severest kind, of these may be mentioned the King of Persia, Amuroth IV. of Turkey, the Emperor Jehan-Gee and Popes Urban VIII. and Innocent XII., the last of whom showed his dislike to many other customs beside that of tobacco taking.

One of the edicts which he issued was against the taking of snuff in St. Peters, at Rome; this was in 1690; it was, however, revoked by Pope Benedict XIV., who himself had acquired the indulgence.

Early in the Seventeenth Century tobacco found its way to Constantinople. To punish the habit, a Turk was seized and a pipe transfixed through his nose.

---

* "King James' violent prejudices against all use of tobacco arose from his aversion to Sir Walter Raleigh, its first importer into England whom he intended a sacrifice to the gratification of the King of Spain."

† The Empress Elizabeth was less severe. She decreed that the snuff-boxes of those who made use of them in church should be confiscated to the use of the beadle.

PUNISHMENT FOR SNUFF-TAKING.

The death of King James, followed by its occupancy of the throne by his son Charles I., did not lessen the persecution against tobacco.* In 1625, the year of his accession, he issued a proclamation against all tobaccos excepting only the growth of Virginia and Somerites. Charles II. also prohibited the cultivation of tobacco in England and Ireland, attaching a penalty of 10£ per rood. Fairholt, in alluding to the Stuarts and Cromwell as persecutors of tobacco, says:

"Cromwell disliked the plant, and ordered his troops to trample down the crop wherever found."

It is an historical fact that both James I. and the two Charleses as well as Cromwell had the strongest dislike against the Indian weed.

With such powerful foes it seems hardly possible that the custom should have increased to such an extent that when William ascended the throne the custom was said to be almost universal.† "Pipes grew larger and ruled by a Dutchman, all England smoked in peace." From this time forward the varieties used served only to increase the demand for the tobacco of the colonies, and as its culture became better understood the leaf grew in favor, until the demand for it was greater than the production.

During the reign of Anne, the custom of smoking appears to have attained its greatest height in England; the consumption of tobacco was then proportionably greater, considering the population, than it is at the present time. Spooner, in his "Looking-Glass for Smokers," 1703, says of the custom:

"The sin of the kingdom in the intemperate use of tobacco, swelleth and increaseth so daily, that I can compare it to nothing but the waters of Noah, that swell'd fifteen cubits above the highest mountains. So that if this practice shall continue to increase as it doth, in an age or two it will be as hard to find a family free, as it was so long time since one that commonly took it."

---

*Tobacco has been able to survive such attacks as these—nay, has raised up a host of defenders as well as opponents. The Polish Jesuits published a work entitled "Anti-Misocapnus," in answer to King James. In 1628, Raphael Thorius wrote a poem "Hymnus Tobaci." A host of names appear in the field: Lesus, Braum and Simon Pauli, Portal, Pia, Vanquelin, Gardaune, Poswelt, Reimann, and De Morveau.

†Says an enthusiastic writer on tobacco, "If judged by the vicissitudes through which it has traveled, it must indeed be acknowledged a hero among plants; and if human pity, respect, or love should be given it for 'the dangers it has passed,' the inspiration of Desdemonia's love for Othello, then might its most eloquent opponent be dumb, or yield it no inconsiderable meed of homage."

OLD CUSTOMS.

When tobacco was first introduced into England its sale was confined to apothecaries, but afterwards it was dealt in by tobacconists, who sold other goods besides tobacco.

About the middle of the Seventeenth Century the culture of tobacco commenced in England; it continued, however, only for a short time, for the rump parliament in 1652 prohibited the planting of it, and two years later Cromwell and his council appointed commissioners for strictly putting this act in execution: and in 1660 it was legally enacted, that from the first of January, 1660–1, no person whatever should sow or plant any tobacco in England, under certain penalties.

In England drinking or smoking tobacco seems to have met with more success (as a mode of use) rather than chewing (now so popular). It was principally confined to the lower classes, and was common among soldiers and sailors.

SILVER SPITTOONS.

When used by gentlemen it was common to carry a silver basin to spit in.

The habit of smoking or using tobacco in any form was

then more constant than now, and its use was common in almost all places of public gathering. It was the custom to smoke in theatres; stools being provided for those who paid for their use and the privilege of smoking on the stage. Tobacco was also sold at some of the play-houses, and proved a source of profit, doubtless, beyond even the representation of the plays. We should infer also from some of the early stage plays, that the "players" used the weed even when acting their parts. Rowlands gives the following poem on tobacco in his "Knave of Clubs," 1611:—

" Who durst dispraise tobacco whilst the smoke is in my nose,
Or say, but fah! my pipe doth smell, I would I knew but those
Durst offer such indignity to that which I prefer.
For all the brood of blackamoors will swear I do not err,
In taking this same worthy whif with valiant cavalier,
But that will make his nostrils smoke, at cupps of wine or beer.
When as my purse can not afford my stomach flesh or fish,
I sop with smoke, and feed as well and fat as one can wish.
Come into any company, though not a cross you have,
Yet offer them tobacco, and their liquor you shall have.
They say old hospitalitie kept chimnies smoking still;
Now what your chimnies want of that, our smoking noses will.
Much vituals serves for gluttony, to fatten men like swine,
But he's a frugal man indeed that with a leaf can dine,
And needs no napkins for his hands, his fingers' ends to wipe,
But keeps his kitchen in a box, and roast meat in a pipe.
This is the way to help down years, a meal a day's enough:
Take out tobacco for the rest, by pipe, or else by snuff,
And you shall find it physical; a corpulent, fat man,
Within a year shall shrink so small that one his guts shall span.
It's full of physic's rare effects, it worketh sundry ways,
The leaf green, dried, steept, burnt to dust, have each their several praise,
It makes some sober that are drunk, some drunk of sober sense,
And all the moisture hurts the brain, it fetches smoking thence.
All the four elements unite when you tobacco take.
For earth and water, air and fire, do a conjunction make.
The pipe is earth, the fire's therein, the air the breathing smoke;
Good liquor must be present too, for fear I chance to choke.
Here, gentlemen, a health to all, 'Tis passing good and strong.
I would speak more, but for the pipe I cannot stay so long.

In 1602 appeared a sweeping tirade entitled, "Work for Chimney Sweepers, or a Warning against Tobacconists." It abounds with threats against all who indulge in tobacco.

The most singular work, however, appeared in 1616, bearing the following singular title: "The Smoking Age, or the Man in the Mist; with the Life and Death of Tobacco. Dedicated to Captain Whiffe, Captain Pipe, and Captain Snuffe." A frontispiece is given representing a tobacconist's shop with shelves, counters, pipes and tobacco; a carved figure of a negro stands upon the counter, which shows how soon such figures were used by dealers in pipes and tobacco.

The title-page contains the following epigram:

"This some affirme, yet yield I not to that,
'Twill make a fat man lean, a lean man fat;
But this I'm sure (howse'ere it be they meane)
That many whiffes will make a fat man lean."

The following effusion resembles many of the verses of the day on the fruitful subject:

"Tobacco's an outlandish weed,
Doth in the land strange wonders breed,
It taints the breath, the blood it dries,
It burns the head, it blinds the eyes;
It dries the lungs, scourgeth the lights,
It numbs the soul, it dulls the sprites;
I brings a man into a maze,
And makes him sit for other's gaze;
It makes a man, it mars a purse,
A lean one fat, a fat one worse;
A sound man sick, a sick man sound,
A bound man loose, a loose man bound;
A white man black, a black man white,
A night a day, a day a night;
The wise a fool, the foolish wise,
A sober man in drunkard's guise;

> A drunkard with a drought or twain,
> A sober man it makes again;
> A full man empty, and an empty full,
> A gentleman a foolish gull;
> It turns the brain like cat in pan,
> And makes a Jack a gentleman."

The well-known song of "Tobacco is an Indian Weed," was written most probably the last half of the Seventeenth Century. Fairholt gives the best copy we have seen of it. It is taken from the first volume of "Pills to Purge Melancholy," and reads thus:

> " Tobacco's but an Indian weed,
> Grows green at morn, cut down at eve,
> It shows our decay, we are but clay;
> Think of this when you smoke tobacco.

> " The pipe, that is so lily white,
> Wherein so many take delight,
> Is broke with a touch—man's life is such;
> Think of this when you smoke tobacco.

> " The pipe, that is so foul within,
> Shews how man's soul is stained with sin,
> And then the fire it doth require;
> Think of this when you smoke tobacco.

> " The ashes that are left behind
> Do serve to put us all in mind
> That unto dust return we must;
> Think of this when you smoke tobacco.

> " The smoke, that does so high ascend,
> Shews us man's life must have an end,
> The Vapor's gone—man's life is done;
> Think of this when you smoke tobacco."

One of the strongest objections against the use of the "Indian novelty" was its ruinous cost at this period. During the reign of James The First and Charles The Second, Spanish tobacco sold at from ten to eighteen shillings per pound while Virginia tobacco sold for a time for three shillings. In no age and by no race excepting perhaps the Indians was the habit so universal or carried to such a length as in the Seventeenth Century—its supposed virtues as a medicine induced many to inhale the smoke constantly. This was one reason why tobacco was condemned by so many of the writers and playwrights of the day yet many of them used the weed in some form from Ben Johnson to Cibber the one fond of his pipe the other of his snuff.

In 1639 Venner published a volume entitled "A Treatise" concerning the taking of the fume of tobacco. His advice is "to take it moderately and at fixed times." Many of the clergy were devoted adherents of the pipe. Lilly says of its use among them:

"In this year Bredon vicar of Thornton a profound divine, but absolutely the most polite person for nativities in that age, strictly adhering to Ptolemy, which he well understood; he had a hand in composing Sir Christopher Heydon's defence of judicial astrology, being that time his chaplain; he was so given over to tobacco and drink, that when he had no tobacco, he would cut the bell-ropes and smoke them."

CHAPTER V.

TOBACCO IN EUROPE. (CONTINUED.)

NEANDER in his work "Tobacologia," (1622) gives a list of the various kinds of tobacco then used and where they were cultivated, among them are the following well known now as standard varieties of tobacco: Brazilian, St. Domingo, Orinoco, Virginia, and Trinidad tobacco. Fairholt says of the latter that it was most popular in England and is frequently named by early authors.* Tobacco when prepared for us was made into long rolls or large balls which often answered for the tobacconist's sign. What we now call cut tobacco was not as popular then as roll. Smokers carried a roll of tobacco, a knife and tinder to ignite their tobacco. At the close of the Sixteenth Century tobacco was introduced into the East. In Persia and Turkey where at first its use was opposed by the most cruel torture it gained at length the sanction and approval of even the Sultan himself. Pallas gives the following account in regard to its first introduction into Asia:

"In Asia, and especially in China, the use of tobacco for smoking is more ancient than the discovery of the New World, I too scarcely entertain a doubt. Among the Chinese, and among the Mongol tribes who had the most intercourse with them, the custom of smoking is so general, so frequent, and become so indispensable a luxury; the tobacco purse affixed to their belt, so necessary an article of dress; the form of the pipes from which the Dutch seem to have taken the model of theirs so original; and, lastly the preparation of the yellow leaves, which are merely rubbed to pieces and

---

* Neander says that Varinas tobacco was the best.

then put into the pipe, so peculiar, that we cannot possibly derive all this from America by way of Europe; especially as India, (where the habit of smoking is not so general,) intervenes between Persia and China. May we not expect to find traces of this custom in the first account of the Voyages of the Portugese and Dutch to China? To investigate this subject, I have indeed the inclination but not sufficient leisure."

We find by research that smoking was the most general mode of using tobacco in England when first introduced. In France the habit of snuffing was the most popular mode

TOBACCO AND THEOLOGY.

and to this day the custom is more general than elsewhere. In the days of the Regency snuff-taking had attained more general popularity than any other mode of using the plant leaves; the clergy were fond of the "dust" and carried the most expensive snuff boxes, while many loved the pipe and indulged in tobacco-smoking. The old vicar restored to his living enjoyed a pipe when seated in his chair musing on the subject of his next Sunday's discourse, "with a jug of sound old ale and a huge tome of sound old divinity on the table before him, for the occasional refreshment as well of the bodily as the spiritual man."

The cultivation of tobacco in Europe was begun in Spain and Portugal. Its culture in these kingdoms as well as by their colonies brought to the crown enormous revenues. In 1626, its culture began in France and is still an important product. A little later it began to be cultivated in Germany where it had already been used as a favorite luxury. From this time its use and cultivation extended to various parts of Europe. The Persecutors whether kings, popes, poets, or courtiers at length gave up their opposition while many of them joined in the use and spread of the custom. It has been said with much truth:

"History proves that persecution never triumphs in its attempted eradications. Tobacco was so generally liked that no legislative measures could prevent its use."

At first the use of tobacco was confined to fops and the hangers on at ale houses and taverns but afterwards by the "chief men of the realm." Soon after the importation of the "durned weed" from Virginia the tobacco muse gave forth many a lay concerning the custom. The following verses describe the method of smoking then in vogue:

> Nor did that time know
> To puff and to blow
> In a peece of white clay,
> As they do at this day
> With fier and coole,
> And a leafe in a hole;
> As my ghost hath late seen,
> As I walked betwene
> Westminister Hall
> And the church of St. Paul,
> And so thorow the citie
> Where I saw and did pitty
> My country men's cases,
> With fiery-smoke faces,
> Sucking and drinking
> A filthie weede stinking,
> Was ne'r known before
> Till the devil and the More
> In th' Indies did meete,
> And each other there greete
> With a health they desire,
> Of stinke, smoke and fier.
> But who e're doth abhorre it.
> The citie smookes for it;
> Now full of fier shop,
> And fowle spitttng chop,
> So sneezing and coughing,
> That my ghost fell to scoffing.
> And to myself said:
> Here's filthie fumes made;
> Good phisicke of force
> To cure a sicke horse."

The Puritans, from the first introduction of the plant, were sincere haters of tobacco, not only in England but in America. Cromwell had as strong a dislike of the plant as King James, and ordered the troopers to destroy the crops by trampling them under foot. Hutton describes a Puritan as one who

> "Abhors a sattin suit, a velvet cloak,
> And sayes tobacco is the Devill's smoke."

Probably no other plant has ever met with such powerful determined opposition, both against its use and cultivation, as the tobacco plant. It was strenuously opposed by all possible means, governmental, legislative, and literary. When tea and coffee were first introduced both were denounced in unmeasured terms, but the opposition was not so bitter or as lasting.

The following verses bearing the *nom de plume* of an "Old Salt," record much of the history of the plant:—

> "Oh muse! grant me the power
> (I have the will) to sing
> How oft in lonely hour,
> When storms would round me lower,
> Tobacco's prov'd a King!
>
> "Philanthropists, no doubt
> With good intentions ripe,
> Their dogmas may put out,
> And arrogantly shout
> The evils of the pipe.
>
> "Kind moralists, with tracts,
> Opinions fine may show:
> Produce a thousand facts—
> How ill tobacco acts
> Man's system to o'erthrow.
>
> "Learn'd doctors have employed
> Much patience, time and skill,
> To prove tobacco cloyed
> With acrid alkaloid,
> With power the nerves to kill
>
> "E'en Popes have curst the plant;
> Kings bade its use to cease;

TOBACCO GLORIFIED.

But all the Pontiff's rant
And Royal Jamie's cant
Ne'er made its use decrease.

"Teetotallers may stamp
And roar at pipes and beer;
But place them in a swamp,
When nights are dark and damp—
Their tune would change, I fear.

"No advocate am I
Of excess in one or t'other,
And ne'er essayed to try
In wine to drown a sigh,
Or a single care to smother.

"Yet, in moderation pure,
A glass is well enough;
But, a troubled heart to cure,
Kind feelings to insure,
Give me a cheerful puff.

"How oft a learn'd divine
His sermons will prepare,
Not by imbibing wine,
But, 'neath th' influence fine
Of a pipe of "baccy" rare!

"How many a pleasing scene,
How many a happy joke,
How many a satire keen,
Or problem sharp, has been
Evolved or born of smoke!

"How oft, amidst the jar
Of storms on ruin bent,
On ship-board, near or far,
To the drenched and shiv'ring tar
Tobacco's solace lent!

"Oh! tell me not 'tis bad,
Or that it shortens life.
Its charms can soothe the sad,
And make the wretched glad,
In trouble and in strife.

"'Tis used in every clime,
By all men, high and low;
It is praised in prose and rhyme,
So let the kind herb grow!

> " 'Tis a friend to the distress'd,
> 'Tis a comforter in need;
> It is social, soothing, blest;
> It has fragrance, force, and zest;
> Then hail the kingly weed!"

While Raleigh * and many of Elizabeth's courtiers indulged frequently in a pipe, some have imagined that even Queen Bess herself tested the rare virtues of tobacco. This is hardly based upon sufficient proof to warrant a very strong belief in it; but the following account of "How to weigh smoke" taken from *Tinsley's Magazine* shows that the Queen was acquainted at least with Raleigh's use of the weed:

"One day it happened that Queen Elizabeth, wandering about the grounds and alleys at Hampton with a single maid of honour, came upon Sir Walter Raleigh indulging in a pipe. Smoking now is as common as eating and drinking, and to smoke amongst ladies is a vulgarity. But not so then: it was an accomplishment, it was a distinction; and one of the feathers in Sir Walter's towering cap was his introduction of tobacco. The all-accomplished hero rose and saluted the Queen in his grand manner, and the Queen, who was in her daintiest humour, gave him her white hand to kiss, and took the seat he had left.

"Now, Sir Walter, I can puzzle you at last." "I suppose I must not be so rude as to doubt your Majesty." "You are bold enough for that, but your boldness will not help you, Sir Walter, this time. You cannot tell me how much the smoke from your pipe weighs." "Your Majesty is mistaken. I can tell you to a nicety. Will your Majesty allow me to call yonder page, and send for a pair of scales and weights?'.
"By my honour," said the Queen, "were any other subject in our realm to make request so absurd, we should very positively deny it. But you are the wisest of our fools, and, though we expect to see but little use made of these weights when brought, your request shall be granted. And, supposing you fail to weigh the smoke, what penalty will you pay?"
"I will be content," said Sir Walter, "to lose my head."
"You may chance to lose it on a graver count than this;" answered the Queen. "If the head shall have done some

---

* It is said that Raleigh in communicating the art to his friends, gave smoking parties at his house, where his guests were treated with nothing but a pipe, a mug of ale, and a nutmeg. Says an English writer: "From the anecdote related respecting the weight of smoke, the vapor of the pipe certainly did not throw a cloud over the brilliant wit of the unfortunate Raleigh."

slight service to your Majesty and the realm," replied the courteous knight, "thee will be well content nevertheless."

"But your Majesty will soon see that I fail not. First, madam, I place this empty pipe in the scales, and I find that it weighs exactly 2 ounces. I now fill it with tobacco, and the weight is increased to 2 1-10th ounce. I must now ask your Majesty to allow me to smoke the pipe out. I shall then turn out the ashes, and place them together with the pipe in the scale once more. The difference between the weight of the pipe with the unsmoked tobacco, and weight of the pipe with the ashes, will be the weight of the smoke."

WEIGHING SMOKE.

"You are too clever for us, Sir Walter. We shall expect you to-night at supper, and if the conversation grow dull, you shall tell our courtiers the story of the pipe."

Many other anecdotes have been told of the adventures of Raleigh with his pipe. One is that while taking a quiet smoke his servant entered and becoming alarmed on seeing the smoke coming from his nose threw a mug of ale in his face.

The same anecdote is also related of others including Tarlton. He gives an account of it in his Jests 1611. it is told in this manner:

"Tarlton as other gentlemen used, at the first coming up of tobacco, did take it more for fashion's sake than otherwise, and being in a roome, sat betweene two men overcome with wine, and they never seeing the like, wondered at it, and seeing the vapour come out of Tarlton's nose, cryed out, 'Fire, fire!' and threw a cup of wine in Tarlton's face. 'Make no more stirre,' quoth Tarlton, 'the fire is quenched; if the sheriffs come, it will turne a fine as the custom is.'

And drinking that againe, 'Fie,' says the other: 'what a stinke it makes. I am almost poysoned.' 'If it offend,' quoth Tarlton, 'let's every one take a little of the smell, and so the savor will quickly go;' but tobacco whiffes made them leave him to pay all."

Rich gives the following account of a similar scene:—

"I remember a pretty jest of tobacco which was this: A certain Welchman coming newly to London, and beholding one to take tobacco, never seeing the like before, and not knowing the manner of it, but perceiving him vent smoke so fast, and supposing his inward parts to be on fire, cried out, 'O Jhesu, Jhesu man, for the passion of Cod hold, for by Cod's splud ty snowt's on fire,' and having a bowle of beere in his hand, threw it at the other's face, to quench his smoking nose."

The following anecdote is equally ludicrous. Before tobacco was much known in Germany, some soldiers belonging to a cavalry regiment were quartered in a German village. One of them, a trumpeter, happened to be a negro. A peasant, who had never seen a black man before, and who knew nothing about tobacco, watched, though at a safe distance, the trumpeter, while the latter groomed and fed his horse. As soon as this business was dispatched, the negro filled his pipe and began to smoke it. Great had been the peasant's bewilderment before; great was his terror now. The terror reached an intolerable point when the negro took the pipe from his mouth, offered it to the peasant, and asked him, in the best language he could command, to take a whiff. "No, no!" cried the peasant, in exceeding alarm; "no, no! Mr. Devil; I do not wish to eat fire."

Henry Fielding, in "The Grub Street Opera" written about a century ago, has the following verses on Tobacco:—

"Let the learned talk of books,
The glutton of cooks,
The lover of Celia's soft smack—O!
No mortal can boast
So noble a toast,
As a pipe of accepted tobacco.

"Let the soldier for fame,
And a general's name,
In battle get many a thwack—O!

  Let who will have most
   Who will rule the rooste,
Give me but a pipe of tobacco.

  "Tobacco gives wit
   To the dullest old cit,
And makes him of politics crack—O!
   The lawyers i' th' hall
   Were not able to bawl,
Were it not for a whiff of tobacco.

  "The man whose chief glory
   Is telling a story,
Had never arrived at the smack—O!
   Between every heying,
   And as I was saying,
Did he not take a whiff of tobacco.

  "The doctor who places
   Much skill in grimaces,
And feels your pulse running tic tack—O!
   Would you know his chief skill?
   It is only to fill
And smoke a good pipe of tobacco.

  "The courtiers alone
   To this weed are not prone;
Would you know what 'tis makes them so slack—O?
   'Twas because it inclined
   To be honest the mind,
And therefore they banished tobacco."

One of the most curious pieces of verse ever written on tobacco is the following by Southey, entitled "Elegy on a Quid of Tobacco:"—

 "It lay before me on the close-grazed grass,
  Beside my path, an old tobacco quid:
 And shall I by the mute adviser pass
  Without one serious thought? now Heaven forbid!

 "Perhaps some idle drunkard threw thee there—
  Some husband spendthrift of his weekly hire;
 One who for wife and children takes no care,
  But sits and tipples by the ale-house fire.

"Ah! luckless was the day he learned to chew!
    Embryo of ills the quid that pleased him first;
Thirsty from that unhappy quid he grew,
    Then to the ale-house went to quench his thirst.

"So great events from causes small arise—
    The forest oak was once an acorn seed;
And many a wretch from drunkenness who dies,
    Owes all his evils to the Indian weed.

"Let no temptation, mortal, ere come nigh!
    Suspect some ambush in the parsley hid;
From the first kiss of love ye maidens fly,
    Ye youths, avoid the first Tobacco-quid!

"Perhaps I wrong thee, O thou veteran chaw,
    And better thoughts my musings should engage;
That thou wert rounded in some toothless jaw,
    The joy, perhaps of solitary age.

"One who has suffered Fortune's hardest knocks,
    Poor, and with none to tend on his gray hairs;
Yet has a friend in his Tobacco-box,
    And, while he rolls his quid, forgets his cares.

"Even so it is with human happiness—
    Each seeks his own according to his whim;
One toils for wealth, one Fame alone can bless,
    One asks a quid—a quid is all to him.

"O, veteran chaw! thy fibres savory, strong,
    While aught remained to chew, thy master chewed,
Then cast thee here, when all thy juice was gone,
    Emblem of selfish man's ingratitude!

"O, happy man! O, cast-off quid! is he
    Who, like as thou, has comforted the poor;
Happy his age who knows himself, like thee,
    Thou didst thy duty—man can do no more."

Another well known song of the Seventeenth Century is entitled "The Tryumph of Tobacco over Sack and Ale:"—

"Nay, soft by your leaves,
    Tobacco bereaves
You both of the garland; forbear it;
    You are two to one,
    Yet tobacco alone
Is like both to win it, and weare it.

TRIUMPH OF TOBACCO.

Though many men crack,
Some of ale, some of sack,
And think they have reason to do it;
Tobacco hath more
That will never give o'er
The honor they do unto it.
Tobacco engages
Both sexes, all ages,
The poor as well as the wealthy;
From the court to the cottage,
From childhood to dotage,
Both those that are sick and the healthy.
It plainly appears
That in a few years
Tobacco more custom hath gained,
Than sack, or than ale,
Though they double the tale
Of the times, wherein they have reigned.
And worthily too,
For what they undo
Tobacco doth help to regaine,
On fairer conditions
Than many physitians,
Puts an end to much griefe and paine;
It helpeth digestion,
Of that there's no question,
The gout and the tooth-ache it easeth:
Be it early, or late,
'Tis never out of date,
He may safely take it that pleaseth.
Tobacco prevents
Infection by scents,
That hurt the brain, and are heady.
An antidote is,
Before you're amisse,
As well as an after remedy.
The cold it doth heate,
Cools them that do sweate,
And them that are fat maketh lean:
The hungry doth feed,
And if there be need,
Spent spirits restoreth again.
The poets of old,
Many fables have told,
Of the gods and their symposia;
But tobacco alone,
Had they known it, had gone

> For their nectar and ambrosia.
>   It is not the smack
>   Of ale or of sack,
> That can with tobacco compare:
>   For taste and for smell,
>   It beares away the bell
> From them both, wherever they are:
>   For all their bravado,
>   It is Trinidado,
> That both their noses will wipe
>   Of the praises they desire,
>   Unless they conspire
> To sing to the tune of his pipe.

The history of the rise and progress of tobacco in England, is one of the most interesting features connected with the use and cultivation of the plant. In Spain, Portugal, Germany and Holland the plant was sustained and encouraged by the throne, and royalty was the strongest and most devoted defender it had. It saw in the encouragement of its use, an income of revenue and a source of profit far greater than that received from any other product. Soon after its cultivation began in France, Spain, and Portugal, the tobacco trade was farmed out.

From its first cultivation in these countries it has been a government monopoly. In 1753, the King of Portugal farmed out the tobacco trade, and from that time until now, the annual amount received has been one of the principal sources of revenue to the crown. In France, as early as 1674, a monopoly of the trade was granted to Jean Breton for six years, for the sum of 700,000 francs.

In 1720 the Indian Company paid for the privilege 1,500,000 francs per annum; and in 1771 the price was increased to 25,000,000 francs. Besides France there are thirteen other European states where the tobacco trade is a government monopoly, namely, Austria, Spain, Sicily, Sardinia, Poland, Papal States, Portugal, Tuscany, Modena, Parma, San Marino, Lichtenstein.

From the first cultivation of the plant, its growers saw in the tobacco trade a vast and constantly increasing source of wealth. They doubtless in some measure comprehended the close relation existing between it and commerce and realized how extensive would be its use.

From the nature of the plant, it affords states and nations an opportunity to engage either in its culture or commerce with the prospect of the largest success. In this respect it is far different from any other tropical plant, and unlike them is capable of being cultivated in portions of the earth far remote from the tropics. In Switzerland and in the Caucassias it attains to a considerable size, but is nevertheless tobacco although it may possess but few of the excellences of some varieties, still it affords some enjoyment to the user, from the fact that it is the Indian weed. Fairholt speaking of the tobacco trade says:

"The progress of the tobacco trade from the earliest introduction of the plant into Europe until now, is certainly one of the most curious that commerce presents. That a plant originally smoked by a few savages, should succeed in spite of the most stringent opposition in church and state, to be the cherished luxury of the whole civilized world; to increase with the increase of time, and to end in causing so vast a trade, and so large an outlay of money; is a statistical fact, without an equal parallel."

The tobacco plant notwithstanding its fascinating powers, has suffered many romantic vicissitudes in its fame and character; having been successively opposed and commended by physicians, condemned and eulogized by priests, vilified and venerated by kings, and alternately proscribed and protected by governments, this once insignificant production of a little island or an obscure district, has succeeded in diffusing itself throughout every clime, and—exhilarating and enriching its thousands—has subjected the inhabitants of every country to its dominion. And every where it is a source of comfort and enjoyment; in the highest grades of civilized society, at the shrine of fashion, in the depths of poverty, in the palace and in the cottage, the fascinating influence of this singular plant demands an equal tribute of devotion and attachment.

CHAPTER VI.

TOBACCO-PIPES, SMOKING AND SMOKERS.

THE implements used in smoking tobacco, from the rude pipe of the Indian to the elaborate hookah of the Turk, show a far greater variety than even the various species of the tobacco plant. The instruments used by the Indians for inhaling the tobacco smoke were no less wonderful to Europeans than the plant itself.

The rude mode of inhaling the smoke and the intoxication produced by its fumes suggested to the Spaniards a better method of "taking tobacco." Hariot, however, found clay pipes in use by the Indians of Virginia, which though having no resemblance to the smoking implements discovered by Columbus, seem to have afforded a model for those afterward manufactured by the Virginia colony. The sailors of Columbus seemed to have first discovered cigar, rather than pipe-smoking, inasmuch as the simple method used by the natives, consisted of a leaf of maize, which enwrapped a few leaves of the plant.

The next instruments discovered in use among the Indians were straight, hollow reeds and forked canes. Their mode of use was to place a few leaves upon coals of fire and by placing the forked end in the nostrils and the other upon the smoking leaves, to inhale the smoke until they were stupified or drunken with the fumes. Their object in inhaling the fumes of tobacco seemed to be to produce intoxication and insensibility rather than a mode of enjoyment, although the enjoyment with them consisted of seeing the most remarkable visions when stupefied by its fumes. Such were the modes of smoking among the Indians when Columbus planted the banner of Spain in America.

A writer in *The Tobacco Plant* has given a very interesting description of Indian pipes in use among the natives of both North and South America. He says:

"In the tumuli or Indian grave mounds of the Ohio and Scioto valleys, large quantities of pipes have been found, bearing traces of Indian ingenuity. That their burial mounds are of great antiquity, is proved by the fact that trees several centuries old are to be found growing upon them. About twenty-five years ago, two distinguished archeologists Squier and Davis—made extensive exploration of these mounds, the results of which were published in an elaborate memoir by the Smithsonian Institution. The mounds indicate that an immense amount of labor has been expended upon them, as the earthworks and mounds may be counted by thousands, requiring either long time or an immense population; and there is much probability in the supposition of Sir John Lubbock that these parts of America were once inhabited by a numerous and agricultural population. It may be asked, have the races who erected these extensive mounds become extinct, or do they exist in the poor uncivilized tribes of Indians whom Europeans found inhabiting the river valleys of Ohio and Illinois? Many of these mounds are in the form of serpents and symbolic figures, and were evidently related to the sacrificial worship of the mound builders."

Squier and Davis are of the opinion that:—

"The mound builders were inveterate smokers, if the great numbers of pipes discovered in the mounds be admitted as evidence of the fact. These constitute not only a numerous, but a singularly interesting class of remains. In their construction the skill of the maker seems to have been exhausted. Their general form, which may be regarded as the primitive form of the implement, is well exhibited in the accompanying sketch. They are always carved from a single piece, and consist of a flat carved bore of variable length and width, with the bowl rising from the centre of the convex side. From one of the ends, and communicating with the hollow of the bowl, is drilled a small hole, which answers the purpose of a tube; the corresponding opposite division being left for the manifest purpose of holding the implement to the mouth.

"The specimen here represented is finely carved from a

beautiful variety of brown porphyry, granulated with various-colored materials, the whole much changed by the action of fire, and somewhat resembling porcelain. It is intensely hard, and successfully resists the edge of the finest-tempered knife.

INDIAN PIPE.

The length of the base is five inches; breadth of the same one inch and a-quarter. The bowl is one inch and a-quarter high, slightly tapering upwards, but flaring near the top. The hollow of the bowl is six-tenths of an inch in diameter. The perforation answering to the tube is one-sixth of an inch in diameter, which is about the usual size. This circumstance places it beyond doubt that the mouth was applied directly to the implement, without the intervention of a tube of wood or metal."

This is an account of a simple pipe, with a small bowl; but most of the pipes found in the mounds are highly ornamented with elaborate workmanship, representing animals such as the beaver, otter, bear, wolf, panther, raccoon, squirrel, wild-cat, manotee, eagle, hawk, heron, swallow, paroquet, etc. One of the most interesting of the spirited sculptures of animal forms to be found on the mound pipes, is the representation of the Lamantin, or Manotee, a cetacean found only in tropical waters, and the nearest place which they at present frequent is the coast of Florida—at least a thousand miles away. According to Sir John Lubbock, these are no rude sculptures, for the characteristics of the animal are all distinctly marked, rendering its recognition complete. Many modern Indians are possessed of a wonderful aptitude for sculpture, and they appear to gladly exchange their work for the necessaries of life.

The material most prized for the purpose of pipe-making is the beautiful red pipe-stone of the Coteau des Prairies, which is an indurated aluminous stone, highly colored with red oxide of iron. It is frequently called "Catlinite," out of compliment to George Catlin, the distinguished collector of Indian traditions, who claims to be the first European that ever visited the Red Pipe-stone Quarry, which is situated amongst the upper waters of Missouri. Catlin gives the following legend as the Indian version of the birth of the mysterious red pipe:—

"The Great Spirit, at an ancient period, here called together the Indian warriors, and standing on the precipice of the red pipe-stone rock, broke from its wall a piece, and made a huge pipe by turning it in his hand, which he smoked over them, and to the north, the south, the east and the west; and told them that this stone was red, that it was their flesh, that they must use it for their pipes of peace, that it belonged to them all, and that the war club and the scalping knife must not be raised on its ground. At the last whiff of his pipe his head went into a great cloud, and the whole surface of the rock for several miles was melted and glazed. Two great ovens were opened beneath, and two women, guardian spirits of the place, entered them in a blaze of fire, and they are heard there yet, answering to the invocations of the priests and medicine-men."

At the pipe-stone quarry there is a row of five huge, granite boulders, which the Indians regard with great reverence, and when they visit the spot to secure some red stone to make pipes, they seek to propitiate the guardian spirits by throwing plugs of Tobacco to them. Some admirable pieces of pipe-sculpture are produced by the Boheen Indians, who are found on the coast of the Pacific to the south of the Russians. These pipes are made from a soft blue clay stone which is found only in slabs, and the sculptures are wrought on both sides, the pipes being generally covered with singular groups of human and animal forms, grotesquely intermingled.

The Chippewas are also celebrated for their pipes, which are cut out of a close-grained stone of a dark color; and Professor Wilson, of Toronto, states that Pobahmesad, or the Flier, one of the famed pipe-sculptors, resides on the Great Manitoulin Island in Lake Huron. The old Chippewa has never deviated from the faith of his fathers, as he still adheres to all their rites and ceremonies. He uses the red pipe-stone and other materials in the production of his pipes, which are ingenious specimens of sculpture. The calumet, or pipe of peace, is still an object of special reverence with the Indian tribes, and the pipe-stem is ornamented with six or eight eagle's

SCULPTURED PIPE.

feathers. Each tribe has an official who takes charge of the calumet, which he keeps rolled up in a bearskin robe; and it's never exposed to view or used, except when the chief enters into a treaty with some neighboring chief. On these occasions the pipe is taken out of its covering by the Indian dignitary, ready charged with the "holy weed," when it is smoked by all the chiefs, each one taking only a single breath of smoke, which is regarded as implementing the treaty. The pipe is then rolled up in its robe of fur, and stowed away in the lodge of its keeper until it is again required. The war pipe is simply a tomahawk, with a perforated handle communicating with the bowl, which is opposite the sharp edge of the weapon. When the Indians joined the British as allies during the American war, they had to be supplied with iron tomahawks of the native pattern, before they could enter the field as allies.

Many tribes of Indians use herbs of various kinds to mix with tobacco to reduce its strength, as they are in the habit of exhaling the smoke from the nostrils, and not from the mouth. By the adoption of this means a much smaller quantity of tobacco suffices to produce the soothing influence on the nervous system so well known to votaries of the weed.

Longfellow, in his great Indian epic of the Song of Hiawatha, has portrayed with graphic power in pleasing verse the mysterious legends describing the birth or institution of the peace-pipe by Gitche Manito, "The Master of Life;" and a few extracts from "Hiawatha" may be interesting to illustrate the deep significance of the ideas which the Indian holds regarding his relations to the Great Spirit of the Universe, and of the esteem with which he views the peace-pipe, which in the words of Catlin "has shed its thrilling fumes over the land, and soothed the fury of the relentless savage."

Longfellow, in the opening of his poem, says:—

> "Ye whose hearts are fresh and simple,
> Who have faith in God and Nature,
> Who believe that in all ages
> Every human heart is human,
> That in even savage bosoms
> There are longings, yearnings, strivings,
> For the good they comprehend not,
> That the feeble hands and helpless,
> Groping blindly in the darkness,
> Touch God's right hand in that darkness
> And are lifted up and strengthened;—
> Listen to this simple story,
> To the song of Hiawatha.

He then describes the making of the pipe from the great Red Pipe-Stone Quarry, as follows:—

> " On the Mountains of the Prairie,
>   On the great Red Pipe-Stone Quarry,
>   Gitche Manito, the mighty,
>   He the Master of Life, descending,
>   On the red crags of the quarry
>   Stood erect, and called the nations,
>   Called the tribes of men together.
>   From his foot-prints flowed a river,
>   Leaped into the light of morning,
>   O'er the precipice plunging downward
>   Gleamed like Ishkoodah, the comet.
>   And the Spirit stooping earthward,
>   With his finger on the meadow
>   Traced a winding pathway for it,
>   Saying to it, 'Run in this way!'

> " From the red stone of the quarry
>   With his hand he broke a fragment,
>   Moulded it into a pipe-head,
>   Shaped and fashioned it with figures;
>   From the margin of the river

MAKING THE "PEACE-PIPES."

Took a long reed for a pipe-stem,
With its dark green leaves upon it;
Filled the pipe with bark of willow;
With the bark of the red willow;
Breathed upon the neighboring forest,
Made its great boughs chafe together,
Till in flame they burst and kindled;
And erect upon the mountains,
Gitche Manito, the mighty,
Smoked the calumet, the Peace-Pipe,
As a signal to the nations."

PIPE OF PEACE.

The next verses describe the assembling of the nations at the call of Gitche Manito, who proceeds to speak to his children words of wisdom and announces that he:

"'Will send a prophet to you,
A Deliverer of the nations,
Who shall guide you and shall teach you,
Who shall toil and suffer with you.
So you listen to his counsels,
You will multiply and prosper;
If his warnings pass unheeded,
You will fade away and perish!

"'Bathe now in the stream before you,
Wash the war-paint from your faces,
Wash the blood-stains from your fingers,
Bury your war-clubs and your weapons,
Break the red stone from this quarry,
Mould and make it into Peace-Pipes,
Take the reeds that grow beside you,
Deck them with your highest feathers,

>   Smoke the calumet together,
>   And as brothers live henceforward!'
>
>   * * * *
>
>   "And in silence all the warriors
>   Broke the red stone of the quarry,
>   Smoothed and formed it into Peace-Pipes,
>   Broke the long reeds by the river,
>   Decked them with their brightest feathers,
>   And departed each one homeward,
>   While the Master of Life, ascending
>   Through the opening of cloud curtains,
>   Through the doorways of the heavens,
>   Vanished from before their faces,
>   In the smoke that rolled around him,
>   The Pukwana of the Peace-Pipe!"

Along the northern parts of America, are to be found the Esquimaux population, estimated to number about 60,000.

They are votaries of the weed, making their pipes either out of driftwood, or of the bones of animals they have used for food.

Tobacco is found growing along the whole western seaboard of South America until we reach the northern boundaries of Patagonia. Far inland on the banks of the Amazon, Rio Niger, and other great rivers, the weed has been found in luxurious abundance, with a delightful fragrance.

Stephens, in his "Travels in Central America," says that "the ladies of Central America generally smoke—the married using tobacco, and the unmarried, cigars formed of selected tobacco rolled in paper or rice straw. Every gentleman carries in his pocket a silver case, with a long string of cotton, steel and flint, and one of the offices of gallantry is to strike a light. By doing it well, he may help to kindle a flame in a lady's heart; at all events, to do it bunglingly would be ill-bred. I will not express my sentiments on smoking as a custom for the sex. I have recollections of beauteous lips profaned. Nevertheless, even in this I have seen a lady show her prettiness and refinement, barely touching the straw on her lips, as it were kissing it gently and taking it away. When a gentleman asks a lady for a light, she always removes the cigar from her lips."

The Rev. Canon Kingsley, in his fascinating novel of "Westward Ho!" has some quaint remarks on the method

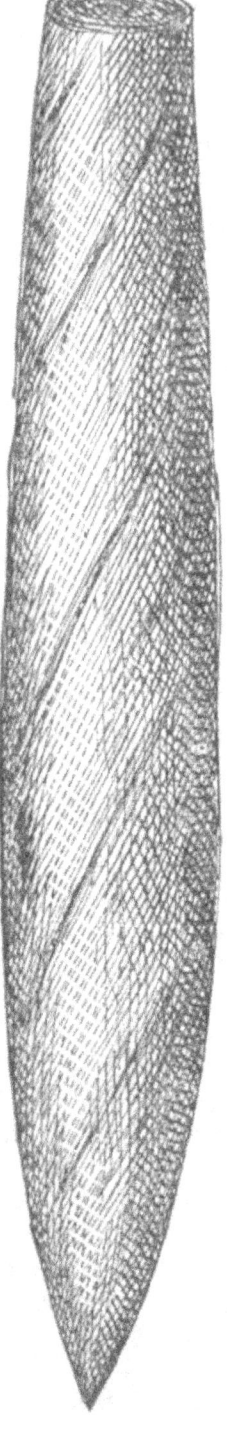

of smoking described by Lionel Wafer, surgeon to Dampier, which are well worth quoting. He says, "When they, (the Darien Indians,) will deliberate on war or policy, they sit round in the hut of the chief; where being placed, enter to them a small boy with a cigarro of the bigness of a rolling-pin, and puffs the smoke thereof into the face of each warrior, from the eldest to the youngest; while they, putting their hands funnel-wise round their mouths, draw into the sinuosities of the brain that more than Delphic vapor of prophecy; which boy presently falls down in a swoon, and being dragged out by the heels and laid by to sober, enter another to puff at the sacred cigarro, till he is dragged out likewise, and so on till the Tobacco is finished, and the seed of wisdom has sprouted in every soul into the tree of meditation, bearing the flower of eloquence, and in due time the fruit of valiant action."

Tobacco in the form of cigarettes, is extensively used by the inhabitants of Nicaragua, Guiana, and the dwellers on the banks of the Orinoco, and the use of the weed is not confined to the male sex, but is freely used both by the female and juvenile portions of the community. Mr. Squier, in his "Travels in Nicaragua," states that the dress of the young urchins consists mainly of a straw hat and a cigar—the cigar when not in use being stuck behind the ear, in the manner in which our clerks place their pens. The natives of Guiana use a tube or pipe not unlike a cheroot, made from the rind of the fruit of a species of palm. This curious pipe is called a "Winna," and the hollow is filled with tobacco, the smoking of which affords much enjoyment to the denizens of the swampy regions of Guiana.

Mr. Cooke, in "The Seven Sisters of Sleep," states that a tube much resembling the "Winna" of Guiana was some years ago to be met with in the Tobacconists' Shops in London. The Indian dwelling in the dense forests in the region of Orinoco has found that tobacco is an excellent solace to relieve the monotony of his life; he uses it "not only to procure an afternoon nap, but also to induce a state of quiescence which they call dreaming with their eyes open." We find from voyagers up the Amazon, that smoking prevails not merely amongst the natives inhabiting the regions which skirt that great river, but also amongst the people on the banks of its numerous tributaries. Mr. Bates the distinguished Naturalist, when making researches far up one of the tributaries of the Amazon, found tobacco extensively cultivated, and some distinguished makers of cigarettes. One maker, Joan Trinidade, was noted for his Tobacco and Tauri cigarettes. This cigar is so named from the bark in which the tobacco is rolled. Some of the tribes inhabiting the district of the lower Amazon indulge in snuff-taking. This snuff is not made from tobacco, it is the produce of a plant of the leguminous order, the seeds being carefully collected and thoroughly dried in the sun before they are pounded in a mortar, when the powder is ready for use. The snuff-making season is quite an event in a Brazilian village, the week or so during which it lasts forming a kind of religious festival mingled with a good deal of indulgence in fermented liquors, chiefly of native origin.

Humboldt, when traveling in South America, found in use among the Ottomac Indians a powder called Niopo, or "Indian snuff." Niopo is a powerful stimulant, a small portion of it producing violent sneezing in persons unaccustomed to its use. Father Gumilla says:—"This diabolical powder of the Ottomacs, furnished by an adolescent tobacco plant, intoxicates them through the nostrils, deprives them of reason for some hours, and renders them furious in battle." Humboldt, however, has shown that this stimulating snuff is not the product of the tobacco plant, but of a species of acacia, Niopo being made from the pods of the plant after they have undergone a process of fermentation. Captain Burton, when traveling in the Highlands of Brazil, found the tobacco plant growing spontaneously, which made him conclude that it is indigenous to Brazil. He found the "Aromatic Brazilian" a kind of tobacco with thin leaves and a pink flower, which is "much admired in the United States, and there found to lose its aroma after the second year." It is usually asserted that the tobacco grown in Brazil contains only two per cent. of nicotine, but Captain Burton is disposed to doubt this, as he states that some varieties of the "holy herb" grown at Sa'a' Paulo and Nimos suggests a larger proportion. In the small towns in the Highlands of Brazil, Captain Burton found that excellent cigars, better than many "Havannas," were retailed at a halfpenny each. In La Plata, Paraguay, and other countries to the south of Brazil, nearly every person smokes, and an American traveler quoted by Mr. Cooke states that women and girls above thirteen years of age use the weed in the form of quids. A magnificent Hebe, arrayed in satin and flashing in diamonds, "puts you back with one delicate hand, while with the fair taper fingers of the other she takes the tobacco out of her mouth previous to your saluting her." A European visiting Paraguay for the first time is rather astonished at the conduct of the fair beauty, but such is the force of custom that the squeamishness of the new-comer is soon overcome, when he finds that he has to kiss every lady to whom he is introduced; and the traveler says that "one half of those you meet are really tempting enough to render you reckless of consequences."

Smoking is practised by the natives of Patagonia, who are a tall and muscular class of men, though not such giants as represented by the early voyagers. Hutchinson, in a valuable paper on the Indians of South America has an account of the Pehuenches, one of the principal tribes of Patagonia, in which he states that "their chief indulgence is smoking. The native pipes are fabricated out of a piece of stone, fashioned into the shape of a bowl, into which is inserted a long brass tube. The latter is obtained by barter at Bohia Blanca. The tobacco in the bowl being lighted, each man of a party takes a suck at the pipe in his turn." Tilston, who witnessed the operation, describes it as a most ludicrous one. "The smoker gives a pull at the pipe, gulping in a quantity of Tobacco vapour, the cubic measurement of which my informant would be afraid to guess at. All the muscles of the body seem in a temporary convulsion whilst it is being taken in, and the neighbour to whom the pipe is transferred follows suit by inhaling as if he were trying to swallow down brass tube, bowl, Tobacco, fire, and all. Meanwhile, there issues from the nose and mouth of the previous smoker such a cumulus of cloud as for a few seconds to render his face quite invisible." Tobacco is more used in Chili than in the other countries on the Pacific side of South America; this is owing to the extensive use of the leaves of the Cocoa plant as a narcotic by the natives of Bolivia, Peru; and Colombia.

We refrain from enlarging on the nature and use of this narcotic, as on some future occasion we may take an opportunity of making some observations on Cocoa, which according to Jonson, holds an undisputed sway over some seven or

SOUTH AMERICANS SMOKING.

eight millions of the inhabitants of South America. The Indians formerly inhabiting the high table-lands of what is now called Peru and Bolivia appear prior to the invasion of the Spaniards to have been much further advanced in civilization than the races occupying the other portions of South America; and there is a strong probability that they are of a different origin from the races occupying Chili, Patagonia, Brazil, and the great district washed by the waters of the West Indian Sea. Science as yet cannot give anything like an accurate idea of the time man has existed in these widely-diversified countries, but we cannot go wrong in accepting the statement of Darwin, who observes that " we must admit that man has inhabited South America for an immensely long period, inasmuch as any change in climate, effected by the elevation of the land must have been extremely gradual."

Another writer says of the pipes of the Indians of North America:

"Great variety of form and material distinguishes the pipes of the modern Indians; arising in part from the local facilities they possess for a suitable material from which to construct them; and in part also from the special style of art and decoration which has become the traditional usage of the tribes. The favorite red pipe-stone of the Coteau des Prairies, has been generally sought after, both from its easiness of working and the beauty of its appearance. A pipe of this favorite and beautiful material, found on the shores of Lake Simcoe, and now in my possession, measures five inches and three-quarters in length, and nearly four inches in greatest breadth, yet the capacity of the bowl hollowed in it for the reception of tobacco is even less than in the smallest of the "Elfin Pipes." In contrast to this, a modern Winnebago pipe recently acquired by me, made of the same red pipe-stone, inlaid with lead, and executed with ingenious skill, has a bowl of large dimensions illustrative of Indian smoking usages modified by the influence of the white man. From the red pipe-stone, as well as from the lime stone and other harder rocks, the Chippeways, the Winnebagos, and the Sioux, frequently make a peculiar class of pipes, inlaid with lead.

"The Chincok and Puget Sound Indians, who evince little taste in comparison with the tribes surrounding them, in ornamenting their persons or their warlike and domestic implements, commonly use wooden pipes. Sometimes these are elaborately carved, but most frequently they are rudely and hastily made for immediate use; and even among these remote tribes of the flat head Indians, the common clay pipe of the fur trader begins to supersede such native arts. Among the Assinaboin Indians a material is used in pipe manufacture altogether peculiar to them. It is a fine marble, much too hard to admit of minute carving, but taking a high polish. This is cut into pipes of graceful form, and made so extremely thin, as to be almost transparent, so that when lighted the glowing tobacco shines through, and presents a singular appearance when in use at night or in a dark lodge. Another favorite material employed by the Assinaboin Indians is a coarse species of jasper also too hard to admit of elaborate ornamentation."

This also is cut into various simple but tasteful designs, executed chiefly by the slow and laborious process of rubbing it down with other stones. The choice of the material for fashioning the favorite pipe is by no means invariably guided by the facilities which the location of the tribe affords. A suitable stone for such a purpose will be picked up and carried hundreds of miles. Mr. Kane informs me that, in coming down the Athabaska River, when drawing near its source in the Rocky Mountains, he observed his Assinaboin guides select the favorite bluish jasper from among the water-worn stones in the bed of the river, to carry home for the purpose of pipe manufacture, although they were then fully five hundred miles from their lodges. Such a traditional adherence to a choice of material peculiar to a remote source, may frequently prove of considerable value as a clue to former migrations of the tribes. Both the Cree and the Winnebago Indians carve pipes in stone of a form now more frequently met with in the Indian curiosity stores of Canada and the States than any other specimens of native carving. The tube, cut at a sharp right angle with the cylindrical bowl of the pipe, is ornamented with a thin vandyked ridge, generally perforated with a row of holes, and standing up somewhat like the dorsal fin of a fish. The Winnebagos also manufacture pipes of the same form, but of a smaller size, in lead, with considerable skill.

Among the Cree Indians a double pipe is occasionally in

use, consisting of a bowl carved out of stone without much attempt at ornament, but with perforations on two sides, so that two smokers can insert their pipe-stems at once, and enjoy the same supply of tobacco. It does not appear, however, that any special significance is attached to this singular fancy. The Saultaux Indians, a branch of the great Algonquin nation, also carve their pipes out of a black stone found in their country, and evince considerable skill in the execution of their elaborate details. But the most remarkable of all the specimens of pipe sculpture executed by the Indians of the north-west are those carved by the Bobeen, or Big-lip Indians,—so called from the singular deformity they produce by inserting a piece of wood into a slit made in the lower lip.

The Bobeen Indians are found along the Pacific coast, about latitude 54°, 40′, and extend from the borders of the Russian dominions eastward nearly to Frazer River. The pipes of the Bobeen, and also of the Clalam Indians, occupying the neighboring Vancouver's Island, are carved with the utmost elaborateness and in the most singular and grotesque devices, from a soft blue clay-stone or slate. Their form is in part determined by the material, which is only procurable in thin slabs, so that the sculptures, wrought on both sides, present a sort of double bas-relief. From this, singular and grotesque groups are carved without any apparent reference to the final destination of the whole as a pipe. The lower side is generally a straight line, and in the specimens I have examined they measure from two or three to fifteen inches long; so that in these the pipe-stem is included. A small hollow is carved out of some protruding ornament to serve as the bowl of the pipe, and from the further end a perforation is drilled to connect with this. The only addition made to it when in use is the insertion of a quill or straw as a mouth-piece. The Indians have both war and peace pipes.

The War pipe is a true tomahawk of ordinary size with a perforated handle the tobacco being placed in the receptacle above the hatchet the handle serving as a pipe-stem and used for either pipe or tomahawk. Many varieties of Indian Pipes have been found not only in the Western and Southern mounds but in Mexico and Central America. Fine specimens are found in Florida and some elaborately carved have been unearthed in Virginia. Wilson says of the pipes used by the Indians: "The pipe stem is one of the characteristics of modern race, if not distinctive of the Northern tribes of Indians." In alluding to the pipes more particularly he says: "Specimens of another class of clay pipes of a larger size, and with a tube of such length as obviously to be designed for use without the addition of a "pipe-stem," most of the ancient clay pipes that have been discovered are stated to have the same form; and this, it may be noted, bears so near a resemblance to that of the red clay pipe used in modern Turkey, with the cherry-tree pipe stem, that it might be supposed to have furnished the model.

A WAR PIPE.

The bowls of this class of ancient clay pipes are not of the miniature proportions which induce a comparison between those of Canada and the early examples found in Britain; neither do the stone pipe-heads of the mound-builders suggest by the size of the bowl either the self-denying economy of the ancient smoker, or his practice of the modern Indian mode of exhaling the fumes of the tobacco, by which so small a quantity suffices to produce the full narcotic effects of the favorite weed. They would rather seem to confirm the indications derived from the other sources, of an essential difference between the ancient smoking usages of Central America and of the mound-builders, and those which are still maintained in their primeval integrity among the Indians of the North West.

Of the mound-builders Foster says:

"The mound-builders were well aware of the narcotic properties of tobacco, a plant which indigenous to America, and which since the discovery of the western continent has been domesticated in every region of the earth where the soil and climate are favorable to its cultivation. No habit at this day, it may be said, is more universal or more difficult to eradicate than that of smoking. With the mound-builder tobacco was the greatest of luxuries; his solace in his hours of relaxations, and the choicest offering he could dedicate to the Great Spirit. Upon his pipe he lavished all the skill he possessed in the lapidary's art.

> "From the red stone of the quarry
> With his hand he broke a fragment
> Moulded it into a pipe head
> Shaped and fashioned it with figures."

Many of these pipes are sculptured from the most obdurate stones and display great delicacy of workmanship. The features of animals are so truthfully cut that often there is no difficulty in their identification, and even the plumage of birds is delineated by curved or straight lines which show a close adherence to nature. The bowl and stem piece wrought from a single block, are as accurately drilled as they could be at this day, by the lapidary's art. Both the War pipe and Peace pipe are the most sacred and the most highly valued of all the various kinds.

"The calumet, or pipe of peace, ornamented with the war eagles quill, is a sacred pipe, and never used on any other occasion than that of peace making, when the chief brings it into treaty, and unfolding the many bandages which are carefully kept around it, has it ready to be mutually smoked by the chiefs, after the terms of the treaty are agreed upon, as the means of solemnizing it; which is done by passing the sacred stem to each chief, who draws one breath of smoke only

PEACE PIPE.

through it. Nothing can be more binding than smoking the pipe of peace and is considered by them to be an inviolable pledge. There is no custom more uniformly in constant use amongst the poor Indians than that of smoking nor any more highly valued. His pipe is his constant companion through life—his messenger of peace; he pledges his friends through its stem and its bowl, and when its care-drowning fumes cease to flow, it takes a place with him in his solitary grave with his tomahawk and war-club companions to his long-fancied 'happy hunting grounds.'"

From specimens of clay pipes found at the South from Virginia to Florida it would seem that the Indians had a great variety of pipes some of which were beautifully carved while others are perfectly plain. Many of them however are of rude workmanship and might have been fashioned by some of the tribe unacquainted with pipe-making.

Dall gives the following account of smoking among the natives of Alaska:

"We broke camp about five o'clock in the morning. Nothing occurred to break the monotony of constant steady plodding. Two Indians in the bow of the boat would row until tired, and then we would stop for a few minutes to rest, and let them smoke. The last operation takes less than a minute; their pipes are so constructed as to hold but a very small pinch of tobacco. The bowl, with ears for tying it to the stem is generally cast out of lead. Sometimes it is made of soft stone, bone or even hard wood. The stem is made of two pieces of wood hollowed on one side, and bound to the bowl and each other by a narrow strip of deerskin. In smoking the economical Indian generally cuts up a little birch wood, or the inner bark of the poplar, and mixes it with his tobacco. A few reindeer hairs pulled from his paska, are rolled into a little ball, and placed in the bottom of the bowl to prevent the contents from being drawn into the stem. A pinch of tobacco cut as fine as snuff is inserted and two or three whiffs are afforded by it.

The smoke is inhaled into the lungs, producing a momentary stupor and the operation is over. A fungus which grows on decayed birch trees, or tinder manufactured from the down of the poplar rubbed up with charcoal is used with flint and steel for obtaining a light. Matches are highly valued and readily purchased. The effect of the Circassian tobacco on the lungs is extremely bad, and among those tribes who use it many die from asthma and congestion of the lungs. This is principally due to the saltpetre with which it is impregnated. The Indian pipe is copied from the Eskimo, as the latter were the first to obtain and use tobacco. Many of the tribes call it by the Eskimo name.

The Kutchin and Eastern Finneh were modeled after the clay pipes of the Hudson Bay Company, but they also carve very pretty ones out of birch knots and the root of the wild rose-bush. The Chukchees use a pipe similar to those of the Eskimo, but with a much larger and shorter stem. This stem is hollow, and is filled with fine birch shavings. After smoking for some months these shavings impregnated with the oil of tobacco, are taken out through an opening in the lower part of the stem and smoked over. The Hudson Baymen make passable pipe-stems by taking a straight-grained piece of willow or spruce without knots, and cutting through the outer layers of bark and wood. This stick is heated in the ashes and by twisting the end in contrary directions the heart-wood may be gradually drawn out, leaving a hollow tube.

The Kutchin make pretty pipe-stems out of goose-quills wound about with porcupine-quills. It is the custom in the English forts to make every Indian who comes to trade, a present of a clay pipe filled with tobacco. We were provided with cheap brown ones, with wooden stems, which were much liked by the natives, and it is probable that small brier-wood pipes, which are not liable to break, would form an acceptable addition to any stock of trading goods". The Tchuktchi of north-eastern Asia are devoted worshipers of tobacco, and is one of the chief articles of trade with them. Their pipes are large, much larger at the stem than the bowl. In smoking, they swallow the fumes of the tobacco which causes intoxication for a time. " The desire to procure a few of its narcotic leaves induces the American Esquimaux from the Ice Cape to Bristol Bay, to send their produce from hand to hand as far as the Guosden Islands in Behrings Straits, where it is bartered for the tobacco of the Tchuktchi, and there again principally resort to the fair of Ostrownoje to purchase tobacco from the Russians.

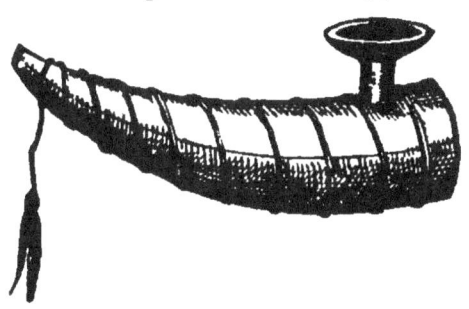

A TCHUKTCHI PIPE.

Generally the Tchuktchi receive from the Americans as money skins for half a pond, or eighteen pounds of tobacco leaves as they afterwards sell to the Russians for two ponds of tobacco of the same quality.

The Russians also are great lovers of the weed. A writer says:—

"Everybody smokes, men, women, and children. They smoke Turkish tobacco, rolled in silk paper—seldom cigars or pipes. These rolls are called parporos. The ladies almost all smoke, but they smoke the small, delicate sizes of parporos, while the gentlemen smoke larger ones. Always at morning, noon and night, comes the inevitable box of parporos, and everybody at the table smokes and drinks their coffee at the same time. On the cars are fixed little cups for cigar ashes in every seat. Ladies frequently take out their part parporos, and hand them to the gentlemen with a pretty invitation to smoke. Instead of having a smoking car as we do, they have a car for those who are so 'pokey' as not to smoke."

Throughout the German States the custom of smoking is universal and tobacco enters largely into their list of expenditures. A writer says of smoking in Austria:—

"We have been rather surprised to find so few persons smoking pipes in Austria. Indeed, a pipe is seldom seen except among the laboring classes. The most favorite mode of using the weed here is in cigarettes, almost every gentleman being provided with a silver box, in which they have Turkish tobacco and small slips of paper, with mucilage on them ready for rolling. They make them as they use them, and are very expert in the handling of the tobacco. The chewing of tobacco is universally repudiated, being regarded as the height of vulgarity. The Turkish tobacco is of fine flavor, and commands high prices. It is very much in appearance like the fine cut chewing tobacco so extensively used at home."

The cigars made by the Austrian Government, which are the only description to be had are very inferior, and it is not to be wondered that the cigarette is so generally used in preference.

The smoking of cigarettes by the ladies is quite common, especially among the higher classes. In no part of the world is smoking so common as in South America; here all classes and all ages use the weed. Smoking is encouraged in the family and the children are early taught the custom. A traveler who has observed this custom more particularly than any other, says of the use of tobacco in Peru:—

"Scarcely in any regions of the world is smoking so common as in Peru. The rich as well as the poor, the old man as well as the boy, the master as well as the servant, the lady as well as the negroes who wait on her, the young maiden as well as the mother—all smoke and never cease smoking, except when eating, or sleeping, or in church. Social distinctions are as numerous and as marked in Peru as anywhere else, and there is the most exclusive pride of color and of blood. But differences of color and of rank are wholly disregarded when a light for a cigar is requested, a favor which it is not considered a liberty to ask, and which it would be deemed a gross act of incivility to refuse. It is chiefly cigarritos which are smoked.

"The cigarrito, as is well known, is tobacco cut fine and dexterously wrapped in moist maize leaves, in paper, or in straw. Only the laborers on the plantations smoke small clay pipes. Dearer than the cigarritos are the cigars, which are not inferior to the best Havanna. Everywhere are met the cigarrito-twisters. Cleverly though they manipulate, cleanliness is not their besetting weakness. But in Peru, and in other parts of South America, cleanliness is not held in more esteem than in Portugal and Spain."

The Turks have long been noted as among the largest consumers of tobacco as well as using the most magnificent of smoking implements. The hookah is in all respects the most expensive and elaborate machine (for so it may be called) used for smoking tobacco. A traveler gives the following graphic description of smoking among them:

"As each man smokes only out of his own pipe, it is not surprising that this instrument is an indispensable accompa-

TURK SMOKING.

niment of every person of rank. Men of the higher classes keep two or three servants to attend to their pipes. While one looks after things at home, the other has to accompany his master in his walks and rides. The long stem is on such occasions packed in a finely embroidered cloth cover, while the bowl, tobacco, and other accessories are carried by the servant in a pouch at his side. A stranger in Constantinople will often regard with curiosity and surprise, a proud Osmanli on foot or horseback, followed by an attendant who, through the long, carefully-packed instrument which he carries, gives one the idea that he is a weapon-bearer of some heroic period following his lord to some dangerous rendezvous. So are the times altered. What the armor-bearer was for the warlike races of old, such is the tchbukdi for their degenerate descendants.

"To smoke from sixty to eighty pipes a day is by no

means uncommon; for whatever be the business, no matter how serious, in which the Turk is engaged, he must smoke at it. In the divan, where the grandees of the empire consult together on the most delicate affairs of State, the question was once mooted whether the tchbukdes should not be excluded from such debates as were of a strictly private nature. There was a great diversity of opinion on the subject. Politics and reason were on opposite sides. At last it was decided that they would not disgrace an ancient national usage, but would allow the harmless attendants to enter the council-room every now and then to change the pipes. In Turkey, pipes and tobacco afford means of distinguishing not only the different classes of the community, but even the several graduates of rank in the same class. A mushir (marshal) would find it derogatory to his dignity to smoke out of a stem less than two yards in length. The artisan or official of a lower rank, would consider it highly unbecoming on his part to use one which exceeded the proper proportions of his class. A superior stretches his pipe before him to his inferior; while the latter must hold his modestly on one side, only allowing the end of the mouth-piece to peep out of his closed fist.

"The pasha has the right to puff out his smoke before him like a steam engine, while his inferiors are only allowed to breathe forth a light curl of smoke, and that must be let off backwards. Not to smoke at all in the presence of a superior, is held the most delicate homage which can be paid him. A son, for instance, acts in this manner in the presence of his father, and only such a one is considered to be well brought up who declines to smoke even after his father has repeatedly invited him to do so. The fair sex in the East is scarcely less addicted to the use of this weed.

"The girl of twelve years old smokes a cigarette of the thickness of pack-thread. When she has attained her fourteenth or fifteenth year, and is already marriageable, she is allowed to indulge her penchant at will, which is forbidden when younger. After this age the diameter of the cigarette increases year by year; and when a lady has reached the mature age of twenty-four, no one sees anything remarkable in her smoking a modest little chibouque as she sits on the lower divan of the harem. Elderly matrons—and in Turkey every lady is an elderly matron in her fortieth year—are passionately devoted to this enjoyment. The pipe-bowls and stems always remain of the size appropriated by etiquette to the use of the harem; but the strongest and most pungent sorts of tobacco are not unseldom smoked, until the mouth, which, according to the assurance of the poet, in the bloom of its youth breathed forth ambergiris and musk, in its fortieth year acquires so strong a smell that the lady can be scented from a distance.

"Like their lords, the hanyrns of rank have also their tchbukdes, of course of their own sex, who accompany them when out walking or on a visit. In this case, however, the cover in which the pipe-stem is made, not of cloth, but of silk. The habit of refreshing oneself with a pipe on some elevated spot which commands a fine view, is common to both sexes. Men can indulge this taste whenever their fancy may suggest, but ladies only in retired spots; for, whenever a Turkish fair one removes the yas mak (veil) from her lips, as she does to smoke, all around her must be harem (sacred).

"Sometimes an eunuch stands guard at a little distance off, and if a stranger of the male sex approaches, gives a signal; the pipe is held aside, while the mouth is kept covered by the veil, until the unexpected Acteon has passed by. But where the pipe plays the most important part is in the bath. It is well known that the Turkish ladies are accustomed to frequent the hommams assiduously, and to remain there for hours together. They enter the bath about eight o'clock in the morning; take their midday meal there, and return home between three and four in the afternoon. During these hours of leisure, the most agreeable in a Mohammedan woman's life, the pipe is their constant resource. In the middle of the warmest room is a round terrace-like elevation, called Gobek-tosh.

"Here are clustered old and young, the snow white daughters of Circassia and the coal-black beauties of Soudan, and beguile the hours with never ending gossip, while around them rise the dense fumes of their pipes. Now one of the elders of the party tells a story, now a learned lady holds a discourse on religion, or extols the beauty and virtue of 'Aisha Fatima.'"

The Fairy, or Dane's pipe is the most ancient form of the tobacco pipe used in Great Britain and of about the same size as the "Elfin pipes" of the Scottish peasantry. A great variety of pipes both in form and size have been found in the British Islands some of which are of ancient origin bearing dates prior to the Seventeenth Century. Some of these ancient pipes are formed of very fine clay and although they held but a small quantity of tobacco were doubtless considered to be fine specimens in their time.

The manufacture of pipes commenced soon after the custom of using tobacco had become fashionable and soon after the Virginians commenced its cultivation. Fairholt says:

"The early period at which tobacco pipes were first manufactured, is established by the fact that the incorporation of the craft of tobacco-pipe makers took place on the 5th of October, 1619. Their privileges extending through the cities of London and Westminster, the kingdom of England

OLD ENGLISH PIPES.

and dominion of Wales. They have a Master, four Wardens, and about twenty-four Assistants. They were first incorporated by King James in his seventeenth year, confirmed again by King Charles I., and lastly on the twenty-ninth of April in the fifteenth year of King Charles II., in all the privileges of their aforesaid charters.

"The London Company of Tobacco Pipe Makers was incorporated in the reign of Charles II (1663); it had no hall and no livery but was governed by a Master two wardens, and eighteen assistants. The first pipes used in the British Islands were made of silver while 'ordinary ones' were made of a walnut shell and a straw. Afterwards appeared the more common clay pipes in various forms and which are in use at the present time."

During the reign of Anne and George I. the pipes assumed a different form and greater length so long were the stems of some of them that they were called yards of clay. The French pipe is one of the finest manufactured and is made of a fine red clay especially those made by Fiolet of St. Omer, one of the best designers of pipes. Many of these like German pipes are made of porcelain, adorned with portraits and landscapes. Others are made of rare kinds of wood turned in the lathe or artistically carved, and lined with clay to resist the action of fire.

The French also make pipes of agate, amber, crystal, carnelian and ivory, as well as the various kinds of pure or mixed metals. Many of the French and German pipes while they are beautiful in design and made of the most costly materials are often exceedingly grotesque, representing often the most ludicrous scenes and all possible attitudes. Many of them have been termed as satirical pipes taking off some public character *a la* Nast.

Fairholt says of satirical pipes:

"England has occasioned the production of one satirical pipe for sale among ourselves. The late Duke of Wellington toward the close of his life, took a strong dislike to the use of tobacco in the army, and made some ineffectual attempts to suppress it. Benda, a wholesale pipe importer in the city employed Dumeril, of St. Omer, to commemorate the event, and the result was a pipe head, in which a subaltern, pipe in hand, quietly 'takes a sight' at the great commander who is caricatured after a fashion that must have made the work a real pleasure to a Frenchman." Many of the French pipes are exceedingly quaint representing all manner of comical scenes. One is formed like a steam-engine the smoke pass-

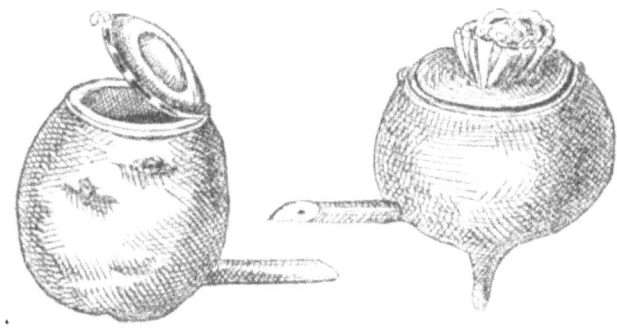

FRENCH PIPES.

ing through the funnel. Another is fashioned after a potato or a turnip while others often represent some military subjects. In England and Ireland also pipes of a whimsical form are common.

CHAPTER VII.

PIPES AND SMOKERS. (CONTINUED.)

IN Russia and Denmark as also in Norway and Sweden the pipes are more simple and are principally formed of wood sometimes tipped with copper but usually of inferior material and work when compared with French and German pipes. The German pipes considered as works of art are doubtless the finest made. Many are made of meerschaum (sea foam). This material is found in various parts of Asia Minor. When first obtained it is capable of forming a lather like soap, and is used by the Tartars for washing purposes. The Turks use it for pipes which are made in the same way that pottery is and afterwards soaked in wax and is then ready for smoking. It heats slowly and is capable of greater absorption than any other material used in pipe making. To properly color a meerschaum is now considered as one of the fine arts and when completed is considered quite a triumph. When the pipe takes on a rich deep brown tint it is considered a valuable pipe and is watched and guarded as a most valuable treasure.

M. Ziegler thus describes the source whence the considerable annual supply of meerschaum for meerschaum pipes is derived:

"Large quantities of this mineral so highly esteemed by smokers, comes from Hrubschitz and Oslawan in Austrian Moravia, where it is found embedded between thick strata of serpentine rock. It is also found in Spain at Esconshe, Vallecos, and Toledo; the best however comes from Asia Minor. The chief places are the celebrated meerschaum mines from six to eight miles southeast of Eskis chehr, on the river Pursak chief tributary to the river Sagarius. They were known to Xenophon, and are now worked principally by Armenian Christians, who sink narrow pits, to the beds of this mineral, and work the sides out until water or imminent danger drives them away to try another place. Some meerschaum comes from Brussa, and in 1869 over 3,000 boxes of raw material were imported from Asia Minor at Trieste, with 345,000 florins. The pipe manufacture and carving is principally carried on in Vienna and in Rhula, Duchy of Saxe-Coburg-Gotha. The commercial value of meerschaum carving at these places may be estimated at $2,000,000 annually. However very large quantities of them are not made from genuine but artificial material. The waste from these carvings is ground to a very fine powder, and then boiled with linseed oil and alum. When this mixture has sufficient cohesion, it is cast in molds and carefully dried and carved, as if these blocks of mineral had been natural. It is said that about one-half of all pipes now sold are made from artificial meerschaum. Meerschaum is one of the lightest of minerals and it is said that in Italy bricks have been made of it so light that they would float on the top of the water. Some pipes (doubtless owing to the quality of meerschaum) take on more color in a given time than others this is owing in a great measure however to the thickness of the bowl."

Pipe-colorers, who go around coloring pipes or meerschaums, pride themselves on the rapidity with which they are enabled to color a pipe. The following, on "Pipe Colorers," is from "The Tobacco Plant":

"There are men who pride themselves upon the skill with which they are able to color the pipes they smoke. Some of these are amateurs, who smoke Tobacco only with the view of gratifying that taste for color which is satisfied when a bowl of clay or meerschaum is sufficiently yellowed, browned, or blacked. There are men who care nothing for Tobacco of itself, and would be much more easily and rationally pleased were they to set their pipes upon an easel and paint them with oils and camel's-hair. Others of the class are professional colorers, who hire themselves to pipe-sellers or connoisseurs by the week, or day, or hour, to smoke so many ounces or pounds of strong Tobacco through such and such pipes in such and such a time, with the view of causing such and such stains of Tobacco-juice to make themselves visible on the bowls or stems of those specified pipes. These are mostly old, well-seasoned smokers, to whose existence the weed has become essential; who smoke their own old pipes, which lack artistic coloring, in the intervals when they lay aside the pipes they are employed to color. Another and much smaller section of the class are those who smoke for smoking's sake, and yet are weak enough to nurse some special pipes for show. To such it is a joy to say, when friends are gathered at the festive board 'Look! is not that well colored? I colored it myself.' In such an age as this, when the learned cannot tell us which of our various branches of knowledge and inquiry are sciences and which are not, it may not seem a great anomaly that this pipe-coloring should, by some, be called 'an art.' Nor is it, when we think that there is such an 'art' as blacking shoes; and when we must perforce admit that he who, barber fashion, cuts our hair—and he who, cook-wise, broils the kindney for our mid-day dinner—is an artist. We have not come as yet to give this title to the weaver who watches the loom that weaves our stockings, or to the hammer-man who beats the red-hot horse-shoe on the anvil in a smithy; but even there we designate 'artisans,' and 'artists' may come next. So, hey! for the art of coloring pipes!

PIPE COLORER.

"It may not be denied that there is beauty in a well-colored meerschaum; but in the admission lies the contradiction of Keats' well-known line—

"A thing of beauty is a joy for ever."

For, your meerschaum is a fragile thing, and eminently frangible. This present writer once did see four beauties break within a single moon. And when they break, what previous joy of coloring can over-top the sorrow of their dire destruction? It is a singular difficulty in the way of those who most desire to beautify utility or utilize the beautiful, or show that beauty is most lovely when made practical, that these artistic colorers of pipes are always those who make least use of Tobacco, save for the immediate purpose of obtaining the clay in which it is smoked. Ask such an artist why he smokes, and he will scarcely tell you. His best reason certainly will be, that others smoke, and, as a custom, it becomes him. And when you find an ardent smoker—one who smokes because he likes Tobacco for itself, or finds it useful—who spends his time in tinting pipes, you will have found a *rara avis*, or a monstrosity. Apart from taste, there are some practical objections to this custom of coloring pipes. Smoking, to be worthy, should be free and unrestrained; while he who colors his pipe is tied by system and confined to rule.

"A pipe to be enjoyable, should be its master's slave; but he who keeps a 'well-colored' pipe is slave thereto. He cannot smoke it as, or when, or where he will. He must not smoke it in a draught, or near a fire; he must not lay it down, or finger it; he must not puff too fast, nor yet too slow. In short, he is the creature of this 'Joss'—this home-made deity—to which he bows down and worships. The pipe-colorers are the Sabbatarians of smoking. Whereas, the pipe was made for man, they treat man as made for the pipe. And thus, as in all cases where the cart is expected to draw the horse, the economy of nature is reversed, and mischief is evolved."

Dibdin, in his "Tour in France and Germany," says of Vienna, that it is a city of smokers,—"a good Austrian thinks he can never pay too much for a good pipe." Many of the Germans use a kind of pipe carved from the root of the dwarf oak; wooden pipes of a similar kind are made of brier root, and are very common, as are also those made from maple and sweet-brier. One of the favorite pipes used by Germans is the porcelain pipe, which consists of a double bowl—the upper one containing the tobacco, which fits into another portion of the pipe, allowing the oil to drain into the lower bowl, which may be removed and the pipe cleaned. The bowls are

GERMAN PORCELAIN PIPES.

sometimes painted beautifully, representing a variety of subjects, and in no way inferior to the painted porcelain for the table.

The Dutch are famous smokers and are constantly "pulling at the pipe." They use those with long, straight stems, and both their clay and porcelain pipes are of the finest form and finish. Irving, in "The History of New York from the Beginning of the World to the End of the Dutch Dynasty," has given a good description of the smoking powers of the Dutch. Speaking of his grandfather's love for the weed, he says:

"My great-grandfather, by the mother's side, Hermanns Van Clattercop, when employed to build the large stone church at Rotterdam, which stands about three hundred yards to your left, after your turn from the Boomkeys; and which is so conveniently constructed that all the zealous Christians of Rotterdam prefer sleeping through a sermon there to any other church in the city. My great-grandfather, I say, when employed to build that famous church, did, in the first place, send to Delft for a box of long pipes; then, having purchased a new spitting-box and a hundred weight of the best Virginia, he sat himself down and did nothing for the space of three months but smoke most laboriously.

"Then did he spend full three months more in trudging on foot, and voyaging in the Trekschuit, from Rotterdam to Amsterdam—to Delft—to Hærlem—to Leyden—to the Hague—knocking his head and breaking his pipe against every church in his road. Then did he advance gradually nearer and nearer to Rotterdam, until he came in full sight of the identical spot whereon the church was to be built. Then did he spend three months longer in walking round it and round it, contemplating it, first from one point of view, and then from another,—now would he be paddled by it on the canal—now would he peep at it through a telescope from the other side of the Meuse, and now would he take a bird's-eye glance at it from the top of one of those gigantic windmills which protect the gates of the city.

"The good folks of the place were on the tip-toe of expectation and impatience. Notwithstanding all the turmoil of my great-grandfather, not a symptom of the church was yet to be seen; they even began to fear it would never be brought into the world, but that its great projector would lie down and die in labor of the mighty plan he had conceived. At length, having occupied twelve good months in puffing and paddling, and talking and walking,—having traveled over all Holland, and even taken a peep into France and Germany,—having smoked five hundred and ninety-nine pipes and three hundred weight of the best Virginia tobacco,—my great-grandfather gathered together all that knowing and industrious class of citizens who prefer attending to anybody's business sooner than their own, and having pulled off his coat and five pair of breeches he advanced sturdily up and laid the corner-stone of the church, in the presence of the whole multitude,—just at the commencement of the thirteenth month."

He also alludes to Hudson whom he says was:

"A seafaring man of renown, who had learned to smoke tobacco under Sir Walter Raleigh, and is said to have been the first to introduce it into Holland, which gained him much popularity in that country, and caused him to find great favor in their High Mightinesses, the lords and states general, and also of the honorable West India Company. He was a short, square, brawny old gentleman, with a double chin, a mastiff mouth, and a broad copper nose, which was supposed in those days to have acquired its fiery hue from the constant neighborhood of his tobacco pipe. * * * As chief mate and favorite companion, the commander chose Master Robert Juet, of Limehouse, in England. By some his name has been spelled Chewit, ascribed to the circumstance of his having been the first man that ever chewed tobacco. * * * * Under every misfortune he comforted himself with a quid of tobacco, and the truly philosophical maxim, 'that it will be all the same a hundred years hence!'"

Further on he alludes to the attempt to subjugate New Amsterdam to the British crown and the effect produced by the burghers lighting their pipes. "When" he says "Captain Argol's vessel hove in sight, the worthy burghers were seized with such a panic, that they fell to smoking their pipes with astonishing vehemence, insomuch that they quickly raised a cloud, which, combining with the surrounding woods and marshes, completely enveloped and concealed their beloved village; and overhung the fair regions of Pavonia:—so that the terrible Captain Argol passed on, totally unsuspicious that a sturdy little Dutch settlement lay snugly couched in the mud, under cover of all this pestilent vapor."

The Persians* are said to be the first to invent the mode of drawing tobacco smoke through water thereby cooling it before inhaling it. Fairholt says "it is to smoking what ice is to Champagne." The *London Review* gives the following description of pipes and smoking apparatus:

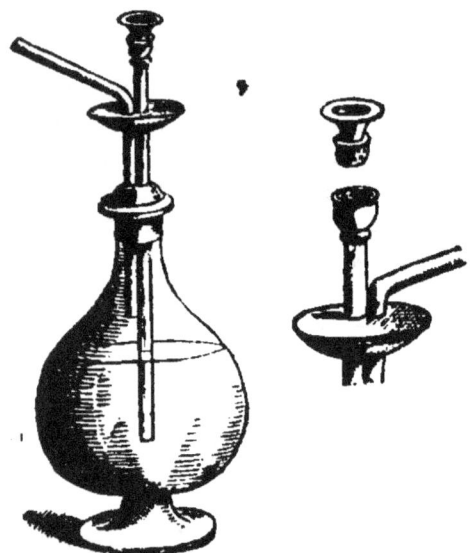

A PERSIAN WATER PIPE.

"The hookah of India is the most splendid and glittering of all pipes; it is a large affair, on account of the arrangements for causing the smoke to pass through water before it reaches the lips of the smoker, as a means of rendering it cooler and of extracting from it much of its rank and disagreeable flavor.

"On the top of an air-tight vessel, half filled with water, is a bowl containing tobacco; a small tube descends from the bowl into the water, and a flexible pipe, one end of which is between the lips of the smoker, is inserted at the other end into the vessel, above the level of the water. Such being the adjustment, the philosophy of the inhalation may be easily understood. The smoke sucks the air out of the vessel, and makes a partial vacuum; the external air, pressing on the burning tobacco, drives the smoke through the small tube into the water beneath; purified from some of its rank qualities, the smoke bubbles up into the vacant part of the vessel above the water, and passes through the flexible pipe to the smoker's mouth. Sometimes the affair is made still more luxurious by substituting rose-water for water *pur et simple*. The tube is so long and flexible that the smoker may sit (or squat) at a small or great distance from the vessel containing the water. In the courts of princes and wealthy natives the vessels and tubes are lavishly adorned with precious metals. One mode of showing hospitality in the

---

*Sandys, writing in 1610 narrates a Persian legend to the effect that Shiraz tobacco was given by a holy man to a virtuous youth, disconsolate at the loss of his loving wife. "Go to thy wife's tomb," said the anchorite, "and there thou wilt find a weed. Pluck it, place it in a reed, and inhale the smoke, as you put fire to it. This will be to you wife, mother, father and brother," continued the holy man, in Homeric strain, "and above all, will be a wise counsellor, and teach thy soul wisdom and thy spirit joy."

East is to place a hookah in the center of the apartment, range the guests around, and let all have a whiff of the pipe in turn; but in more luxurious establishments a separate hookah is placed before each guest. Some of the Egyptians use a form of hookah called the narghile or nargeeleh—so named because the water is contained in the shell of a cocoanut of which the Arabic name is nargeeleh. Another kind, having a glass vessel, is called the sheshee—having, like the other, a very long tube. Only the choicest tobacco is used with the hookah and nargeeleh; it is grown in Persia.

"Before it is used, the tobacco is washed several times, and put damp into the pipe-bowl, two or three pieces of live charcoal are put on the top. The moisture gives mildness to the tobacco, but renders inhalation so difficult that weak lungs are unfitted to bear it. The dry tobacco preferred by the Persians does not involve so much difficulty in 'blowing a cloud.'"

TURKISH CHIBOUQUES AND WOOD PIPES.

"The stiff-stemmed Turkish pipes, quite different from the flexible tube of the hookah and narghile, are of two kinds, the kablioun or long pipe, and the chibouque or short pipe. Some of the stems of the kablioun, made of cherry tree, jasmine, wild plum, and ebony, are five feet in length, and are bored with a kind of gimlet. The workman, placing the gimlet above the long, slender branchlet of wood, bores half the length, and then reverses the position to operate upon the other half. The wild cherry tree wood, which is the most frequently employed, is seldom free from defects in the bark, and some skill is exercised in so repairing these defective places that the mending shall be invisible."

The tubes or pipe-bowls used with these stems are mostly a combination of two substances—the red clay of Nish and the white earth of Rustchuk; they are graceful in form and sometimes decorated with gilding. It is characteristic of some of the Turks that they estimate the duration of a journey, and with it the distance traveled, by the number of pipes smoked, a particular size of pipe-bowl being understood. Dodwell, in his "Tour through Greece," says that "a Turk is generally very clean in his smoking apparatus, having a small tin dish laid on the carpet of his apartment, on which the bowl of the pipe can rest, to prevent the tobacco from burning or soiling the carpet. The tubes of the kabliouns are often as much as seven or eight feet long. Some of the gardens of Turkey and Greece contain jasmine trees purposely cultivated to produce straight stems for these pipes."

Of those Turkish pipes which are used in Egypt, Mr. Lane, after mentioning the narghile and the chibouque or "shibuk," says:—

"The most common kind used in Egypt is made of wood called garmashak (I believe it is maple). The greater part of the stick, from the mouth-piece to three-fourths of its length, is covered with silk, which is confined at each extremity by gold thread, often intertwined with colored silks, or by a tube of gilt or silver; and at the lower extremity of the covering is a tassel of silk. The covering was originally designed to be moistened with water in order to cool the pipe, and consequently the smoke by evaporation; but this is only done when the pipe is old or not handsome. These stick pipes are used by many persons, particularly in winter; in summer the smoke is not so cool from them as from the kind before mentioned. The bowl is of baked earth, colored red or brown."

AUSTRIAN AND HUNGARIAN PIPE STEMS.

Before passing to the subject of the costly mouth-pieces of Oriental pipes, we must say a few words concerning the extraordinary care bestowed on some kinds of plain wood sticks for stems or tubes. Cherry-tree stems, under the name of agriots, constitute a specialty of Austrian manufacture. The fragrant cherry (prunus makaleb) is a native of that country; and the young trees are cultivated with special reference to this application. They are all raised from seed. The seedlings, when two years old, are planted in small pots, one in each; as they grow, every tendency to branching is choked by removing the bud; and as they increase in size from year to year, they are shifted into larger pots or into boxes. Great care is taken to turn them round daily, so that every part shall be equally exposed to sunshine. When the plants have attained a sufficient height they are allowed to form a small bushy head; but the daily care is continued until the stems grow to a proper thickness. They are then taken out of the ground, the roots and branches removed, and the stem bored through after being seasoned for some time. The care shown in rearing insures a perfect straightness of stem, and an equable diameter of about an inch or an inch and a half. The last specimens, when cut from the tree, are as much as eight feet in length, dark purple-brown in color, and highly fragrant. At Pesth are made pipes about eighteen inches in length, of the shoots of the mock orange, remarkable for their quality in absorbing the oil of tobacco, they are flexible without being weak. The French make elegant pipe-bowls of the root of the tree-heath, but their chief attention is directed, as far as concerns wood pipes, to those of brier-root, which are made by them in large quantities. The bowl and the short stems are carried out of one piece, and the wood is credited with absorbing some of the rank oil of tobacco.

Amber—the only kind of resin that rises to the dignity of a gem—is unfitted for the bowl of a tobacco-pipe, because it cannot well bear the heat; but it is largely used for mouth-pieces, especially by wealthy Oriental smokers. The Turks have a belief that amber wards off infection; an opinion which, whether right or wrong, tells well for the amber workers. There has always been a mystery connected with this remarkable substance. So far back as the Phenicians, amber was picked up on the Baltic shore of what is now called Prussia; and the same region has ever since been the chief store-house for it. Tacitus was not far wrong when he conjectured that amber is a gum or resin exuded from certain trees, although other authorities have preferred a theory that it is a kind of wax or fat which has undergone slow petrifaction. At any rate, it must at one time have been liquid or semi-liquid; for insects, flies, detached wings and legs, and small fragments of various kinds, are often found imbedded in it—those odds and ends of which Pope said:—

> "The things, we know, are neither rich nor rare;
> The wonder's how the devil they got there!"

Whether new stores of amber are now being formed, or whether, like coal, it was the result of causes not now in operation, is an unsolved problem. The specimens obtained differ considerably; some are pale as primrose, some deep orange or almost brown; some nearly as transparent as crystal, some nearly opaque. Large pieces, uniform in color and translucency, fetch high prices; and there are fashions in this matter for which it is not easy to account,—seeing that the Turks and other Orientals buy up, at prices which Europeans are unwilling to give, all the specimens presenting a straw-yellow color and a sort of cloudy translucency. The Russians, on the contrary, prefer orange-yellow transparent specimens. The amber is seldom obtained by actual mining.

It is usually found on sea-coasts, after storms, in rounded nodules; or, if scarce on shore, it is sought for by men clad in leather garments, who wade up to their necks in the sea, and scrape the sea-bottom with hooped nets attached to the end of long poles; or (rather dangerous work) men go out in boats, and examine the faces of precipitous cliffs, picking off, by means of iron hooks, the lumps of amber which they may see here and there. Sometimes a piece weighing nearly a pound is found, and a weight of even ten pounds is recorded. As small pieces can easily be joined by smoothing the surfaces, moistening them with linseed oil, and pressing them together over a charcoal fire, and as gum copal is sometimes very like amber, there is much sophistication indulged in, which none but an expert can guard against. In fashioning the nodules of amber, whether genuine or

SEARCHING FOR AMBER.

fictitious, into pipe mouth-pieces, they are split on a leaden plate in a turning lathe, smoothed into shape by whet-stones, rubbed with chalk and water, and polished with a piece of flannel. It is an especially difficult kind of work; for unless the amber is allowed frequent intervals for cooling, it becomes electrically excited by the friction and shivers into fragments; the men, too, are put into nervous tremors if kept too long at work at one time. Amber is one of the most electrically excitable of all known substances; in fact, the name electricity itself was derived from *electron*, the Greek name for amber. Hookahs, chibouques, narghiles, meerschaums, all are largely adorned with amber mouth-pieces. The mouth-piece often consists of two or three pieces of amber, interjoined with ornaments of gold and gems; it is in such case the most costly part of the pipe.

At one of the greater industrial exhibitions four Turkish amames, or amber mouth-pieces, were shown, illustrating clearly enough the value attached to choice specimens; two of them were worth £350 each, two £200 each, diamond studded. The Turkish and Persian pipes have often a small wooden tube inside the amber mouth-piece. They require frequent cleaning with a long wire and a bit of tow, and in some large towns there are professional pipe-cleaners.

The natives of British Guiana have a curious kind of pipe, made of the rind of the fruit of the areca-palm, coiled up into a kind of cheroot, with an internal hollow to hold the tobacco. The poorer Hindoos make a simple pipe of two pieces of bamboo,—one cut close to a knot for the bowl, and a more slender piece for the tube. A lower class of natives in India make two holes of unequal length, with a piece of stick, in a clay soil; the holes are unequally inclined so as to meet at the bottom; the tobacco is placed in the shorter hole, and the smoker, applying his mouth to the longer, inhales the fumes in this primitive fashion. The pipes used for opium-smoking in various parts of the East have small bowls; the drug is too costly to be used otherwise than in small portions at a time, and too powerful to need more than a few whiffs to produce the opium-smoker's dreary delirium.

The Tunisians use reeds for pipes. Stone pipes are found among the natives of Vancouver; while Strong Bow, the North American Indian chief, has his long wooden pipe of peace, decked out with tassels and fringes, but with an ominous-looking sharp steel cutting instrument near the end most remote from the bowl.

Chinese, Japanese, Phillipine Islanders, Madagascans, Central Africans, Algerine Arabs, Mexicans, Paraguayans, Siamese, Tahitians, South American Indians, Mongols, Malays, Tartars, Turcomans, as well as the nations of Europe and the chief nations of Southern Asia, all have their smoking-pipes, plain or ornate, as the case may be, and made of wood, reeds, bamboo, bone, ivory, stone, earthenware, glass, porcelain, amber, agate, jade, precious metals and common metals, according to the civilization of the country and the pecuniary means of the smoker.

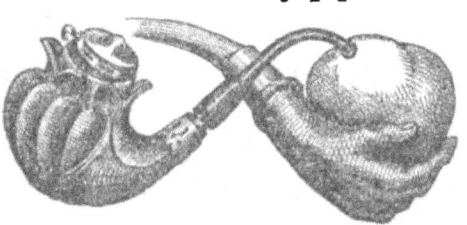

FANCY PIPES.

"The French clay pipes have quite a special character; they are well made, and great ingenuity is shown in the preparation of the moulds in which they are pressed; but being mostly intended for a class of purchasers who prefer grotesque ideas to refined taste, the bowls are often ornamented with queer shaped heads, having bead-like eyes; sometimes imaginary beings, sometimes caricature portraits of eminent persons. Where more than the head is represented, license is given to a certain grossness of idea; but this is not a general characteristic. The clay of which these French pipes are made is admitted to be superior to that of England, due to the careful mixture of different kinds, and to skilful manipulation.

"We need not say much about Dutch pipes as distinct articles of manufacture, because the process adopted in their production are pretty much like those in use elsewhere. The Dutch are famous clay-pipe smokers, not countenancing the cigar so much as their neighbors the Belgians, nor the meerschaum so largely as their German neighbors on the Rhine frontier. A notable bit of sharp practice is on record in connexion with the pipe-smokers of Holland—a dodge only to be justified on the equivocal maxim that all is fair in trade provided it just keeps within the margin we need not speak. A pipe manufactory was established in Flanders about the middle of the last century.

"The Dutch makers, alarmed at the competition which this threatened, cunningly devised a stratagem for nipping it in the bud. They freighted a large worn-out ship with an enormous quantity of pipes of their own make, sent it to Ostend, and wrecked it there. By the municipal laws of that city the wreck became public property; the pipes were sold at prices so ridiculously low that the town was glutted with the commodity; the new Flemish factory was thereby paralyzed, ruined, and closed.

The Turks (especially those of the lower orders) use a kind of clay pipe made of red earth decorated with gilding. The stem of the pipe is made from a branch of jasmine, cherry tree or maple and is sufficiently long to rest on the floor when used by the smoker. A writer in the *Tobacco Plant* says of Old English Clay pipes:

"Of all the various branches of the subject of tobacco, that of the history of pipes is one of the most interesting, and one that deserves every attention that can possibly be given. Whether considered ethnographically, historically, geographically, or archæologically, pipes present food for speculation and research of at least equal importance to any other set of objects that can be brought forward. Some branches of the subject have already been treated in these columns, and others, in what is intended shall follow, will hereafter be discussed. The present article will be devoted to 'Fairy Pipes' and the history of the earliest pipes of this country. Smoking is an old and venerable institution in this kingdom of ours, and dates far back beyond the introduction of tobacco to our shores. Long before Sir Walter Raleigh was thought of, there is reason to believe herbs and leaves of one kind or other—coltsfoot, yarrow, mouse-lax, sword-grass, dandelion, and other plants, and even dried cow-dung—were smoked for one ailment or other, and in some instances for relaxation and pleasure, and thus, no doubt, became habitually used. These are still, in some of our rural districts, smoked by people as cures for various ailments, and are considered not only highly efficacious but very pleasant. I have known these or other herbs smoked through a stick from which the pith had been removed, the bowl being formed of a lump of clay moulded by the fingers at the time, and baked in the household fire.

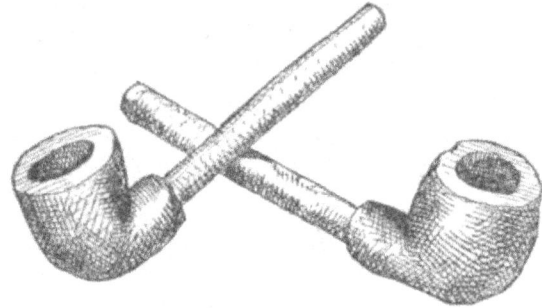
CLAY AND REED PIPES.

"The small branches of the elder tree, or sometimes the stem of the briar and bramble, are what I have seen used, but even the stem of the hemlock and keckse are sometimes brought into requisiton for the purpose.

"I believe that long before the time Dr. Wilson states on the authority of Sharpe, that it was common within memory, for the old wives of Annandale to smoke a dried white moss gathered on the neighboring moors, which they declared to be much sweeter than tobacco, and to have been in use long before the American weed was heard of; before Sir Walter Raleigh wooed and won Elizabeth Throgmorton, or Sir Richard Granville voyaged to Virginia with Masters Ralph Layne, Thomas Candish, John Arundell, Master Stukely, Bremize, Vincent, Heryot, and John Clarke; before Sir Francis Drake made his first voyage, or the Spanish Armada was dreamed of; before Sir John Hawkins, Captain Price, Coft, Keat or others for whom the honor of the introduction of tobacco has been claimed, drew breath—smoking was to some extent indulged in by our forefathers and (still medicinally, of course) in this country. In mediæval times, when the Ceramic art was but little practiced, and when all the domestic vessels that were produced were of the rudest and coarsest character both in material, form, and decoration, it is not to be expected that pipes for the smoking of herbs would be manufactured as a matter of sale, and those of the people who wished for such an indulgence would naturally be thrown on their own primitive resources such as I have described, for instruments for the purpose.

"A portion of a very rude pipe-head, formed of common red clay—a lump of clay moulded by hand, and ornamented with small circles pressed into it as from the end of a stick—has come under my notice, as have also others of an equally primitive character, found in different parts of this kingdom. These I have no hesitation in ascribing to a pre-Raleigh period. It is not to these, however, but to the small pipes formerly used in this kingdom for smoking tobacco, and tobacco alone, that I wish to draw attention. Most people, especially in the Midland and Northern counties of England, as well as in Scotland and Ireland, will have heard the name of Fairy Pipes applied to the small, old-fashioned, and sometimes oddly-shaped tobacco pipes which are not infrequently turned up in digging and plowing and other operations. To these and the general forms of old English pipes, I purpose confining myself in the present article. Many years ago I collected together a large number of these 'Fairy Pipes' from all parts of the kingdom. Since then, my own researches have, with the aid of inquiries carried on for me, enabled me to bring forward many interesting points, so as to verify dates of manufacture and more fully to carry out their classification. Like their Irish brethren and sisters, English people were formerly apt to ascribe everything unusually small to the fairies, and anything out of the common way to the people of very remote ages.

"Thus, these small pipes are commonly in England called 'fairy pipes,' or 'Carl's pipes,' or 'old man's pipes;' in Ireland, where they are likewise known as 'fairy pipes,' they are also called 'Dane's pipes;' and in Scotland, where their common name is 'elf pipes,' or 'elfin pipes,' they are, in like manner, known as 'Celtic pipes.' They are also sometimes named 'Mab pipes,' or 'Queen's pipes,' from the same fairy majesty, Queen Mab. Thus, while in each country they are ascribed to the elfin race—the 'small people' of Cornish folk-lore—their secondary names attach to them a popular belief in their extreme antiquity. Anything apparently old is at once, by the Irish, set down to the 'Danes;' by the Scots to the 'Celts;' and by people in the rural districts of our own country to the 'carls,' or 'old men'—carl being indicative of extreme antiquity. In Ireland, the pipes are believed to have belonged to the *cluricaunes*—a kind of wild, ungovernable, mischievous fairy-demon—who were held in awe by the 'pisantry;' and whenever found, these pipes were, with much superstitious feeling, immediately broken up, so as to destroy and break up the spell their finding might have cast around the finder. But it was not only among the peasantry that this belief in the extreme antiquity of tobacco pipes existed.

"Serious essays were written to prove their pre-historic origin, and to claim for them a history that in our day reads as arrant nonsense. In 1784, a short pipe was asserted to have been found between the jaws of the skull of an ancient Milesian exhumed at Bannockstown, county Kildare. Upon this discovery, an elaborate and learned paper was written in the 'Authologia Hibernica,' setting forth this pipe as a proof of the use of tobacco in Ireland long before that country was invaded by the Danes. This pipe has been proved by comparison to be probably quite late in the reign of Elizabeth. They also have a more modern pipe, the stem of which describes one or more circles, while another is tied in a knot, yet allows a free passage of air. At another time,

FAIRY PIPES.

in opening an Anglo-Saxon grave mound, some of the men employed came across a fairy pipe which evidently had rolled down from among the surface-soil, and, being turned out in juxtaposition with undoubted Anglo-Saxon remains, was immediately set down by the learned director of the proceedings as a relic of that period. At another time I had brought to me, as a great curiosity, two 'Roman pipes,' as I was informed—the finders jumping to the conclusion that because they had dug them up at little Chester (the Roman station Derventio), they must be Roman pipes! I believe they expected to receive a large sum from these relics: how grievously they were disappointed I need not tell. Instances of this kind are far from rare.

"I remember a man once bringing me some fragments of Roman pottery and other things of the same period, which he had turned up in the course of excavations, and among them was a Tobacco stopper formed of a Sacheverell medal! and a George II. half-penny, all of which he was ready to swear he had found "all of a heap together," inside a hypocaust tile, which, on examination, certainly had remained *in situ* from Romano-British times! The cupidity of a man had evidently led him to collect together these odds and ends, and try to turn them to profitable account. Some twenty years ago, a large number of "elfin pipes" were dug up at Bomington, near Edinburgh, along with a quantity of placks or bodles of James VI., which thus gave trustworthy evidence of their true date. Others were found in the ancient cemetery at North Berwick, adjoining to which is a small Romanesque building of the Twelfth Century, close upon the shore. Within the last half-century, the sea has made very great inroads upon this ancient burial-place, carrying off a considerable ruin, and exposing the skeletons, and bringing to light many interesting relics at almost every spring-tide. Among these, many pipes have been washed down. A similar circumstance has occurred on the seashore at Hoy Lake, Cheshire, where several "fairy pipes" have been found.

"Notices of several discoveries occur. Dr. Wilson says, in the statistical accounts of Scotland, many of which are suggestive of a pre-Raleigh period. Thus, 'in an ancient British encampment in the parish of Kirk Michael, Dumfriesshire, on the farm of Gilrig, a number of pipes of burnt clay were dug up, with heads smaller than the modern tobacco-pipes, swelled at the middle and straighter at the top. Again, in the vicinity of a group of standing stones at Cairney Mount, in the parish of Carluke Lanarkshire, a celt or stone hatchet, elfin bolts (flint and bone arrow-heads), elfin pipes, numerous coins of the Edwards and of later date, and other things are all stated to have been found.' An example is also recorded of the discovery of a tobacco-pipe in sinking a pit for coal, at Misk, in Ayrshire, after digging through many feet of sand. All these notes are pregnant with significant warnings of the necessity for cautious discrimination in determining the antiquity of such buried relics."

In Turkey the jasmine is cultivated for the purpose of pipe smoking. Barillet describes the growing of the common jasmine near Constantinople. He says:

"The object sought is a long straight stem, free from leaves and side branches. For this purpose the plants are grown quickly in a rich soil, and drawn up by being grown in a sheltered situation, to which the sun has little access at the sides, but only at the top. Pinching is resorted to, and during the second year's growth one end of a thread is attached to the top of the jasmine stem. This thread passes over a pulley attached to the post to which this jasmine is trained, and from it is suspended a weight, the effect of which is to keep the stem always in a vertical direction. When the jasmine stem is about two centimeters (say three quarters of an inch) in diameter a cloth is wrapped around it to prevent access of dust and of the sun's rays. Twice or thrice in the year the stem is washed with citron-water, which is said to give the clear color so much esteemed. When the stem has acquired a length of some fifteen feet, it is cut down and perforated by the workmen, and fitted with a terra-cotta bow and an amber mouth-piece."

Blackburn, in his work entitled "Artists and Arabs," gives the following picture of life and manners in Algiers:—

"There is one difficulty here, however, for the artist—that of finding satisfactory models. You can get one at last, and here is her portrait. Her costume, when she throws off her haik (and with it a tradition of the Mohammedan faith, that forbids her to show her face to an unbeliever), is a rich, loose, crimson jacket embroidered with gold, a thin white bodice, loose silk trousers reaching to the knee and fastened round the waist by a magnificent sash of various colors, red morocco slippers, a profusion of rings on her little fingers, and bracelets and anklets of gold filagree work. Through

FEMALE SMOKING IN ALGIERS.

her waving black hair are twined strings of coins and the folds of a silk handkerchief, the hair falling at the back in plaits below the waist. She is not beautiful, she is scarcely interesting in expression, and she is decidedly unsteady. She seems to have no more power of keeping herself in one position or of remaining in one part of the room, or even of being quiet, than a humming-top. The whole thing is an unutterable bore to her, for she does not even reap the reward—her father, or husband, or other male attendant always taking the money. She is petite, constitutionally phlegmatic, and as fat as her parents can manage to make her; she has small hands and feet, large rolling eyes—the latter made to appear artificially large by the application of henna or antimony black; her attitudes are not ungraceful, but there is a want of character about her, and an utter abandonment to the situation, peculiar to all her race. In short, her movements are more suggestive of a little caged animal that had better be petted and caressed, or kept at a safe distance, according to her humor. She does one thing—she smokes incessantly, and makes cigarettes with a skill and rapidity which are wonderful. Her age is thirteen, and she has been married six months; her ideas appear to be limited to three or four, and her pleasures, poor creature, are equally circumscribed. She had scarcely ever left her father's house, and had never spoken to a man until her marriage. There seems to be in the Moorish nature a wonderful sense of harmony and contrasts of color. Two Orientals will hardly walk down a street side by side unless the colors of their costumes harmonize. You find a negress selling oranges or citrons; an Arab boy with red fez and white turban, carrying purple fruit in a basket of leaves—always the right juxtaposition of colors. The sky furnishes them a superb background of deep blue, and the repose of these solemn Orientals, who sit here like bronze statues, save that they smoke incessantly, inspires you with a curious respect. They are men who believe in fate—what need that they should make haste?"

In Africa the pipes are made of clay and horn, and are mostly rude affairs, but well suited to their ideas of implements used for holding tobacco. King gives the following description of smoking among them:—

"A party of headmen and older warriors, seated cross-legged in their tents, ceremoniously smoked the daghapipe, a kind of hookah, made of bullock's horn, its downward point filled with water, and a reed stem let into the side, surmounted by a rough bowl of stone, which is filled with the dagha, a species of hemp, very nearly, if not the same, as the Indian bang. Each individual receives it in turn, opens his jaws to their full extent, and placing his lips to the wide mouth of the horn, takes a few pulls and passes it on. Retaining the last draught of smoke in his mouth, which he fills with a decoction of bark and water from a calabash, he squirts it on the ground by his side through a long ornamented tube in his left hand, performing thereon, by the aid of a reserved portion of the liquid, a sort of boatswain's whistle, complacently regarding the soap-like bubbles, the joint production of himself and neighbor. It appeared to be a sign of special friendliness and kindly feeling to squirt into the same hole."

We give an engraving of a kind of pipe used by the natives of interior Africa. It is made of clay, and holds but a small portion of the weed. The natives are great smokers and indulge in it almost constantly, but their love for it can hardly exceed that of the more hardy Laplanders, who are described as "passionately fond of the plant."

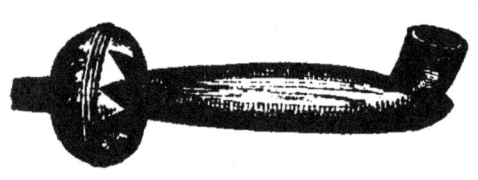

AFRICAN PIPE.

Nothing is so indispensable as tobacco to their existence. A Laplander who cannot get Tobacco sucks chips of a barrel or pieces of anything else which has contained it. Tobacco gives the Laplanders a pleasure which often rises to ecstacy. They both chew and smoke, and they are certainly the dirtiest chewers in the world. When they chew they spit in their hands, then raise them to their nose that they may inhale from the saliva the irritating principles of the plant. Thus they satisfy two senses at the same time. They regularly smoke after their meals. If their supply of Tobacco falls short, they sit down in a circle and pass the pipe round, so that every one in his turn may have a whiff.*

"A Painter's Camp in the Highlands" defends the custom of smoking in the following well chosen words:

"People who don't smoke—especially ladies—are exceedingly unfair and unjust to those who do. The reader has, I daresay, amongst his acquaintances ladies who, on hearing any habitual cigar-smoker spoken of, are always ready to exclaim against the enormity of such an expensive and useless indulgence; and the cost of Tobacco-smoking is generally cited by its enemies as one of the strongest reasons for its general discontinuance. One would imagine, to hear these

---

*Reynard, in his "Travels in Lapland," says of the use of tobacco: "We interrogated our Laplander upon many subjects. We asked him what he had given his wife at their marriage. He told us that she had been very expensive to him during his courtship, having cost him two pounds weight of tobacco and four or five pints of brandy."

people talk, that smoking was the only selfish indulgence in the world. When people argue in this strain, I immediately assume the offensive. I roll back the tide of war right into the enemy's intrenched camp of comfortable customs; I attack the expensive and unnecessary indulgences of ladies and gentlemen who do not smoke. I take cigar-smoking as an expense of, say, half-a-crown a-day, and pipe-smoking at threepence.

"I then compare the cost of these indulgences with the cost of other indulgences not a whit more necessary, which no one ever questions a man's right to if he can pay for them. There is luxurious eating, for instance. A woman who has got the habit of delicate eating will easily consume dainties to the amount of half-a-crown a-day, which cannot possibly do her any good beyond the mere gratification of the palate. And there is the luxury of carriage-keeping, in many instances very detrimental to the health of women, by entirely depriving them of the use of their legs. Now, you cannot keep a carriage a-going quite as cheaply as a pipe. Many a fine meerschaum keeps up its cheerful fire on a shilling a-week. I am not advocating a sumptuary law to put down carriages and cookery; I desire only to say that people who indulge in these expensive and wholly superfluous luxuries, have no right to be so hard on smokers for their indulgence.

"Nearly every gentleman who drinks good wine at all will drink the value of half-a-crown a-day. The ladies do not blame him for this. Half-a-dozen glasses of good wine are not thought an extravagance in any man of fair means, but women exclaim when a man spends the same amount in smoking cigars. The French habit of coffee-drinking and the English habit of tea-drinking are also cases in point. They are quite as expensive as ordinary Tobacco-smoking, and, like it, defensible only on the ground of the pleasurable sensation they communicate to the nervous system. But these habits are so universal that no one thinks of attacking them, unless now and then some persecuted smoker in self-defence.

"Tea and tobacco are alike seductive, delicious, and deleterious. The two indulgences will, perhaps, become equally necessary to the English world. It is high treason to the English national feeling to say a word against tea, which is now so universally recognized as a national beverage that people forget it comes from China, and that it is both alien and heathen. Still, I mean no offence when I put tea in the same category with Tobacco. Now, who thinks of lecturing us on the costliness of tea? And yet it is a mere superfluity. The habit of taking it as we do is unknown across the Channel, and was quite unknown amongst ourselves a very little time ago, when English people were no less proud of themselves and their customs than they are now, and perhaps with equally good reason. A friend of mine tells me that he smokes every day, at a cost of about sixpence a-week. Now, I would like to know in what other way so much enjoyment is to be bought for sixpence. Fancy the satisfaction of spending sixpence a-week in wine! It is well enough to preach about the selfishness of this expenditure; but we all spend more selfishly, and we all love pleasure, and I should very much like to see that cynic whose pleasures cost less than sixpence a-week."

The Egyptian pipes, especially those of modern date are

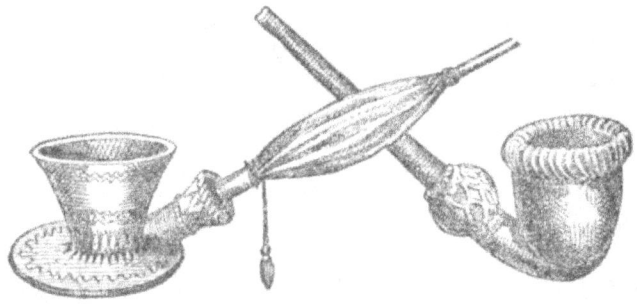

EGYPTIAN PIPES.

exceedingly fanciful in shape and resemble somewhat the pipes used by the Persians. Many of them are made of clay and are sold very cheap.* The Chinese use a variety of pipes but all of them have small bowls for the tobacco. Some of their pipes are made of brass and attached to the pipe is a receptacle for water, so as to cool the smoke before it passes into the mouth. The Japanese use both copper and silver pipes, most of them similar in shape and size to those used by the Chinese.

A writer says of smoking among the Japanese:

---

* Watlin says of smoking in Egypt: Tobacco is tolerated, and seems to become more common again, though a smoker is generally disliked and not allowed to perform the part of Imam or rehearse, of the prayers, before a congregation. The greater part of the people, however, detest and condemn still the use of tobacco, and I remember a Shaumar Bedawry who assured me that he would not carry that abominable herb on his Camel, even if a load of gold were given him."

SMOKING IN JAPAN.

"Let us sit down to a good Japanese dinner—down on the floor. Food on the floor. Fire and cigars or pipes on the floor. Sit on your heels, waiting. Enter first course—Fish-skin soup. Smoke. Third—Fish, cake and bean-cheese. Smoke. Fourth—Row fish and horse-radish. Smoke. Fifth—Broiled fish. Smoke again, Sixth—Custard soup. Smoke. Seventh—Chicken stew, turnips and onions. Smoke a little. Eighth—Cuttle-fish, wafer cakes, Nipon tea. Here, if tired you can stop at the end of about two hours' ankle-ache. All is cleanly, well spiced with talk, and served with the utmost politeness. Sipping tea may be substituted for the infinitesimal whiffs of polite smoking. A grand dinner is much more elaborate; at least, so far as the variety of smokes is concerned. After dinner, rest and smoke."

JAPANESE PIPES.

An English writer could very appropriately call this a cloud of smoke as he has another scene herein described.

"'Tis all smoke, possibly, but what cannot we discern, through a cloud of smoke? Objects dim, but

'Thick as autumnal leaves that strew the brooks
In Vallambrosa.'

Be the medium of the smoke an honest 'churchwarden,' a short clay, or a costly meerschaum; does the smoke emanate from a refined Havana, a neat Manilla, or a dainty cigarette, such as we are at this moment enjoying as a sequel to a modest breakfast, 'tis all smoke."

We have thus given a somewhat lengthy description of the custom and implements used in smoking, from the first discovery of the plant until now, and turn to other implements used in connection with the pipe. We, however, give the following from Cop's "Tobacco Plant," descriptive of the part played by tobacco on the stage two centuries ago:

"The 'Return from Parnassus' was published anonymously, and the copy I have used is dateless. It was 'publicly acted by the students of St. John's College in Cambridge.' In Act I., Scene 2d, characters are given of Spenser, Ben Jonson, Marlow, Drayton, Marston and Shakespeare, together with some other of the known poets and dramatists of the Elizabethan age. It contains many references to tobacco. In 'Act IV., Scene 1st,' the characters are thus placed: 'Sir Rodericke and Prodigo at one corner of the stage, Recorder and Amaretto at the other. Two pages scouring of Tobacco pipes.' Actual smoking from tobacco-pipes was introduced on the stage afterwards; and instances from the early dramas have been given by the writers on tobacco history. In the second scene of Act III. smoking is alluded to as one of the marks of the current man of fashion, and is coupled with that of wearing love-locks, which was to prove such a scandal to the Puritans. 'He gins to follow fashions. He wore thin sireduelt in a smooky roofe, must take tobacco and must weare a locke.' 'Work for Chimney Sweepers, or a Warning against Tobacconists, by J. H.,' was published in quarto in the year 1602.

"It was answered in the same year by the anonymous 'Defence of Tobacco,' a quarto of seventy pages. The author of the attack followed the line of King James, or, I should rather say, showed him the line to take, for the King's 'Counterblast' did not appear until he had been King of England for some years. The book is divided into sections, each section being called 'A Reason.' The seventh 'Reason' against the use of tobacco is, that the devil is the discoverer and suggester of smoking. 'It was first used and practised,' says J. H., 'by devils, priests, and, therefore, not to be used by us Christians. That the devil was the first author hereof. Monardus, in his 'Treatise of Tabaco,' dooth sufficiently witnesse, saying: The Indian priests, who, no doubt, were instruments of the devil, whom they serve, even before they answer to questions propounded to them by their princes, drinke of this tobacco-fume, with the vigour and strength whereof they fall suddenly to the ground as dead men, remaining so according to the quantity of smoke that they had taken. And when the hearbe hath done his worke, they revive and wake, giving answers according to the vissions and illusions which they saw while they were wrapt in that order.' It is not unlikely that J. H.'s authority had confused opium with tobacco.

"It was the opinion of the age that every Pagan deity had a real existence in the world of evil spirits. After further quotations of Monardus, to prove that the devil is 'the author of Tobacco, and of the knowledge thereof,' J. H. concludes his seventh reason by declaring, 'Wherefore in mine opinion this practice is more to be excluded of us

Christians, who follow Veritie and Truth, and detest and abhor the devil as a lyar and deceiver of mankind.' In the first year of this century, pipes were not only exhibited, but were used upon the stage. They seem at first to have been smoked, not during 'the induction.' In the induction to Ben Jonson's 'Cynthia's Revels' (1601), the Third Child says: 'Now, sir, suppose I am one of your genteel auditors, that am come in, having paid my money at the door, with much ado; and here take my place, and sit down, I have my three sorts of tobacco in my pocket, my light by me, and thus I begin.' The Third Child thereupon smokes; but it seems as if the smoking on the stage was a kind of protest against a prior smoking in the pit. In John Webster's 'Malcontent,' as augmented by John Marston in 1604, Sly says in the introduction: 'Come, coose, (coz or goose!) let's take some tobacco.'

"In 'The Puritan, or the Widow of Watling Street,' published in 1607, and attributed by some to Shakespeare, tobacco-taking or tobacco-drinking (as smoking was then usually called) appears no longer in the induction, but in the play itself, Idle, the highwayman, says to the old soldier, Skirmish, 'Have you any tobacco about you?' Idle being supplied, smokes a pipe on the stage. These extracts, however, may have been cited before, together with others of like character in the great days of the English Drama. Pipes continued to appear upon the stage until its abolition (in company with the Prayer Book) by the Puritan rulers. They reappeared on the stage of the Restoration. In Thomas Shadwell's 'Virtuos' (1676),—to take one instance,—Mirando and Clarinda fling away Snarl's cane, hat and periwig, and break his pipes, because he 'takes nasty tobacco before ladies.'"

There is printed evidence, however, in this same period to show not only that all the English ladies of the time were not enemies to tobacco, but that some of them were themselves smokers. In 1674 an anonymous quarto appeared under the title of "The Women's Petition against Coffee." It was a protest against the growing influence of the coffee-houses in seducing men away from their homes to sit together making mischief and drinking "this boiled soot." It was answered in the same year by "The Men's Answer to the Women's Petition." After speaking of the providential introduction of coffee into England in the midst of the Puritan epoch, when Englishmen wanted some kind of drink which would "at once make them sober and merry," the writer glorifies the coffee-house.

John Taylor, "the Water Poet," made a kind of compromise when he attributed the introduction of tobacco, not to the devil, but to Pluto,—"Pluto's Proclamation concerning his Infernal Pleasure for the Propagation of Tobacco." It appears in the folio collection of his works of the year 1628. The confusion of tobacco with opium and such destructive drugs seems to have been common with the travelers of the Sixteenth and Seventeenth centuries. Camerarius, in his "Historical Meditations," translated into English by John Malle (folio, 1621), speaks of tobacco as to be seen growing in many gardens throughout Europe. He quotes Jerome Benzo as saying that in Hispaniola "there be among them some that take so much of it, as their senses being all overcome and made drunke with the same, they fell down flat to the ground as if they were dead, and there lie without sense or feeling most part of the day or of the night."

The tobacco-box, during the reign of Elizabeth, was no unimportant part of a dandy's outfit; sometimes a pouch or bag was used. Tobacco-boxes came into general use in England soon after the introduction of tobacco, and were much sought after by all who "drank" tobacco. Marston, the Duke of New Castle, and other dramatists, alluded to the tobacco-box as a part of the smoker's outfit; thus in the play of "The Man in the Moone" (1609), one character, in answer to an inquiry who one of the company is, answers: "I know not certainly, but I think he cometh to play you a fit of mirth, for I behelde pipes in his pocket; now he draweth forth his tinder-box and his touchwood, and falleth to his tacklings; sure his throate is on fire, the smoke flyeth so fast from his mouth; blesse his beard with a bason of water, lest he burn it; some terrible thing he taketh, it maketh him pant and look pale, and hath an odious taste, he spitteth so after it.

The tobacco boxes of the Seventeenth Century were much larger than those of the present. Some of them held a pound of tobacco besides space for a number of pipes.

Many of them were made of brass while others were fashioned from horn:

"There is also a simple and ingenious tobacco-box used frequently in ale-houses, 'which keeps its own account,' with each smoker and acts also as a money-box. It is kept on parlor tables for the use of all comers; but none can obtain a pipe-full, till the money is deposited through a hole in the lid. A penny dropped in, causes a bolt to unfasten, and allow the smoker to help himself from a drawer full of tobacco. His honor is trusted so far as not to take more than his pipe-full, and he is reminded of it by a verse engraved on the lid:—

'The custom is, before you fill,
'To put a penny in the till.'"

Some of the tobacco boxes were made of silver and beautifully engraved with fancy sketches, historical scenes, or

ENGRAVED BOXES.

representations of personages, landscapes, flowers, etc. The late Duke of Sussex had a large collection of pipes and tobacco boxes.

A journal describing them says of the collection: "The Duke of Sussex had a wonderful collection of these, the values attached to some of them being almost fabulous. One example from the work-shop of Vienna—long celebrated for this description of art,—represented the combat of Hector and Achilles, the cover of the pipe being a golden hemlet cristatus of the Grecian type." Swiss and Tyrolean artists also produce exquisite carving, but use wood as a material; and in the famous collection of Baron de Watteville will be found a marvelous piece of carving representing Bellerophon overturning the Chimera. But French pipes are the most interesting of all to collectors, from the fact that tobacco was introduced into that country long before it was known in England, and also from the ingenuity of a people who can give interest of various kinds to what might seem a simple and prosaic branch of manufacture. In the sentiment of the following lines on "A pipe of Tobacco" by John Usher, all lovers of the plant will heartily join:

> "Let the toper regale in his tankard of ale,
> Or with alcohol moisten his thropple,
> Only give me I pray, a good pipe of soft clay,
> Nicely tapered, and thin in the stopple;
> And I shall puff, puff, let who will say enough,
> No luxury else I'm in lack o',
> No malice I hoard, 'gainst Queen, Prince, Duke or Lord,
> While I pull at my pipe of Tobacco.
>
> "When I feel the hot strife of the battle of life,
> And the prospect is aught but enticin',
> Mayhap some real ill like a protested bill,
> Dims the sunshine that tinged the horizon;
> Only let me puff, puff,—be they ever so rough,
> All the sorrows of life I lose track o',
> The mists disappear, and the vista is clear,
> With a soothing mild pipe of Tobacco.
>
> "And when joy after pain, like the sun after rain,
> Stills the waters, long turbid and troubled,
> That life's current may flow, with a ruddier glow,
> And the sense of enjoyment be doubled,—
> Oh! let me puff, puff, till I feel *quantum suff*,
> Such luxury still I'm in lack o',
> Be joy ever so sweet, it would be incomplete,
> Without a good pipe of tobacco.
>
> "Should my recreant muse,—Sometimes apt to refuse
> The guidance of bit and of bridle,
> Still blankly demur, spite of whip and of spur,
> Unimpassioned, inconstant, or idle;
> Only let me puff, puff, till the brain cries enough,

Such excitement is all I'm in lack o',
And the poetic vein soon to fancy gives reign,
Inspired by a pipe of Tobacco.

"And when with one accord, round the jovial board,
In friendship our bosoms are glowing;
While with toast and with song we the evening prolong,
And with nectar the goblets are flowing;
Still let us puff, puff—be life smooth, be it rough,
Such enjoyment we're ever in lack o';
The more peace and goodwill will abound as we fill
A jolly good pipe of Tobacco."

The tobacco jar is another accessory of more recent date than tobacco pipes but interesting from the varieties of style

TOBACCO JARS.

and shapes. The finest are made of porcelain and are lavish in design and enrichment. Of all the articles of the smokers' paraphernalia none however exhibit more fanciful designs than Tobacco-stoppers used by smokers for crowding the tobacco into the pipe while smoking. The author of "A Paper of Tobacco" says:

"This was the only article on which the English smoker prided himself. It was made of various materials—wood, bone, ivory, mother-of-pearl, and silver: and the forms which it assumed were exceedingly diversified. Out of a collection of upwards of thirty tobacco-stoppers of different ages, from 1688 to the present time, the following are the most remarkable: a bear's tooth tipped with silver at the bottom, and inscribed with the name of Captain James Rogers of the

Happy Return whaler, 1688; Dr. Henry Sacheverel in full canonicals, carved in ivory, 1710; a boat, a horse's hind leg, Punch, and another character in the same Drama, to wit: his Satanic majesty; a countryman with a flail; a milkmaid; an emblem of Priopus; Hope and Anchor; the Marquis of Granby; a greyhound's head and neck; a paviour's rammer; Lord Nelson; the Duke of Wellington; and Bonaparte. The tobacco-stopper was carried in the pocket or attached to a ring worn on the finger."

In Butler's Hudibras it is alluded to in connection with the astronomer's sign.

> "——Bless us! quoth he,
> It is a planet now I see;
> And if I err not, by his proper
> Figure that's like tobacco-stopper,
> It should be Saturn!"

In James Boswell's "Shrubs of Parnassus" (1760) a description in verse of the various kinds of tobacco-stoppers is given:

> "O! let me grasp thy waist, be thou of wood
> Or levigated steel, for well 'tis known
> Thy habit is disease. In iron clad
> Sometimes thy feature roughen to the sight,
> And oft transparent art thou seen in glass,
> Portending frangibility. The son
> Of laboring mechanism here displays
> Exuberance of skill. The curious knot,
> The motley flourish winding down the sides,
> And freaks of fancy pour upon the view
> Their complicated charms, and as they please,
> Astonish. While with glee thy touch I feel,
> No harm my fingers dread. No fractured pipe
> I ask, or splinters aid, wherewith to press
> The rising ashes down. Oh! bless my hand,
> Chief when thou com'st with hollow circle crowned
> With sculptured signet, bearing in thy womb
> The treasured Cork-screw. Thus a triple service
> In firm alliance may'st thou boast."

Tobacco-stoppers were often made of wood from some relic like a celebrated tree or mansion which gave additional value by its historic associations. Taylor alludes to several made from the well known Glastonbury thorn. He says:—

"I saw the sayd branch, I did take a dead sprigge from it,

TOBACCO STOPPERS.

wherewith I made two or three tobacco-stoppers, which I brought to London."

Pipes and tobacco-stoppers have often been favorite testimonials of friendship and reward. Fairholt says:—

"It was the custom during the last century to present country churchwardens with tobacco-boxes, after the faithful discharge of their duties."

The following lines from "The Tobacco Leaf," penned by some favored one on receiving a rare pipe, are no doubt as neat as the object that called them forth:—

> "I lifted off the lid with anxious care,
> Removed the wrappages, strip after strip,
> And when the hidden contents were laid bare,
> My first remark was: "Mercy, what a pipe!"
>
> A pipe of symmetry that matched its size,
> Mounted with metal bright—a sight to see—
> With the rich umber hue that smokers prize,
> Attesting both its age and pedigree.
>
> A pipe to make the royal Freidrich jealous,
> Or the great Teufelsdrockh with envy gripe!
> A man should hold some rank above his fellows
> To justify his smoking such a pipe!

MUSINGS OVER A PIPE.

What country gave it birth? What blest of cities
Saw it first kindle at the glowing coal?
What happy artist murmured "*Nunc dimittis*,"
When he had fashioned this transcendent bowl!

Has it been hoarded in a monarch's treasures?
Was it a gift of peace, or price of war?
Did the great Khalif in his "Houre of Pleasures,"
Wager and lose it to the good Zaafar?

It may have soothed mild Spenser's melancholy,
While musing o'er traditions of the past,
Or graced the lips of brave Sir Walter Raleigh,
Ere sage King Jamie blew his "Counterblast."

Did it, safe hidden in some secret cavern,
Escape that monarch's pipoclastic ken?
Has Shakespeare smoked it at the Mermaid Tavern,
Quaffing a cup of sack with rare old Ben?

Ay, Shakespeare might have watched his vast creation
Loom through its smoke—the spectre-haunted Thane,
The Sisters at their ghostly invocations,
The jealous Moor and melancholy Dane.

Round its orbed haze and through its mazy ringlets,
Titania may have led her elfin rout,
Or Ariel fanned it with his gauzy winglets,
Or Puck danced in the bowl to put it out.

Vain are all fancies—questions bring no answer;
The smokers vanish, but the pipe remains;
He were indeed a subtle necromancer,
Could read their records in its cloudy stains.

Nor this alone: its destiny may doom it
To outlive e'en its use and history—
Some ploughman of the future may exhume it
From soil now deep beneath the eastern sea.

And, treasured by some antiquarian Stultus,
It may to gaping visitors be shown,
Labelled: "The symbol of some ancient Cultus,
Conjecturally Phallic, but unknown."

Why do I thus recall the ancient quarrel
'Twixt Man and Time, that marks all earthly things?

"PUFFS FROM A PIPE."

Why labor to re-word the hackneyed moral,
Ὡς φύλλωνγενεή, as Homer sings?

For this: Some links we forge are never broken:
Some feelings claim exemption from decay;
And Love, of which this pipe was but the token,
Shall last, though pipes and smokers pass away."

The verse that has been written in praise as well as dispraise of the "Indian Novelty" would of itself fill a volume of no "mean pretentions." The following clever lines from The Tobacco Plant entitled "Puffs from a Pipe," convey much advice to all smokers of tobacco.

Sage old friend! with judgment ripe;
Come and join me in a pipe.

Brother student! brother joker,
Thee I greet, O! brother smoker.

Smoke, O! men of every station,
Every climate, every nation.

East and West, and South and North,
Recognize Tobacco's worth.

Red man! let thy warfare cease:
Smoke the calumet of peace.

Chinaman! shun opium-grief:
Use the pure Tobacco leaf.

Frenchmen! no more foes provoke:
Follow arts of peace—and smoke!

German victors! crowned with laurel,
Smoke, content; and seek no quarrel.

Americans no one needs bid
To blow a cloud, or take a quid.

Though rows shake Dame Europa's school,
Johnny Bull smokes, calm and cool.

Toffy, it will ease thy brain, man!
Smoke and snuff, and smoke again, man!

A GOOD THING.

Paddy, light of heart and gay,
Smoke thy dhudeen: short black clay.

Sawney, on thy Hielen' hill,
Tak' thy sneishin'; tak' thy gill!

Tourist, thou hast journey'd far;
Rest, and light a mild cigar.

Sailor, from the stormy seas,
Take a quid, and take thine ease.

"Soldier tired," put off thy shako;
Prepare to fire, and burn tobacco.

Workman, prize thine honest labor;
Burn thy weed, and love thy neighbor!

Evil-doers, when ye burn
The weed; think how soon 'twill be your turn.

Artist, let thy " coloring " be
Of a pipe; thy " drawing," free!

Miser, moderate thy greed!
Mend thy life, and take a weed.

Lawyer, loose thy bitter gripe!
Burn thy writ—to light a pipe.

Statesman, harassed night and day,
Blow a cloud; puff care away!

Hardy tiller of the soil!
Light a pipe; 'twill lighten toil.

Usurer, we surely know
Thou wilt have thy *quid pro quo*.

Merchant, smoke thy pipe; hang care!
Draughts are always honored there.

Gentle friend, whom troubles fret!
Smoke a soothing cigarette.

Preacher! take a pinch with me:
Snuff is dust, and so are we.

A WARNING.

Hence with moralizings musty!
I say life is "not so dusty."

Smoke in gladness; smoke in trouble;
Soothe the last, the former double!

Teach the Fiji Indians, then,
To chew their quids, instead of men.

Pain from heart and brain to wipe,
Pass the weed, and fill your pipe!

LORD AND LACKEY.

Prince and peasant, lord and lackey,
All in some form take their 'Baccy."

The evil effects occasioned by man's indulging too frequently in tobacco have been the subject of many a fierce debate between the friends and foes of the "great plant." Many, however, are not aware of the fatality attending its use by the brute creation. A modern Englisn poet on hearing of the result produced on a cow from chewing tobacco, penned the following sad lines which he entitles—"An elegy on somebody's Cow."

Weep! weep, ye chewers! Lowly bend, and bow;
Here lieth what was once a happy cow.
No more her voice she'll raise, now low, now high,
In amber fields, beneath an autumn sky;

SAD FATE OF A "CHEWER."

No more she'll wander to the milking-pail,
While swine stand by to see her chew " pig-tail;"
No more round her the bees, a busy crew,
Shall linger, eager after " honey-dew;"
No more for her shall smoking grains be spread:
All bellowless remains her empty shed.

Sad was her fate. Reflect, all ye who read:
Life's flower destroyed by the accursed weed.
When first the yellow juice streamed o'er her lip,
One might have said, " This is a sad cow-slip."
To chew the peaceful cud by nature bid,
Degraded man taught her to chew a quid.
Sad the effect on body and on mind:
Her coat grew " shaggy," her milk nicotined;
Over her head shall naught but clover grow,
While o'er her peaceful grave the clouds shall blow.

No invalid shall ask for her cow-heel,
To heal his ailments with the simple meal;
Her whiskful tail into no soup shall go;
Mother of " weal " that would but bring us woe.
Her tripe shall honor not the festive meal.
Where smoking onions all their joys reveal;
Nor shall those shins that oft lagged on the road,
Be sold in cheap cook-shops as " *a la mode*,"
Her tongue must soon be sandwiched under ground,
Nor at pic-nics with cheap champagne go round;
Yea, even her poor bones are past all hope—
Not fit to be boiled down for scented soap.

Ah! hide her hide, poor beast. Her stomachs five
Dyed with the chewing she could not survive;
The very worms from her will turn away,
To seek some anti-chewer for their prey.
Ye chewers! be ye pilgrims to her tomb;
Lament with us o'er her untimely doom.
Awhile she stood the anti-chewer's butt,
Till scythe-arm'd Time gave her an " ugly cut."
She stagger'd to her death, and feebly cried,
And sneezed, " Achew! achew! " and chewing died.

There are many parodies of popular poems written in praise of the weed; of which the following in imitation of Tennyson's " Charge of the Light Brigade," entitled " The Charge of the Tobacco Jar Brigade," is one of the best.

"Epigrams, epigrams,
Pour'd in, and numbered—
Good, bad, indifferent—
More than Six Hundred.
"Epigrams potters want,"
Quoth The Tobacco Plant:
Write! you for fame who pant;
Write! we'll three prizes grant."
Wrote for Tobacco-Jars,
Over Six Hundred.

Postmen, ere morning's light;
Postmen, whilst day was bright;
Postmen, as closed in night,
Ran—tan'd and thunder'd
Loud at our office door;
Brought letters, many score—
Contents of bags—to pour
Table and desk all o'er:
Handfuls and armfuls bore,
Casting them on the floor.
Then through the town they tore,
Hastening back for more—
More than Six Hundred.

Letters to right of us,
Letters to left of us,
Letters in front of us,
Seeming unnumbered!
Envelopes every size
Met our astonish'd eyes.
Writer with writer vies!
Which wins the chiefest prize
Out of Six Hundred.

How did each writer strain
After a happy vein!
Pegasus, spurning rein,
Shied, jibb'd, and blunder'd.
Reverend writers, then
Took up the winged pen;
Suff'rers on beds of pain
Sought the bright muse again;
Lawyer and barrister
Courted and harassed her;
M. D.s and editors;
Debtors and creditors;
Artists and artisans,
Nicotine's partisans;
Nurses and gentle dames
Call'd it endearing names;
Poets, ship-masters, too;
Ay! poetasters, too;
Wooing fair Nicotine,
Six hundred scribes were seen.
Anti-Tobacco cant,
Bigoted, bilious rant,
Bursting to vent their spleen,
Joined the Six Hundred.

Flash'd many fancies rare;
Flash'd like Aurora's glare;
Quick jotted down with care;
Some the reverse of fair;
Some that we well could spare;
Some that were made to bear
Blunders unnumbered.
Plunging in metaphor,
Not a bit better for—
Pardon the Cockney rhyme!—
Similies plunder'd.
Praising Tobacco smoke,
Heeding not grammar's yoke,
Prosody's rules they broke.
Many a rhyming moke,
Sense from rhyme sundered:
Many wrote well, but not—
Not the Six Hundred.
Honour Tobacco! roll'd,
Cut, press'd, however sold.
Alpha and Beta, bold,
Ye shall be tipp'd with gold.
Omega shall be sold,
Others in type behold
Nearly Six Hundred."

The following poem entitled "Weedless," after Byron's "Darkness," gives a vivid description of the world without tobacco.

A BAD DREAM.

"I had a dream, and it was all a dream:
Tobacco was abolish'd, and cigars
Were flung by "Antis" fearsome space—
The foreign and the British fared alike—
And the blue smoke was blown beyond the moon.
Night came and went and came, and brought no "weed,"
And men forgot their suppers, in the dread
Of the dire desolation; and all tongues
Were tingling with the taste of empty pipes;
And they did live all wretched; old hay bands,
And street-door mats, and clover brown and dry;
Carpets, rope-yarn, and such things as men sell,
Were burnt for 'bacca; haystacks were consumed,
And men were gathered round each blazing mass,
To have another makeshift sniff.
Happy were those who smoked, with smould'ring logs,
The harmless Yarmouth bloater after death—
Another pipe not all the world contain'd;
The furze was set on fire, but, hour by hour,
The stock diminish'd; all the prickly points
Quivered to death, and soon it all was gone.
The lips of men by the expiring stuff
Drew in and out, and all the world had fits.
The cinders fell upon them; some sprang up,
And blew their noses loud, and some did stand
Upon their heads, and sway'd despairing feet;
And others madly up and down the world
With "two-pence" hurried, shouting out for "Shag;"
And wink'd and blink'd at th' unclouded sky,
The "Anti's" smokeless banner—then again
Flung all their halfpence down into the dust,
And chewed their tainted pockets; snuffers wept,
And, flatt'ning noses on the dreary ground,
Inhaled the useless dust; the biggest "rough"
Came mild, tobacco-begging; p'licement came,
And mix'd themselves among the multitude,
Run in" forgotten; uniforms were chew'd,
And teeth which for a moment had had rest,
Did move themselves again; old beaver hats
Fetch'd little fortunes; they were torn in bits,
And smok'd or chew'd at will; no bits were left.
All earth was but one thought, and that was smoke,
Immediate and glorious; and a pang
Of horror came at intervals, and men
Cried; and the boys were restless as themselves,
Till by degrees their stockings were devour'd;

E'en pipes were dropp'd despairing—all, save one,
One man was faithful to his pipe, and kept
Despair and deeper misery at bay,
By seeking ever for a " topper," dropped
From some spurned pipe, but that he could not find;
So, with a piteous and perpetual glare,
And a quick dissolute word, sucking the pipe,
Which answer'd never with a whiff, he slept;
The crowd dispersed by slow degrees, but two
Of all the dreary company remain'd,
And they kept 'bacca shops; they sat upon
The scented lid of a tobacco tub,
Wherein was heap'd a mass of coined bronze—
Profits of 'bacca sold—they were sold out;
They, grinning, scraped with their warm, eager hands
The little halfpence and the bigger pence,
Counted a little time, and cried " Haw! haw!"
Like a whole rookery; then lifted up
The tub as it grew lighter, and beheld
Each other's profits; saw, and smiled, and winked,
Uncaring that the world was poor indeed,
So they were rich in pence. The world was mad,
The populace and peerage both alike
Birds—Eyeless, Shagless, and returnless, too—
Oh! day of death, oh! chaos of hard times!—
And princes, dukes, and lords, they all stood still,
Feeling within their pockets' silent depths;
And sailors went a-moaning out to sea,
And chew'd their cables piecemeal: then they wept,
And slept on the abyss without a quid.
All quids were gone, cigars were in their graves;
The plant, their mother, had been rooted up;
Pawnbrokers had a ton of pipes apiece,
And " Antis " triumph'd. Then they had no need
To keep a " Sec.," so Reynolds got the " sack."

One of the best of all parodies is one in imitation of Longfellow's " Excelsior," entitled " Tobacco." It is from " Copis' Tobacco Plant."

" The summer blight was falling fast,
 When straight through dirty London passed
 A youth, who bore, through road and street,
 A packet, thereon written neat;
                    "Tobacco!"

THE TRAVELER.

His brow was glad, his laughing eye
Flashed like a gooseberry in a pie;
And like a penny whistle rung
The piping notes of that strange tongue—
     "Tobacco!"

In dusty homes he saw the light
Of supper fires gleam warm and bright;

THE STRANGE YOUTH.

Above, the ruddy chimneys smoked:
He from his lips the word evoked—
     "Tobacco!"

"Try not the weed," good Reynolds said;
"I've smoked it 'till I'm nearly dead:
Take not the juice in thy inside;"
But loud the jovial voice replied—
     "Tobacco!"

"Oh! stay," the maiden said, "and rest;
I have got on my Sunday best:"
A wink stood in his bright blue eye,
And answered he, without a sigh—
     "Tobacco!"

"Beware the briar's poison'd root;
Beware the birds-eye put into 't."

This was the Anti's latest greet.
A voice replied, far up the street—
    "Tobacco!"

At break of day, on Clapham Rise,
A pot-boy opened both his eyes,
And to himself did gently swear,
To hear a voice call through the air—
    "Tobacco!"

A traveler up a tree he found,
Who smoked and spat upon the ground;
And then among the blossoms ripe
He cried, while puffing at his pipe—
    "Tobacco!"

There in the grayish twilight, "What's
That you say?" cried eager Pots,
And from the branch so green and far,
A voice fell like a broken jar—
    "Tobacco."

The following lines from the same source have been very appropriately called "The Smoker's Calendar."

When January's cold appears,
A glowing pipe my spirit cheers;
And still it glads the length'ning day,
'Neath February's milder sway.
When March's keener winds succeed,
What charms me like the burning weed?
When April mounts the solar car,
I join him, puffing a cigar;
And May, so beautiful and bright,
Still finds the pleasing weed a-light.
To balmy zephyrs it gives zest,
When June in gayest livery's drest.
Through July Flora's offspring smile,
But still Nicotia's can beguile;
And August, when its fruits are ripe,
Matures my pleasure in a pipe.
September finds me in the garden,
Communing with a long churchwarden.
Ev'n in the wane of dull October,
I smoke my pipe and sip my "robur,"
November's soaking show'rs require
The smoking pipe and blazing fire:

> The darkest day in drear December's—
> That's lighted by their glowing embers.

The Hon. "Sunset" Cox in his lecture on American Humor alluded to the national characteristics of the French, Spanish, German, and other nationalities, says:—

"The highest enjoyment of a Frenchman is to hear the last cantatrice, the Spaniard enjoys the most skillful thrust of the matador in the bull arena, the Neapolitan the taste of the maccaroni, the German his beer and metaphysics, the darkey his banjo, and the American—

> 'To the American there's nothing so sweet
> As to sit in his chair and tilt up his feet,
> Enjoy the Cuba, whose flavor just suits,
> And gaze at the world through the toes of his boots.'"

This would seem to be a feature of the Dutch according to a late traveler, who says:—

"I like Holland—it is the antidote of France. No one is ever in a hurry here. Life moves on in a slow, majestic stream, a little muddy and stagnant, perhaps, like one of their own canals; but you see no waves, no breakers; not an eddy, nor even a froth bubble, breaks the surface. Even a Dutch child, as he steals along to school, smoking his short pipe, has a mock air of thought about him."

The following epigrams for tobacco jars from "The Tobacco Plant" evince much "taste, wit, and ingenuity."

> Fill the bowl, you jolly soul,
> And burn all sorrow to a coal.
> *Henry Clay.*

> That man is frugal and content indeed,
> Who finds food, solace, pleasure in a weed.
> *The " Weed.*

> Behold! this vessel hath a moral got,
> Tobacco-smokers all must go to pot.
> *Epigrammatic.*

> A weed you call me, but you'll own
> No rose was e'er more fully blown.
> *Sic Itur ad Nostra.*

> Great Jove, Pandora's box with jars did fill
> This Jar alone has power those jars to still.
> *In Nubilus.*

EPIGRAMS. 193

Tobacco some say, is a potent narcotic,
That rules half the world in a way quite despotic;
So to punish him well for his wicked and merry tricks,
We'll burn him forthwith, as they used to do heretics.
Zed.

SMOKER READING EPIGRAMS.

No use to draw upon a bank if no effects are there,
But a draw of this Tobacco is quite a safe affair;
And a pipe with fragrant weed (such as I hold) neatly stuffed,
Is just the only thing on earth that ought to be well puffed.
R. S. V. P.

Poor woman "pipes her eye,"
When in affliction's gripe;
But man, far wiser grown,
Just eyes his pipe.
In Nubilus.

Sir Walter Raleigh! name of worth,
How sweet for thee to know
King James, who never smoked on earth,
Is smoking down below.
Ex Fumo dare Lucem.

Travelers say Tobacco springs
From the graves of Indian kings:
Fill your pipe, then—smoke will be

EPIGRAMS.

Incense to their memory.
Though the weed's nor rich nor rare,
'Tis a balm for every care.
Peter Piper.

Give me the weed, the fragrant weed,
My wearied brain to calm;
In a wreath of smoke, while I crack my joke,
I'll find a healing balm.

Day after day, let come what may,
The pipe of peace I'll fill;
I readily pay for briar or clay,
To save a doctor's bill.
Pompons.

Great men need no pompous marble
To perpetuate their name;
Household gear and common trinkets
Best remind us of their fame.

Raleigh's glory rests immortal
On ten thousand thousand urns,
Every jar is *in memoriam*,
Every fragrant pipe that burns.
At an Ash.

There are jars of jelly, jars of jam,
Jars of potted-beef and ham;
But welcome most to me, by far,
Is my dear old Tobacco-Jar.
There are pipes producing sounds divine,
Pipes producing luscious wine;
But when I consolation need,
I take the pipe that burns the weed.
Jars.

Friend of my youth, companion of my later days,
What needs my muse to sing thy various praise?
In country or in town, on land or sea,
The weed is still delightful company.
In joy or sorrow, grief or racking pain,
We fly to thee for solace once again.
Delicious plant, by all the world consumed,
'Tis pity thou, like man, to ashes too art doom'd.
Erutxim.

EPIGRAMS.

Hail plant of power, more than king's renown,
Beloved alike in country and in town;
In hotter climes oft mingled with the jet
Of falling fountains; whilst the cigarette
Kisses the fair one's lips, and by thy breath
Redeems the wearied heart from ennui's death.
Theta.

If e'er in social jars you join,
Seek this, and let them cease:
Let all your quarrels end in smoke,
And pass the pipe of peace.
Fumigator.

Many a jar of old outbroke
Into fire and riot;

THE EXPLOSION.

This will yield, with fragrant smoke,
Happy thought, and quiet.
41,911.

The moralist, philosopher, and sage.
Have sought by every means, in every age,
That which should cause the strife of men to cease,
And steep the world in fellowship and peace;

But all their toil and diligence were vain,
'Till Raleigh, noble Raleigh! crossed the main,
And brought to Britain's shores the wish'd-for prize,
The sovereign balm of life—within it lies.
Dum Spiro Fumigo.

To rich men a pastime, to poor men a treat,
To all a true tonic most bracing and sweet,
To talent a pleasure, to genius a joy,
To workmen a comfort, to none an alloy,
The tyrant it softens; it soothes him if mad,
The king who may rule if he smokes not, is sad.
Kit.

Sacred substance! sweet, serene;
Soothing sorrow's saddest scene:
Scent-suffusing, silv'ry smoke,
Softly smoothing suffering's stroke;—
Solacing so silently—
Still so swift, so sure, so sly:
Smoke sublimated soars supreme,
Sweetest soul-sustaining stream!
Similia Similibus.

Why should men reek, like chimneys, with foul smoke,
Their neighbors and themselves to nearly choke?
Avoid it, ye John Bulls, and eke ye Paddies!
Avoid it, sons of Cambria, and Scottish laddies!
Let reason convince you that it very sad is,
   And far too bad is,
   And enough to make one mad is
To be smoked like a red herring or rank Finedon haddies.
J. S.

No punishment save hanging's too severe
For those who'd rob the poor man of his beer;
But for the wretch who'd take away his pipe,
I think he's fully execution ripe!
Pipe Clay.

Weeds are but cares! Well, what of that!
There's one weed bears a goodly crop;
And this exception, then, 'tis flat,
Doth give that rule a firmer prop.
Tobacco brings the genial mood,
Warm heart, shrewd thought, and while we reap
From this poor weed such harvest good,
We'll hold more boasted harvests cheap.
Festus.

EPIGRAMS.

To poets give the laurel wreath, let heroes have their lay,
Of roses twine for lovely youth the garland fresh and gay;
But we poor mortals, quite content, life's fev'rish way pursue,
Can we but crown our foolish pates with wreaths of fragrant blue,
Convinced that all terrestrial things which please us or provoke,
Of ashes come, to ashes go, and only end in smoke.
Pocosmipo,

Whilst cannon's smoke o'erwhelms with deadly cloud
The soldier's comrades in a common shroud,
And whilst the conflagration in the street,
With crushing roar the ruin makes complete,
Tobacco's smoke like incense seeks the skies—
Blesses the giver, and in silence dies!
Theta.

Use me well, and you shall see
An excellent servant I will be;
Let me once become your master,
And you shall rue the great disaster!

As coin does to he who borrows,
I'll soothe your cares and ease your sorrows;
Abuse me, and your nerves I'll shatter,
Your heart I'll break, your cash I'll scatter,
Use, not Abuse.

The savage in his wild estate,
When feuds and discords cease,
Soothes with the fragrant weed his hate,
And smokes the pipe of peace.

Long may the plant good-will create,
And banish strife afar:
Our only cloud its incense sweet,
And this our only jar.
Scire Facias.

Breathes there a man with soul so dead,
Who never to himself hath said;
I'll have to smoke, or I'll be dead?
If so, then let the caitiff dread!
My wrath shall fall upon his head.
'Tis plain he ne'er the Plant hath read;
But "goody" trash, perchance, instead.
Dear Cope, good night!—Yours, Master Fred.

DOCTOR PARR AS A SMOKER.

That tobacco in one form or another has been patronized from the cottage to the throne, no one will deny who is at all acquainted with the history of the plant. And while it has had many a royal hater, it can also boast of having many a kingly user. A favorite of king and courtier, its use was alike common in the palace and the courtyard. It can claim, also, many celebrated physicians who have been its patrons, and among them the noted Dr. Parr. We give an anecdote of him showing his love of weed and wit.

The partiality this worthy Grecian always manifested for smoking is well known. Whenever he dined he was always indulged with a pipe. Even His Majesty, when Dr. Parr was his guest at Carlton Palace, condescended to give him a smoking-room and the company of Colonel C———, in order that he might suffer no inconvenience. "I don't like to be smoked myself, doctor," said the royal wit, "but I am anxious that your pipe should not be put out." One day, Dr. Parr was to dine at the house of Mr. ———, who informed his lady of the circumstance, and of the doctor's passion for the pipe. The lady was much mortified by this intimation, and with warmth said, "I tell you what, Mr. ———, I don't care a fig for Dr. P.'s Greek; he shan't smoke here." "My dear," replied the husband, "he must smoke; he is allowed to do so everywhere." "Excuse me, Mr. ———, he shall not smoke here; leave it to me, my dear, I'll manage it." The doctor came; a splendid dinner ensued; the Grecian was very brilliant. After dinner, the doctor called for his pipes. "Pipes!" screamed the lady. Pipes! For what purpose?" "Why, to smoke, madam!" "Oh! my dear doctor, I can't have pipes here. You'll spoil my room; my curtains will smell of tobacco for a week." "Not smoke!" exclaimed the astonished and offended Grecian. "Why, madam, I have smoked in better houses." "Perhaps so, sir," replied the lady, with dignity; and she added with firmness, "I shall be most happy, doctor, to show you the rights (rites?) of hospitality; but you cannot be allowed to smoke." "Then, madam," said Dr. Parr, looking

THEORY AGAINST EXPERIENCE.

at her ample person; "then, madam,—I must say, 'madam,—'" "Sir, sir, are you going to be rude?" "I must say, madam," he continued, ".you are the greatest tobacco-stopper in all England." Of the clergy, Whatley was one of the greatest in intellect, and, as a smoker, was devotedly attached to tobacco; his pipes, when out, served him for a book-marker. In summer-time he might be seen, of an evening, sitting on the chains of Stephen's Green, thinking of "that," as the song says, and of much more, while he was "smoking tobacco." In winter he walked and smoked, vigorously in both cases, on the Donnybrook road; or he would be out with his dogs, climbing up the trees to hide amid the branches a key or a knife, which, after walking some distance, he would tell the dogs he had lost, and bid them look for it and bring it to him.

Of many warriors, none have been more devoted to the plant than Napoleon, Frederick of Prussia and Blücher the Bold. The following anecdote of the latter is one of the best of its kind: "As is well-known, Field-Marshal Blücher, in addition to his brave young 'fellows' (as he called his horsemen), loved three things above all, namely, wine, gambling, and a pipe of Tobacco. With his pipe he would not dispense, and he always took two or three puffs, at least, before undertaking anything. 'Without Tobacco, I am not worth a farthing,' he often said. Though so passionately fond of Tobacco, yet old 'Forwards' was no friend of costly smoking apparatus; and he liked best to smoke long, Dutch clay pipes, which, as everybody knows, very readily break. Therefore, from among his 'young fellows' he had chosen for himself a Pipe-master, who had charge of a chest well packed with clay pipes; and this chest was the most precious jewel in Blücher's field baggage. If one of the pipes broke, it was, for our hero, an event of the greatest importance. On its occurrence, the 'wounded' pipe was narrowly examined, and if the stem was not broken off too near the head, it was sent to join the corps of Invalids, and was called 'Stummel' (Stump, or Stumpy). One of these Stumpies the

Field-Marshal usually smoked when he was on horseback, and when the troops were marching along or engaged in a reconnoissance, and eye-witnesses record that many a Stumpy was shot from his mouth by the balls of the enemy—nothing but a piece of the stem then remaining between his lips. Blücher's Pipe-master, at the time of the Liberation War, was Christian Hennemann, a Mecklenburg and Rostock man, like Blücher himself, and most devotedly attached to the Field-Marshal. He knew all the characteristic peculiarities of the old hero, even the smallest, and no one could so skillfully adapt himself to them as he. His duties as Pipemaster, Hennemann discharged with great fidelity; yea, even with genuine fanatical zeal. The contents of the pipe-chest he thoroughly knew, for often he counted the pipes. Before every fierce fight, Prince Blücher usually ordered a long pipe to be filled. After smoking for a short time, he gave back the lighted pipe to Hennemann, placed himself right in the saddle, drew his sabre, and with the vigorous cry, 'Forward, my lads!' he threw himself into the fierce onset on the foe.

On the ever-memorable morning of the battle of Belle-Alliance (Waterloo), Hennemann had just handed a pipe to his master, when a cannon-ball struck the ground near, so that earth and sand covered Blücher and his gray horse. The horse made a spring to one side, and the beautiful new pipe was broken before the old hero had taken a single puff. 'Fill another pipe for me,' said Blücher; 'keep it lighted, and wait for me here a moment, till I drive away the French rascals. Forwards, lads!' Thereupon there was a rush forwards; but the chase lasted not only 'a moment,' but a whole hot day. At the Belle-Alliance Inn, which was demolished by shot,—the battle having at last been gained,—the victorious friends, Blücher and Wellington, met and congratulated each other on the grand and nobly achieved work, each praising the bravery of the other's troops. 'Your fellows slash in like the very devil himself!' cried Wellington. Blücher replied, 'Yes; you see, that is their business. But brave as they are, I know not whether one of them would stand as firmly and calmly in the midst of the shower of balls and bullets as your English.' Then Wellington asked Blücher about his previous position on the field of battle, which had enabled him to execute an attack so fatal to the enemy.

Blücher, who could strike tremendous blows, but was by no means a consummate orator, and could not paint his deeds in words, conducted Wellington to the place itself. They found it completely deserted; but on the very spot where Blücher had that morning halted, and from which he had galloped away, stood a man with his head bound up, and with his arm wrapped in a handkerchief. He smoked a long, dazzling white clay pipe. 'Good God!' exclaimed Blücher, 'that is my servant, Christian Hennemann. What a strange look you have, man! What are you doing here?' 'Have you come at last?' answered Christian Hennemann, in a grumbling tone; 'here I have stood the whole day, waiting for you. One pipe after another have the cursed French shot away from my mouth. Once even a blue bean (a bullet) made sad work with my head, and my fist has got a deuce of a smashing. That is the last whole pipe, and it is a good thing that the firing has ceased; otherwise, the French would have knocked this pipe to pieces, and you must have stood there with a dry mouth.' He then handed the lighted pipe to his master, who took it, and after a few eagerly-enjoyed whiffs, said to his faithful servant, 'It is true, I have kept you waiting a long time; but to-day the French fellows could not be forced to run all at once.' With astonishment, Wellington listened to the conversation. Amazed, he looked now at the Field-Marshal, now at the 'Pipe-master,' and now at the branches of trees and the balls scattered all round, which made it only too evident what a dangerous post this spot must have been during the battle. The wound in Hennemann's head proved to be somewhat serious; his hand was completely shattered; and yet, in the midst of the tempest

THE FAITHFUL ATTENDANT.

of shot, he had stood there waiting for his beloved master."*

Tobacco smoking, however, can boast of many patrons besides warriors, physicians and statesmen, some of the finest writers of the last three centuries have indulged in the weed. The following extract from the "Australasian" entitled, "Tobacco Smoking" refers to many literary smokers.

"Burke felt himself precluded from 'drawing an indictment against a whole community.' The critical moralist pauses before the formidable array of the entire social world, civilized and savage. The Cockney, leaving behind him the regalias and meerschaums of the Strand, finds the wax-tipped clay-pipe in the parlors of Yorkshire: finds dhudeen and cutty in the wilds of Galway and on the rugged shores of Skye and Mull. The Frenchman he finds enveloped in clouds of Virginia, and the Swede, Dane, and Norwegian, of every grade or class, makes the pipe his travelling companion and his domestic solace. The Magyar, the Pole and the Russian rival the Englishman in gusto, perhaps excel him in refinement; the Dutch boor smokes finer Tobacco than many English gentlemen can command, and more of it than many of our hardened votaries could endure; but all must yield, or rather, all must accumulate, ere our conceptions can approach to the German. America and the British colonies round off the picture, adding Cherokees, Redmen and Mongolians *ad libitum*. The Jew whether in Hounds ditch, Paris Hamburgh, or Constantinople, ever inhales the choicest growths, and the Mussulman's 'keyf' is proverbial. India and Persia dispute with us the palm of refinement and intensity, but the philosopher of Australia is embarrassed when he asks himself to whom shall I award that of zealous devotion?

"Dr. Adam Clarke, who could never reconcile himself to the practice, deemed it due to his piety to find a useful purpose in the creation of tobacco by all-seeing Wisdom, and as that discovered by the instincts of all the nations of the planet, and practiced by mankind for three centuries, is wrong, the benevolent Wesleyan of Heydon, applied himself diligently and generously to correct the world, and to vindicate its Author. "In some rare cases of internal injury tobacco may be used but not in the customary way.' Be it

---

*During the conquest of Holland, Louvais paid more attention to furnishing tobacco than provisions; and even at this day, as well as in former times, more care is taken to procure tobacco than bread to the soldier. Every soldier was obliged to have his pipe and his matches.

known, then, that the Creator has not created it in vain. Dr. Clarke must have been a very good-natured man. He tortured his brains to find a hope of pardon for Judas Iscariot, and held that the creature (Nachash) who tempted Eve was not a serpent but a monkey cursed by the forfeiture of *patella* and *podex;* therefore doomed to crawl! But I fear, if the present form of using tobacco be not the true one, we must despair of ever finding it, and people will go on smoking and 'hearing reason' as long as the world goes round. Robert Hall received a pamphlet denouncing the pipe. He read it, and returned it. 'I cannot, sir, confute your arguments, and I cannot give up smoking,' was his comment. It is loosely asserted that smoking is more prevalent among scholars, intellectualists, and men who live by their brains, than among artisans and subduers of the soil. This is an error. Tobacco is less a fosterer of thought than a solace of mental vacuity. The thinker smokes in the intervals of work, impatient of *ennui* as well as of lassitude, and the ploughman, the digger, the blacksmith or the teamster, lights his cutty for the same reason. No true worker, be he digger, or divine, blends real work with either smoking or drinking. Whenever you see a fellow drink or smoke during work, spot him for a gone coon; he will come to grief, and that right soon. Sleep stimulates thought, and sometimes a pipe will bring sleep, but trust it not of itself for either thought or strength. It combats *ennui,* lassitude, and intolerable vacuity, soothing the nerves and diverting attention from self. Sam Johnson came very near the mark: 'I wonder why a thing that costs so little trouble, yet has just sufficient semblance of doing something to break utter idleness, should go out of fashion. To be sure, it is a horrible thing blowing smoke out; but every man needs something to quiet him—as, beating with his feet.'

"Life is really too short for moralists and medici who have read Don Quixote, to attack a verdict arrived at and acted upon by the combined nations of the entire world, during the experience of three centuries, and apparently deepened by their advancing civilization. Give us rules and modifications, give us guides and correctives, give us warnings against excess, precipitancy, and neglect of other enjoyments, or of important duties, if you will. The urbane æstheticism that regulates pleasure also limits it; and true refinement ever modifies the indulgence it pervades. But it is emulating Mrs. Partington and her mop to attempt to preach down a world. When they do agree, their unanimity is irresistible. Prohibition may give zest to enjoyment, and provocation to curiosity, but can never overcome the instincts of nature or cravings of nervous irritability, and he who rises in rebellion against her absolute decree will respect the limits and study the laws of a recognized and regulated enjoyment.

"Let, then, the moralist point out what social duties may be imperilled; let the physician apprise us of the disorders to be guarded against; and let the lover of elegance see that no neglect or slight awaits her. Of abstract arguments we have seen the futility, of moral and medical crusades even the most patient are weary, and we gladly turn to something real in the suffrages of a by-gone great man of acknowledged fame—Ben Jonson. Ben Jonson loved the 'durne weed,' and describes its every accident with the gusto of a connoisseur. Hobbes smoked, after his early dinner, pipes innumerable. Milton never went to bed without a pipe and a glass of water, which I cannot help associating with his:

'Adam waked,
So custom'd, for his sleep was æry light, of pure digestion bred
And temperate vapors bland!'

"Sir Isaac Newton was smoking in his garden at Woolsthorpe when the apple fell. Addison had a pipe in his mouth at all hours, at 'Buttons.' Fielding both smoked and chewed. About 1740 it became unfashionable, and was banished from St. James' to the country squires and parsons. Squire Western, in *Tom Jones*, arriving in town, sends off Parson Supple to Basingstoke, where he had left his Tobaccobox! The snuff-box was substituted. Lord Mark Kerr, a brave officer who affected the *petit maitre* (*à la* Pelham, in Lord Lytton's second novel), invented the invisible hinges, and it was this 'going out of fashion' that Jonson alluded to in 1774.

"We next find Tobacco rearing its head under the auspices of Paley and Parr. Paley had one of the most orderly minds ever given to man. A vein of shrewd and humorous sarcasm, together with an under-current of quiet selfishness, made him a very pleasant companion. 'I cannot afford to keep a conscience any more than a carriage,' was worthy of Erasmus, perhaps of Robelais. 'Our delight was,' said an old Jonsonian to the writer, 'to get old Paley, on a cold winter's night, to put up his legs, wrap them well up, stir the fire, and fill him a long Dutch pipe; he would talk away, sir,

NEWTON AND HIS PIPE.

like a being of a higher sphere. He declined any punch, but drank it up as fast as we replenished his glass. He would smoke any given quantity of Tobacco, and drink any *given* quantity of punch.'

"Parr smoked ostentatiously and vainly, as he did everything. He used only the finest Tobacco, half-filling his pipe with salt. He wrote and read, and smoked and wrote, rising early, and talking fustian. He was a sort of miniature Brummagem Johnson. Except his preface to *Bellendenus*, you might burn all he has written. His 'Life of Fox' is beneath contempt. His letters are simply laughable, especially his characters of contemporaries. He, however, was an amiable and good-natured man, and had sufficient humanity to regard dissent as an impediment to his recognition of intellectual or moral worth. Parr was an arrogant old coxcomb, who abused the respectful kindness he received, and took his pipe into drawing-rooms. I pass over the Duke of Bridgewater, because he was early crossed in love by a most beautiful girl, could not bear the sight of a flower even growing, and passed life in a pot-house with a pipe, listening to Brindley, whose intellect and dialect must have been alike incomprehensible to him.

"The cigar appeared about 1812; it received the countenance of the Regent, who had hitherto confined himself to macobau snuff, scented with lavender and the tonquin bean. Porson smoked many bundles of cheroots, which nabobs began to import. After 1815 the continental visits were resumed, and the practice of smoking began steadily to increase. The German china bowl with globular receiver of the essential oil, the absorbent meerschaum, the red Turkish bell-shaped clay, the elaborate hookah,—a really elegant ornament, and perhaps the most healthful and rational form of smoking,—pipes of all shapes, began to fill the shops of London. Coleridge, when cured of opium, took to snuff. Byron wrote dashingly about 'sublime Tobacco,' but I do not think he carried the practice to excess. Shelley never smoked, nor Wordsworth, nor Keats. Campbell loved a pipe. John Gibson Lockhart was seldom without a cigar. Sir Walter Scott smoked in his carriage, and regularly after dinner, loving both pipes and cigars. Professor Wilson smoked steadily, as did Charles Lamb. Carlyle, now somewhat past seventy, has been a sturdy smoker for years. Goethe did not smoke, neither did Shakespeare. I cannot recall a single allusion to Tobacco in all his plays; even Sir Toby Belch does not add the pipe to his burnt sack. But

Shakespeare hated every form of debauchery. The penitence of Cassio is more prominent than was his fun. 'What! drunk? and talk fustian and speak parrot, and discourse with one's shadow?' Shakespeare held drunkenness in disgust. Even Falstaff is more an intellectual man than a sot. What actor could play Falstaff after riding forty miles and being well thrashed? Yet, when Falstaff sustains the evening at the Boar's Head, he has ridden to Gadshill and back, forty-four miles! No palsied sot, he. Hamlet's disgust at his countrymen is well known. 'Grim death, how foul and loathsome is thine image!' is the comment on the drunken Kit Sly. In short, when you look at the smooth, happy, half-feminine face of Shakespeare, you see one to whom all forms of debauchery were ungenial. A courtier certainly, and a lover of money. The king had written against Tobacco, and Will Shakespeare set his watch to the time. Raleigh and Coliban Jonson might smoke at the Mermaid— Will kept his head clear and his doublet sweet.

"Alfred Tennyson is a persistent smoker of some forty years. Dickens, Jerrold and Thackeray all puffed. Lord

TENNYSON, SMOKING.

Lytton loves a long pipe at night and cigars by day. Lord Houghton smokes moderately. The late J. M. Kemble, author of 'The Seasons in England,' was a tremendous smoker. Moore cared not for it; indeed, I think that Irish gentlemen smoke much less than English. Wellington shunned it; so did Peel. D'Israeli loved the long pipe in his youth, but in middle age pronounced it 'the tomb of love.' While I am writing, it is not too much to aver that 99 persons out of 100, taken at random, under forty years of age, smoke habitually every day of their lives. How many in Melbourne injure health and brain, I leave to more skilled and morose critics. But my mind misgives me. Paralysis is becoming very frequent.

"I have seen stone pipes from Gambia, shaped like the letter U consisting each of one solid flint, hollowed through, also hookahs made by sailors with cocoanut shells. All, however, now agree that it is impossible to have either comfortable, cool, or safe smoking, unless through a substance like clay, porous and absorbent, especially as portable pipes are the mode. Those of black charcoal are not handsome; indeed, I always feel like a mute at a funeral while smoking one, but they are delightfully cool, absorbing more essential oil of nicotine, and more quickly than any meerschaum. I caution the smoker to have an old glove on; as these pipes 'sweat,' the oil comes through, and nothing is more pertinacious than oil of tobacco when it sinks into your pores, or floats about hair or clothes. My own taste inclines to the German receiver, long cherry tube and amber, and to my own garden, for all street smoking is unæsthetic, and the traveller by coach, boat, or rail has the tastes of others to consult. Surely it is not urbane to throw on another the burden of saying that he likes not the smell or the inhaling of burning tobacco. Better postpone your solace to more fitting time and place—the close of day and your own veranda. Indoor smoking is detestable. Life has few direr disenchanters than the morning smells of obsolete tobacco, relics though they be of hesternal beatitude. Give me, in robe or jacket, a hookah, or German arrangement, Chinese recumbency in matted and moistened veranda, and the odors of fresh growing beds of flowers wafted by the southern breeze. Nor be wanting the fragrant perfume of coffee. 'Meat without salt,' says Hafiz, 'is even as tobacco without coffee.' The tannin of the coffee corrects the nicotine. And it may not be amiss to learn that a plate of watercress, salt, and a large glass of cold water should be at hand to the smoker by day; the watercress corrects any excess, and is at hand in a garden. Smoke not before breakfast, nor till an hour has elapsed after a good meal. Smoke not with or before wine, you destroy the wine-palate. If you love tea, postpone pipe till after it; no man can enjoy fine tea who has smoked. In short, smoke not till the day is done, with all its tasks and duties.

"I have seen men of pretension and position treat carpets most contumeliously, trampling on the pride of Plato with a recklessness that would bring a blush to the cheek of Diogenes himself. Can they forget the absorbent powers of carpet tissues, and the horrors of next morning to non-smokers, perhaps to ladies? Surely this is unæsthetic and illiberal: it is in an old man most pitiable, in a young one intolerable, in a scholar inexcusable, from an uncleanness that seems willful. Let the young philosopher avoid such practice, and give a wide berth to those who follow them. Take the following rules, tyro, *meo periculo*:—

1. Never smoke when the pores are open: they absorb, and you are unfit for decent society. Be it your study ever to escape the noses of strangers. First impressions are sometimes permanent, and you may lose a useful acquaintance.

2. Learn to smoke slowly. Cultivate 'calm and intermittent puffs.'—*Walter Scott.*

3. On the first symptom of expectoration lay down the pipe, or throw away the cigar; long-continued expectoration is destructive to yourself and revolting to every spectator.

4. Let an interval elapse between the filling of succeeding pipes.

5. Clean your tube regularly, and your amber mouthpiece with a feather dipped in spirits of lavender. Never suffer the conduit to remain discolored or stuffed.

6. A German receiver can be washed out like a teacup, and the oil collected is of value, but a meerschaum should never be wetted. A small sponge at the end of a wire dipped in sweet oil should be used carefully and persistently round and round, coaxing out any hard concretions, till the inside be smooth in its dark polished grain, of a rich mahogany tint. The outside, also, well polished with sweet oil and stale milk, then enveloped in chamois leather. The rich dark coloring is the pledge of your safety—better there than darkening your own brains.

" The pale gold c'noster and Turkey have now given way to the splendid varieties of America, and my knowledge halts behind the age. The black sticks resembling lollipops are said to be compounds of rum, bullocks' blood and tobacco lees. A taste for them, when once contracted, is abiding. Fine volatile tobacco, with aromatic delicacy, requires a long tube; used in a short pipe of modern fashion, they parch and shrivel the tongue. In short, what is true of all other pleasures is also true of tobacco-smoking. Fruition is sometimes too rapid for enjoyment, as the dram-drinker is less wise than the calm imbiber of the fragrant vintage of the Garonne. With Burke's common sense I began, and with it I end. Depurate vice of all her offensiveness, and you prune her of half her evil. Let not your love of indulgence be so inordinate as to purchase short pleasure by impairing health, neglecting duty, or, while promoting your own self complacency, allow yourself to become permanently revolting to society, by offending more senses as well as more principles than one.'"

Mantegazza, one of the most brilliant of all writers on tobacco, in alluding to the enchantment of the "weed," says:—

"If a winged inhabitant of some remote world felt the impulse to traverse space, and, with an astronomical map, to fly round our planetary system, he would at once recognize the earth by the odor of tobacco which it exhales, forasmuch

MODERN SMOKERS.

as all known nations smoke the nicotian herb. And thousands and thousands of men, if compelled to limit themselves to a single nervous aliment, would relinquish wine and coffee, opium and brandy, and cling fondly to the precious narcotic leaf. Before Columbus, tobacco was not smoked except in America; and now, after a lapse of a few centuries in the furthest part of China and in Japan, in the island of Oceanica as in Lapland and Siberia, rises from the hut of the savage and from the palace of the prince, along with the smoke of the fireplace, where man bakes his bread and warms his heart, another odorous smoke, which man inhales and breathes forth again to soothe his pain and to vanquish fatigue and anxiety.

"In the early times of the introduction of tobacco, smokers in many countries were condemned to infamous and cruel punishments; had their noses and their lips cut off, and with blackened faces and mounted on an ass, exposed to the coarse jests of the vilest vagabonds and the insults of the multitude. But now the hangman smokes, and the criminal condemned to death smokes before being hanged. The king in his gilt coach smokes; and the assassin smokes who lies in wait to throw down before the feet of the horses the murderous bomb. The human family spends every year two thousand six hundred and seventy millions of francs (about a hundred millions in English money) on tobacco, which is not food, which is not drink, and without which it contrived to live for a long succession of ages.

"In the discomfitures and disasters which befell the Army of Lavalle, in the civil wars of the Argentine Republic, the poor fugitives had to suffer the most horrible privations, which can be imagined. By degrees the tobacco came to an end, and the Argentines smoked dry leaves. One man, more fortunate than his comrades, continued to use with much economy the most precious of all his stores—tobacco. A fellow soldier begged to be allowed to put the economist's pipe in his own mouth, and thus to inhale at second-hand the adored smoke, paying two dollars for the privilege. What is more striking still, when, in 1843, the convicts in the prison of Epinal, France, who had for some time been deprived of tobacco, rose in revolt, their cry was 'tobacco or death!' When Col. Seybourg was marching in the interior of Surinam against negro rebels, and the soldiers had to bear the most awful hardships, they smoked paper, they chewed leaves and leather, and found the lack of tobacco the greatest of all their trials and torments."

Elsewhere, inquiring what nervous aliments harmonize the one with the other, he says:—

"The only, the true, the legitimate companion of coffee is the nicotian plant; and wisely and well the Turkish epicures declare that for coffee—the drink of Heaven—tobacco is the salt. The smoke of a puro, of a manilla, or of real Turkish tobacco, which passes amorously through the voluptuous tip of amber, blends magnificently with the austere aroma of the coffee, and the inebriated palate is agitated between a caress and a rebuke."

From a Southern paper we extract these whimsical lines. "On the Great Fall in the Price of Tobacco in 1801," by Hugh Montgomery, Lynchburgh, Va.,

> "Lately a planter chanced to pop
> His head into a barber's shop—
> Begged to be shaved; it soon was done,
> When Strap (inclined oft-times to fun,)
> Doubling the price he'd asked before,
> Instead of two pence made it four.
> The planter said, 'You sure must grant,
> Your charge is most exhorbitant.'
> 'Not so,' quoth Strap, 'I'm right and you are wrong,
> For since tobacco fell, your face is twice as long.'"

Another quaint whim in the form of an advertisement for a lost meerschaum is from an Australian paper:

"To Honest men and others,—Driving from Hale Town to Bridgetown, on Sunday, last, the advertiser lost a cigar holder with the face of a pretty girl on it. The intrinsic value of the missing article is small, but as the owner has been for the last few months converting the young lady from a blonde into a brunette, he would be glad to get it back again. If it was picked up by a gentleman, on reading this notice, he will, of course, send it to the address below. If it was picked up by a poor man, who could get a few shillings by selling it, on his bringing it to the address below, he shall be paid the full amount of its intrinsic value. If it was picked up by a thief, let him deliver it, and he shall be paid a like amount, and thus for once can do an honest action, without being a penny the worse for it."

A humorous writer thus discourses on man, who he denominates as "common clays": "Yet we are all common clays! There are long clays and short clays, coarse clays and refined clays, and the latter are pretty scarce, that's a fact. To follow out the simile, life is the tobacco with which we are loaded, and when the vital spark is applied we live; when that tobacco is exhausted we die, the essence of our life ascending from the lukewarm clay when the last fibre burns out, as a curl of smoke from the ashes in the bowl of the pipe, and mingling with the perfumed breeze of heaven, or the hot breath of—well, never mind; we hope not. Then the clay is cold, and glows no more from the fire within; the pipe is broken, and ceases to comfort and console. We say,

'A friend has left us,' or 'Poor old Joe; his pipe is out.' We have all a certain supply of life, or, if we would pursue the comparison, a share of tobacco. Some young men smoke too rapidly, even voraciously, and thus exhaust their share before their proper time,—then we say they have 'lived too fast," or 'pulled at their pipes too hard.' Others, on the contrary, make their limited supply go a long way, and when they are taking their last puffs of life's perfumed plant their energy is unimpaired; they can run a race, walk a mile with any one, and show few wrinkles upon their brow,

"A delicate person is like a pipe with a crack in the bowl,

THE ARTIST.

for it takes continued and careful pulling to keep his light in; and to take life is like willfully dashing a lighted pipe from the mouth into fragments, and scattering the sparks to the four winds of heaven. An artist is a good coloring pipe;

an attractive orator is a pipe that draws well; a communist is a foul pipe; a well-educated woman whose conversation is attractive is a pipe with a nice mouthpiece; a girl of the period is a fancy pipe, the ornament of which is liable to chip; a female orator on woman's rights is invariably a plain pipe; an old toper is a well-seasoned pipe; an escaped thief is a cutty pipe, and the policeman in pursuit is a shilling pipe, for is he not a Bob?"

From these ingenious "conceits" we turn to a few thoughts on the present condition and history of the plant.

The calumet or pipe of peace, decorated with all the splendor of savage taste, is smoked by the red man to ratify good feeling or confirm some treaty of peace. The energetic Yankee bent upon the accomplishment of his ends, puffs vigorously at his cigar and with scarcely a passing notice, strides over obstacles that lie in his path of whatever nature they may be. The dancing Spaniard with his eternal castanets whispers but a word to his dark-eyed senorita as he hands her another perfumed cigarette. The lounging Italian hissing intrigues under the shadow of an ancient portico, smokes on as he stalks over the proud place where the blood of Cæsar dyed the stones of the Capitol, or where the knife of Virginius flashed in the summer sun. The Turk comes forth from the Mosque only to smoke. The priest of Nicaragua with solemn mien strides up the aisle and lights the altar candles with the fire struck from his cigar. The hardy Laplander invites the stranger to his hut and offers him his pipe while he inquires,

THE YANKEE SMOKER.

if he comes from the land of tobacco. The indigent Jakut exchanges his most valuable furs and skins for a few ounces of the "Circassian weed." Its charms are recognized by the gondolier of Venice and the Muleteer of Spain. The Switzer lights his pipe amid Alpine heights. The tourist climbing Ætna or Vesuvius' rugged side, puffs on though *they* perchance have long since ceased to smoke. Tobacco, soothed the hardships of Cromwell's soldiers and gave novelty to the court life of the daughters of Louis XIV, delighted the courtiers of Queen Elizabeth and bidding defiance to the ire of her successors, the Stuarts, has never ceased to hold sway over court and camp, as well as over the masses of the people.

In nothing cultivated has there been so remarkable a development. Originally limited to the natives of America, it attracted the attention of Europeans who by cultivation increased the size and quality of the plant. But not alone has the plant improved in form and quality, the rude implements once used by the Indians have given away (even among themselves) to those of improved form and modern style. These facts are without a doubt among the most curious that commerce presents. That a plant primarily used only by savages, should succeed in spite of the greatest opposition in becoming one of the greatest luxuries of the civilized world, is a fact without parallel. It can almost be said, so universally is it used, that its claims are recognized by all. Though hated by kings and popes it was highly esteemed by their subjects. Their delight in the new found novelty was unbounded and doubtless they could sing in praise as Byron did in later times of:

> "Sublime tobacco which from East to West
> Cheers the tar's labor and the Turkman's rest."

CHAPTER VIII.

SNUFF, SNUFF-BOXES AND SNUFF-TAKERS.

THE custom of snuff-taking is as old at least as the discovery of the tobacco plant. The first account we have of it is given by Roman Pane, the friar who accompanied Columbus on his second voyage of discovery (1494), and who alludes to its use among the Indians by means of a cane half a cubit long. Ewbank says:

"Much has been written on a revolution so unique in its origin, unsurpassed in incidents and results, and constituting one of the most singular episodes in human history; but next to nothing is recorded of whence the various processes of manufacture and uses were derived. Some imagine the popular pabulum* for the nose of translantic origin. No such thing! Columbus first beheld smokers in the Antilles. Pizarro found chewers in Peru, but it was in the country discovered by Cabral that the great sternutatory was originally found. Brazilian Indians were the Fathers of snuff, and its best fabricators. Though counted among the least refined of aborigines, their taste in this matter was as pure as that of the fashionable world of the East. Their snuff has never been surpassed, nor their apparatus for making it."

Soon after the introduction and cultivation of tobacco in Spain and Portugal its use in the form of snuff came in vogue and from these notions it spread rapidly over Europe, particularly in France and Italy. It is said to have been used

---

* Dr. John Hill in his tract "Cautions against the immoderate use of snuff" gives the following definition of it. "The dried leaves of tobacco, rasped, beaten, or otherwise reduced to powder, make what we call snuff." This tract was published in 1761. The author, afterwards Sir John Hill, was equally celebrated as a physician and a writer of farces, as denoted by the following epigram by Garrick:

"For physic and farces his equal there scarce is;
His farces are physic, his physic a farce is."

first in France* by the wife of Henry II., Catherine de Medici, and that it was first used at court during the latter part of the Sixteenth Century. The Queen seemed to give it a good standing in society and it soon became the fashion to use the powder by placing a little on the back of the hand and inhaling it. The use of snuff greatly increased from the fact of its supposed medicinal properties and its curative powers in all diseases, particularly those affecting the head, hence the wide introduction of snuff-taking in Europe. Fairholt says of its early use:

"Though thus originally recommended for adoption as a medicine, it soon became better known as a luxury and the gratification of a pinch was generally indulged in Spain, Italy and France, during the early part of the Seventeenth Century. It was the grandees of the French Court who 'set the fashion' of snuff, with all its luxurious additions of scents and expensive boxes. It became common in the Court of Louis le Grand, although that monarch had a decided antipathy to tobacco in any form."

Says an English writer "Between 1660 and 1700, the custom of taking snuff, though it was disliked by Louis XIV., was almost as prevalent in France as it is at the present time. In this instance, the example of the monarch was disregarded; *tobac en poudre* or *tobac rape*† as snuff was sometimes called found favor in the noses of the French people; and all men of fashion prided themselves on carrying a handsome snuff-box. Ladies also took snuff; and the belle whose grace and propriety of demeanour were themes of general admiration, thought it not unbecoming to take a pinch at dinner, or to blow her pretty nose in her embroidered *mouchoir* with the sound of a trombone. Louis endeavored to discourage the use of snuff and his valets-de-chambre were obliged to renounce it when they were appointed to their office. One of these gentlemen, the Duc d' Harcourt, was supposed to have died of apoplexy in consequence of having, in order to please the king, totally discontinued the habit which he had before indulged to excess."

Other grandees were less accommodating: thus we are

---

* An English writer gives a different account—"The custom of taking snuff as a nasal gratification does not appear to be of earlier date than 1620, though the powdered leaves of tobacco were occasionally prescribed as a medicine long before that time. It appears to have first become prevalent in Spain, and from thence to have passed into Italy and France.
† Grated tobacco.

told that Marechal d' Huxelles used to cover his cravat and dress with it. The Royal Physician, Monsieur Fagon, is reported to have devoted his best energies to a public oration of a very violent kind against snuff, which unfortunately failed to convince his auditory, as the excited lecturer in his most enthusiastic moments refreshed his nose with a pinch.

Although disliked by the most polished prince of Europe, the use of snuff increased and soon spread outside the limits of the court of France and in a short time became a favorite mode of using tobacco as it continues to be with many at this day.* The snuff-boxes of this period were very elegant and were decorated with elaborate paintings or set with gems. It was the custom to carry both a snuff-box and a tobacco grater, which was often as expensive and elegant as the snuff-box itself. Many of them were richly carved and ornamented in the most superb manner. Others bore the titles and arms of the owner and it was considered as part of a courtier's outfit to sport a magnificent box and grater. The French mode of manufacturing snuff was to saturate the leaves in water, then dry them and color according to the shade desired. The perfume was then added and the snuff was prepared for use. The kind of tobacco used was "Tobac de Virginie." Spanish snuff was perfumed in the same manner with the additional use of orange-flower water. Carver gives the mode of manufacturing snuff in America (1779).

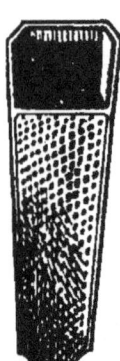

A TOBACCO GRATER.

"Being possessed of a tobacco wheel, which is a very simple machine, they spin the leaves, after they are properly cured, into a twist of any size they think fit; and having

---

* The Rev. S. Wesley speaking of the abuses of tobacco, intimates that the human ear, will not long, remain exempted from its affliction.

"To such a height with some is fashion grown
They feed they very nostrils with a spoon,
One, and but one degree is wanting yet.
To make their senseless luxury complete;
Some choice regale, useless as snuff and dear,
To feed the mazy windings of the ear.

folded it into rolls of about twenty pounds weight each, they lay it by for use. In this state it will keep for several years, and be continually improving, as it every hour grows milder. When they have occasion to use it, they take off such a length as they think necessary, which, if designed for smoking, they cut into small pieces, for chewing into larger, as choice directs; if they intend to make snuff of it they take a quantity from the roll, and laying it in a room where a fire is kept, in a day or two it will become dry, and being rubbed on a grater will produce a genuine snuff. Those in more improved regions who like their snuff scented, apply to it such odoriferous waters as they can procure, or think most pleasing."

Dutch snuff was only partially ground, and was therefore coarse and harsh in its effects when inhaled into the nostrils. The Irish, according to Everards, used large quantities of snuff "to purge their brains." Snuff-taking became general in England* at the commencement of the Seventeenth Century, and scented snuffs were used in preference to the plain. Frequent mention is made in the plays of this time of its use and varieties. In Congreve's "Love for Love," one of the characters presents a young lady with a box of snuff, on receipt of which she says, "Look you here what Mr. Tattle has given me! Look you here, cousin, here's a snuff-box; nay, there's snuff in't: here, will you have any? Oh, good! how sweet it is!"

Portuguese snuff seemed to be in favor and was delicately perfumed. It was made from the fibres of the leaves, and was considered among many to be the finest kind of the "pungent dust." Some varieties of snuff were named after the scents employed in flavoring them. In France many kinds became popular from the fact of their use at court, and by the courtiers throughout the kingdom. Pope notes the use of the snuff-box by the fops and courtiers of his time in this manner:—

---

*"The custom of taking snuff was probably brought into England by some of the followers of Charles II., about the time of the Restoration. During his reign, and that of his brother, it does not appear to have gained much ground; but towards the end of the Seventeenth Century it had become quite the "rage" with beaux, who at that period, as well as in the reign of Queen Anne, sometimes carried their snuff in the hollow ivory head of their canes."—*A Paper of Tobacco.*

"Sir Plume of amber snuff-box justly vain,
And the nice conduct of a clouded cane;
With earnest eyes, and round, unthinking face,
He first the snuff-box open'd, then the case."

The mode of "tapping the box" before opening was characteristic of the beaux and fops of this period, and is commented on in a poem on snuff:—

"The lawyer so grave, when he opens his case,
In obscurity finds it is hid,
Till the bright glass of knowledge illumines his face,
As he gives the three taps on the lid."

Spain, Portugal, and France early in the Seventeenth Century became noted as the producers of the finest kinds of snuff. In Spain and Portugal it was the favorite mode of using tobacco, and rare kinds were compounded and sold at enormous prices. Its use in France by the fair sex is thus commented on by a French writer:—

"Everything in France depends upon *la mode*; and it has

DEMI-JOURNÉES.

pleased *la mode* to patronize this disgusting custom, and carry about with them small boxes which they term *demi-journées*."

The most expensive materials were employed in the manufacture of snuff-boxes, such as agate, mosaics, and all kinds of rare wood, while many were of gold, studded with diamonds. Some kinds were made of China mounted in metal, and were very fanciful. In "Pandora's Box," a "Satyr against Snuff," 1719, may be found the following description of the snuff-boxes then in vogue:

> "For females fair, and formal fops to please,
> The mines are robb'd of ore, of shells the seas,
> With all that mother-earth and beast afford
> To man, unworthy now, tho' once their lord:
> Which wrought into a box, with all the show
> Of art the greatest artist can bestow;
> Charming in shape, with polished rays of light,
> A joint so fine it shuns the sharpest sight;
> Must still be graced with all the radiant gems
> And precious stones that e'er arrived in Thames.
> Within the lid the painter plays his part,
> And with his pencil proves his matchless art;
> There drawn to life some spark or mistress dwells,
> Like hermits chaste and constant to their cells."

Some of the more highly perfumed snuffs sold for thirty shillings a pound, while the cheaper kinds, such as English Rappee and John's Lane, could be bought for two or three shillings per pound. There are at least two hundred kinds of snuff well known in commerce. The Scotch and Irish snuffs are for the most part made from the midribs; the Strasburgh, French, Spanish, and Russian snuffs from the soft parts of the leaves. An English writer gives the following account of some of the well-known snuffs and the method of manufacturing:—

"For the famous fancy snuff known as Maroco, the recipe is to take forty parts of French or St. Omer tobacco, with twenty parts of fermented Virginia stalks in powder; the whole to be ground and sifted. To this powder must be added two pounds and a half of rose leaves in fine powder; and the whole must be moistened with salt and water and thoroughly incorporated. After that it must be 'worked up' with cream and salts of tartar, and packed in lead to preserve its delicate aroma. The celebrated 'gros grain Paris snuff' is composed of equal parts of Amersfoort and James River tobacco, and the scent is imported by a 'sauce,' among the ingredients of which are salt, soda, tamarinds, red wine, syrup, cognac, and cream of tartar."

The mode of manufacture of snuff now is far different than that employed in the Seventeenth Century. Then the leaves were simply dried and made fine by rubbing them together in the hands, or ground in some rude mill; still later the tobacco was washed or cleansed in water, dried, and then ground. Now, however, the tobacco undergoes quite a process, and must be kept packed several months before it is ground into snuff. One of the most celebrated manufacturers of snuff was James Gillespie, of Edinburgh, who compounded the famous variety bearing his name. The following account of him we take from "The Tobacco Plant:"—

"In the High Street of Edinburgh, a little east from the place where formerly stood the Cross,—

"'Dun-Edin's Cross, a pillar'd stone,
Rose on a turret octagon,'

was situated the shop of James Gillespie, the celebrated snuff manufacturer.

JAMES GILLESPIE.

The shop is still occupied by a tobacconist, whose sign is the head of a typical negro, and in one of the windows is exhibited the effigy of a Highlander, who is evidently a competent judge of 'sneeshin.' Not much is known regarding the personal history of James Gillespie, but it is understood that he was born shortly after the Jacobite rebellion of 1715, at Roslin, a picturesque village about six miles from Edinburgh. He became a tobacconist in Edinburgh, along with his brother John, and by the exercise of steady industry and frugality, he was enabled to purchase Spylaw, a small estate in the parish of Colinton, about four miles from Edinburgh, where he erected a snuff-mill on the banks of the Water of Leith, a small stream which flows through the finely-wooded grounds of Spylaw. The younger brother, John, attended to the shop, while the subject of our notice resided at Spylaw, where he superintended the snuff-mill. Mr. Gillespie was able to continue his industrious habits through a long life, and having made some successful speculations in tobacco during the war of American Independence, when the 'weed' advanced considerably in price, he was enabled to increase his Spylaw estate from time to time by making additional purchases of property in the parish.

"Mr. Gillespie remained through life a bachelor. His establishment at Spylaw was of the simplest description. It is said that he invariably sat at the same table with his servants, indulging in familiar conversation, and entering with much spirit into their amusements. Newspapers were not so widely circulated at that period as they are now, and on the return of any of his domestics from the city, which one of them daily visited, he listened with great attention to the 'news, and enjoyed with much zest the narration of any jocular incident that had occurred. Mr. Gillespie had a *penchant* for animals, and their wants were carefully attended to. His poultry, equally with his horses, could have testified to the judicious attention which he bestowed upon them. A story is told of the familiarity between the laird and his riding horse, which was well-fed and full of spirit.

"The animal frequently indulged in a little restive curvetting with its master, especially when the latter was about to get into the saddle. 'Come, come,' he would say, on such occasions, addressing the animal in his usual quiet way, 'hae dune, noo, for ye'll no like if I come across your lugs (ears) wi' the stick.'

"Even in his old age Mr. Gillespie regularly superintended the operations in the mill, which was situated in the rear of his house. On these occasions he was wrapped in an old blanket ingrained with snuff. Though he kept a carriage he very seldom used it, until shortly before his death, when increasing infirmities caused him occasionally to take a drive. It was of this carriage, plain and neat in its design, with nothing on its panel but the initials 'J. G.' that the witty Henry Erskine proposed the couplet—

> 'Who would have thought it
> That noses had bought it?'

as an appropriate motto. In those days snuff was much more extensively used than at present, and Mr. Gillespie was in the habit of gratuitously filling the 'mulls' of many of the Edinburgh characters of the last century. Colinton appears to have been a great snuff-making centre. About thirty years ago there were five snuff mills in operation in the parish, the produce of which was sold in Edinburgh. Even now a considerable quantity of snuff is made in the district, chiefly by grinders to the trade.

Murray, alluding to the popularity of the custom in

England during the reign of the House of Brunswick, says:—

"The reigns of the four Georges may be entitled the snuffing period of English history. The practice became an appanage of fashion before 1714, as it has continued after 1830, to be the comfort of priests, literary men, highlanders, tailors, factory hands, and old people of both sexes. George IV. was a nasute judge of snuffs, and so enamoured of the delectation, that in each of his palaces he kept a jar chamber, containing a choice assortment of tobacco powder, presided over by a critical superintendent. His favorite stimulant in the morning was violet Strasburgh, the same which had previously helped Queen Charlotte to 'kill the day'—after dinner Carrotte — named from his *penchant* for it. King's Carrotte, Martinique, Etrenne, Old Paris, Bureau, Cologne, Bordeaux, Havre, Princeza, Rouen, and Rappee, were placed on the table, in as many rich and curious boxes."

FOPS TAKING SNUFF. (*From an old print*).

Sterne, in his "Sentimental Journey," gives a pleasing description of snuff-taking with the poor monk. He writes:

"The good old monk was within six paces of us, as the idea of him crossed my mind; and was advancing towards us a little out of the line, as if uncertain whether he should break in upon us or no. He stoop'd, however, as soon as he came up to us with a world of frankness; and having a horn snuff-box in his hand, he presented it open to me.

"'You shall taste mine,' said I, pulling out my box (which was a small tortoise one), and putting it into his hand.

"''Tis most excellent,' said the monk.

"'Then do me the favor,' I replied 'to accept of the box and all, and when you take a pinch out of it, sometimes recollect it was the peace-offering of a man who once used you unkindly, but not from his heart.'

"The poor monk blushed as red as scarlet, 'Mon Dieu!' said he, pressing his hands together, 'you never used me unkindly.'

"'I should think,' said the lady, 'he is not likely.'

I blushed in my turn; but from what motives, I leave to the few who feel to analyze. 'Excuse me, madam,' replied I, 'I treated him most unkindly, and from no provocations.'

"''Tis impossible,' said the lady.

"'My God!' cried the monk, with a warmth of asseveration which seemed not to belong to him, 'the fault was in me, and in the indiscretion of my zeal.'

"The lady opposed it, and I joined with her in maintaining it was impossible, that a spirit so regulated as his could give offence to any. I knew not that contention could be rendered so sweet and pleasurable a thing to the nerves as I then felt it. We remained silent, without any sensation of that foolish pain which takes place when, in a circle, you look for ten minutes in one another's faces without saying a word.

"Whilst this lasted, the monk rubb'd his horn box upon

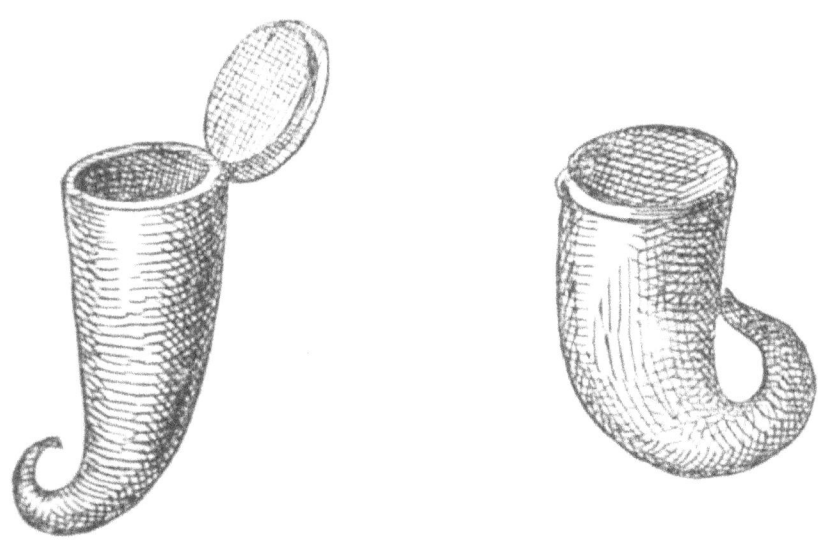

HORN SNUFF-BOXES.

the sleeve of his tunic; and as soon as it had acquired a little air of brightness by the friction, he made a low bow and said, 'twas too late to say whether it was the weakness or goodness of our tempers which had involved us in this contest, but be it as it would, he begg'd we would exchange boxes. In saying this, he presented this to me with one, as he took mine from me in the other; and having kissed it, with a stream of good nature in his eyes, he put it into his bosom, and took his leave. I guard this box as I would the instrumental parts of my religion, to help my mind on to something better: in truth I seldom go abroad without it;

and oft and many a time have I called up by it the courteous spirit of its owner, to regulate my own in the jostlings of the world; they had found full employment for his, as I learnt from his story, till about the forty-fifth year of his age, when upon some military services ill-requited, and meeting at the same time with a disappointment in the tenderest of passions, he abandoned the sword and the sex together, and took sanctuary, not so much in his convent as in himself. I feel a damp upon my spirits, as I am going to add, that in my last return through Calais, upon inquiring after Father Lorenzo, I heard he had been dead near three months, and was buried, not in his convent, but according to his desire, in a little cemetery belonging to it about two leagues off. I had a strong desire to see where they had laid him, when, upon pulling out his little horn box, as I sat by his grave, and plucking a nettle or two at the head of it, which had no business to grow there, they all struck together so forcibly upon my affections that I burst into a flood of tears—but I am as weak as a woman; I beg the world not to smile, but pity me."

Many pleasing effusions have been written promoted doubtless by a sneeze among which the following on "A pinch of Snuff" from "The Sportsman Magazine," exhibits the custom and the benefits ascribed to its indulgence.

"With mind or body sore distrest,
Or with repeated cares opprest,
What sets the aching heart at rest?
    A pinch of snuff!

"Or should some sharp and gnawing pain
Creep round the noddle of the brain,
What puts all things to rights again?
    A pinch of snuff!

"When speech and tongue together fail,
What helps old ladies in their tale,
And adds fresh canvass to their sail?
    A pinch of snuff!

"Or when some drowsy parson prays,
And still more drowsy people gaze,
What opes their eyelids with amaze?
    A pinch of snuff!

"A comfort which they can't forsake,
What is it some would rather take,
Than good roast beef, or rich plum cake?
A pinch of snuff!

"Should two old gossips chance to sit,
And sip their slop, and talk of it,
What gives a sharpness to their wit?
A pinch of snuff!

"What introduces Whig or Tory,
And reconciles them in their story,
When each is boasting in his glory?
A pinch of snuff!

"What warms without a conflagration
Excites without intoxication,
And rouses without irritation?
A pinch of snuff!

"When friendship fades, and fortune's spent,
And hope seems gone the way they went,
One cheering ray of joy is sent—
A pinch of snuff!

"Then let us sing in praise of snuff!
And call it not such 'horrid stuff,'
At which some frown, and others puff,
And seem to flinch.

"But when a friend presents a box,
Avoid the scruples and the shocks
Of him who laughs and he who mocks,
And take a pinch!"

From "Pandora's Box" from which we have already quoted, we extract the following in which the use of snuff is deprecated by the author:

—"now, 'tis by every sort
And sex adored, from Billingsgate to court.
But ask a dame 'how oysters sell?' if nice,
She begs a pinch before she sets a price.
Go thence to 'Change, inquire the price of Stocks;
Before they ope their lips they open first the box.
Next pay a visit to the Temple, where
The lawyers live, who gold to Heaven prefer;
You'll find them stupify'd to that degree,

> They'll take a pinch before they'll take their fee.
> Then make a step and view the splendid court,
> Where all the gay, the great, the good resort;
> E'en they, whose pregnant skulls, though large and thick,
> Can scarce secure their native sense and wit,
> Are feeding of their hungry souls with pure
> Ambrosial snuff. * * * *
> But to conclude: the gaudy court resign,
> T' observe, for once, a place much more divine,
> When the same folly's acted by the good,
> And is the sole devotion of the lewd;
> The church, more sacred once, is what we mean,
> Where now they flock to see and to be seen;
> The box is used, the book laid by, as dead,
> With snuff, not Scripture, there the soul is fed;
> For where to heaven the hands by one of those,
> Are lifted, twenty have them at the nose;
> And while some pray, to be from sudden death
> Deliver'd, others snuff to stop their breath."

Paolo Mantegazza, one of the most brilliant and witty of Italian writers on tobacco, says of its use and "some of the delights that may be imagined through the sense of smell:"—

"Human civilization has not yet learned to found on the sense of smell aught but the moderate enjoyment derived from snuffing, which, confined within the narrow circle of a few sensations, renders us incapable of entering into the most delicate pleasures of that sense.

"Snuff procures us the rapture of a tactile irritation, of a slight perfume; but, above all, it furnishes the charm of an intermittent occupation which soothes us by interrupting, from time to time, our labor. At other times it renders idleness less insupportable to us, by breaking it into the infinite intervals which pass from one pinch of snuff to another. Sometimes our snuff-box arouses us from torpor and drowsiness; sometimes, it occupies our hands when in society we do not know where to put them or what to do with them. Finally, snuff and snuffing are things which we can love, because they are always with us; and we can season them with a little vanity if we possess a snuff-box of silver or of gold, which we open continually before those who humbly content themselves with snuff-boxes of bone or of wood. We gladly concede the pleasures of snuffing to men of all conditions, and to ladies who, having passed a certain age, or who, being deformed, have no longer any sex; but we solemnly and resolutely refuse the snuff-box to young and beautiful women, who ought to preserve their delicate and pretty noses for the odors of the mignonette and the rose."

With royalty snuff has been a prime favorite. Charles III. of Spain had a great predilection for rappee snuff, but only indulged his inclination by stealth, and particularly while shooting, when he imagined himself to be unnoticed. Frederick the Great and Napoleon[*] both loved and used large quantities of the "pungent dust." Of the former the following anecdote is related:—

"The cynical temper of Frederick the Great is well known. Once when his sister, the Duchess of Brunswick, was at Potsdam, Frederick made to the brave Count Schwerin the present of a gold snuff-box. On the lid inside was painted the head of an ass. Next day, when dining with the king, Schwerin, with some ostentation, put his snuff-box on the table. Wishing to turn the joke against Schwerin, the king called attention to the snuff-box. The Duchess took it up and opened it. Immediately she exclaimed, 'What a striking likeness! In truth, brother, this is one of the best portraits I have ever seen of you.' Frederick, embarrassed, thought his sister was carrying the jest too far. She passed the box to her neighbor, who uttered similar expressions to her own. The box made the round of the table, and every one was fervently eloquent about the marvelous resemblance. The king was puzzled what to make of all this. When the box at last reached his hands, he saw, to his great surprise, that his portrait was really there. Count Schwerin had simply, with exceeding dispatch, employed an artist to remove the ass's head, and to paint the king's head instead. Frederick could not help laughing at the Count's clever trick, which was really the best rebuke of his own bad taste and want of proper and respectful feeling."

"As Frederick William I., of Prussia, was eminently the Smoking King, so his son Frederick the Great was eminently the Snuffing King. Perhaps smoking harmonizes best with action; and it might, without much stretch of fancy, be shown that as the Prussian monarchy was founded on tobacco smoke, it flourished on snuff. Possibly, if Napoleon the Great, who like Frederick the Great, was an excessive snuffer,

---

[*]Napoleon, having been unable to undergo the ordeal of a first pipe, stigmatized it as a habit only fit to amuse sluggards. What he renounced in smoking, however, he compensated in snuff.

had smoked as well as snuffed, he might have preserved his empire from overthrow, seeing that smoking steadies and snuffing impels. The influences of smoking and snuffing on politics and war are ascertainable. What the effect of chewing is on political and military affairs, it is not so easy to discover. We recommend the subject for meditation to the profoundest metaphysicians. How many of the American politicians and generals have been chewers as well as snuffers and smokers? Is there to be some mysterious affinity between chewing and the revolutions, especially the social revolutions of the future? May not apocalyptic interpreters be able to show that chewing is the symbol of anarchy and annihilation?"

When first used in Europe snuff was made ready for use by the takers—each person being provided with a box or

SCOTCH SNUFF-MILLS.

"mill," as they were termed, to reduce the leaves to powder.

In connection with this, the following may not be irrelevent:—

The following anecdote of Huerta the celebrated Spanish guitarist, is taken from one of M. Ella's programmes:—

"In the year 1826 the famous Huerta, who astonished the English by his performances on the guitar, was anxious to be introduced to the leader of the Italian Opera Band—a warm-hearted and sensitive Neapolitan—Spagnoletti. The latter had a great contempt for guitars, concertinas, and other fancy instruments not used in the orchestra. He was fond of snuff, had a capacious nose, and, when irritated, would ejaculate '*Mon Dieu!*' On my presenting the vain Spaniard to Spagnoletti, the latter inquired, 'Vat you play?' Huerta—'De guitar-r-r, sare.' Spagnoletti—'De guitar! humph!' (takes a pinch of snuff.) Huerta—'Yeas, sare, de guitar-r-r, and ven I play my *adagio*, de tears shall run down both side your pig nose.' 'Vell den,' (taking snuff,) said Spagnoletti, 'I vill not hear your *adagio*.'"

The anecdote related of Count de Tesse, a celebrated courtier of France, is one of the best of its kind:—

"Count de Tesse, Marshal of France, was an eminent man during the reign of Louis XIV. Though he was a brave soldier and by no means an incompetent general, yet he was more remarkable as a skillful diplomatist and a pliant and prosperous courtier. During the War of Succession in Spain, he besieged Barcelona with a considerable army, in the spring of 1705. Terrible was the assault, and terrible was the resistance. At the end of six weeks the arrival of the British fleet, and reinforcements thrown into the place, forced Marshal Tesse to retire. Besides immense losses in dead and wounded, he had to abandon two hundred and twenty cannon and all his supplies. Incessantly fighting for fifteen days in his retreat towards the Pyrenees, he lost three thousand more of his men. It ought to be said, in vindication of Tesse, that he undertook the siege by express and urgent command of the French King, and contrary to his own judgment; for in writing to a friend, he said: 'If a Consistory were held to decide the infallibility of the King, as Consistories have been held to decide the infallibility of the Pope, I should by my vote declare His Majesty infallible. His orders have confounded all human science.'

"Soon after the siege of Barcelona, a lady at a fashionable party took out her snuff-box and offered a pinch to any one who wished it. Marshal Tesse approached to take a pinch; but suddenly the lady drew her snuff-box back, saying, 'For you, Marshal, the snuff is too strong—it is Barcelona.'"

In Scotland the dry kinds of snuff are in favor and are esteemed as highly as the moister snuffs. Robert Leighton gives the following pen picture of the snuff-loving Scotchman; it is entitled "The Snuffie Auld Man:"—

"By the cosie fire-side, or the sun-ends o' gavels,
　The snuffie auld bodie is sure to be seen.
Tap, tappin' his snuff-box, he snifters and sneevils,
　And smachers the snuff frae his mou' to his een.
Since tobacco cam' in, and the snuffin' began,
　There hasna been seen sic a snuffie auld man.

"His haurins are dozen'd, his een sair bedizzened
　And red round the lids as the gills o' a fish;
His face is a' bladdit, his sark-breest a' smaddit.
　As snuffie a picture as ony could wish.

He makes a mere merter o' a' thing he does,
Wi' snuff frae his fingers an' drops frae his nose.

"And wow but his nose is a troublesome member—
Day and nicht, there's nae end to its snuffle desire:
It's wide as the chimlie, it's red as an ember,
And has to be fed like a dry-whinnie fire.
It's a troublesome member, and gi'es him nae peace,
Even sleepin', or eatin', or sayin' the grace.

"The kirk is disturbed wi' his hauckin' and sneezin'
The dominie stoppit when leadin' the psalm;
The minister, deav'd out o' logic and reason,
Pours gall in the lugs that are gapi' for balm.
The auld folks look surly, the young chaps jocose,
While the bodie himsel' is bambazed wi' his nose.

"He scrimps the auld wife baith in garnal and caddy
He snuffs what wad keep her in comfort and ease;
Rapee, Lundyfitt, Prince's Mixture, and Toddy,
She looks upon them as the worst o' her faes.
And we'll see an end o' her koosbian nar
While the auld carle's nose is upheld like a Czar.

Sharp has written some verses founded upon the following singular anecdote in Dean Ramsay's "Reminiscences of Scottish Life and Character:"

"The inveterate snuff-taker, like the dram-drinker, felt severely the being deprived of his accustomed stimulant, as in the following instance: A severe snow-storm in the Highlands, which lasted for several weeks, having stopped all communication betwixt neighboring hamlets, the snuff-boxes were soon reduced to their last pinch. Borrowing and begging from all the neighbors within reach were first resorted to, but when these failed all were alike reduced to the longing which unwillingly abstinent snuff-takers alone know. The minister of the parish was amongst the unhappy number; the craving was so intense that study was out of the question, and he became quite restless. As a last resort, the beadle was dispatched through the snow, to a neighboring glen, in the hope of getting a supply; but he came back as unsuccessful as he went. 'What's to be done, John?' was the minister's pathetic inquiry. John shook his head, as much as to say that he could not tell, but immediately thereafter started up, as if a new idea occurred to him. He came back in a few minutes, crying, 'Hae!' The minister, too eager to be scrutinizing, took a long deep pinch, and then said, 'Whour did you get it?' 'I soupit (swept) the poupit,' was John's expressive reply. The minister's accumulated superfluous Sabbath snuff now came into good use."

"Near the Highlands,
  Where the dry lands
  Are divided into islands,
  And distinguish'd from the mainland
  As the Western Hebrides.

"Stormy weather,
  Those who stay there,
  Oftentimes for weeks together
  Keep asunder from their neighbors,
  Hemm'd about by angry seas.

"For, storm-batter'd,
  Boats are shattered,
  And their precious cargoes scatter'd
  In the boist'rous Sound of Jura,
  Or thy passage, Colonsay;

"While the seamen,
  Like true freemen,
  Battle bravely with the Demon
  Of the storm, who strives to keep them
  From their harbor in the bay.

"For this reason
  One bad season,
  (If to say so be not treason,)
  In an island town the people
  Were reduced to great distress.

"Though on mainland
  They would fain land,
  They were storm-bound in their ain land,
  Where each luxury was little,
  And grew beautifully less.

"But whose sorrow,
  That sad morrow,
  When no man could beg or borrow
  From a friend's repository,
  Equall'd theirs who craved for snuff.

"But, most sadden'd,
  Nearly madden'd
  For the lack of that which gladden'd
  His proboscis, was the parson,
  Hight the Rev'rend Neil Macduff.

"If a snuffer,
  Though no puffer,
  You may guess what pangs he'd suffer
  In his journey through a snow-drift,
  Visiting a neighboring town.

"From his rushing
  For some sneishing;
  But his choring and his fishing
  Could procure no Toddy's Mixture,
  Moist Rappee, or Kendal Brown.

"In his trouble—
  Now made double,
  Since his last hope proved a bubble—
  To his aid came Beadle Johnnie,
  In his parish right-hand man.

"With a packet,
Saying, Tak' it,
It's as clean as I can mak' it,
If ye'd save yer snuff on Sabbath
A toom box ye needna scan.

"Being lusty
(Though 'twas musty)
To his nose the snuff so dusty
Put the minister, too much in want,
The gift to scrutinize.

"An idea
He could see a
Blessing in this panacea;
So he took such hearty pinches as brought
Tears into his eyes.

"Then to Johnnie,
His old cronie,
Cried—'I fear'd I'd ne'er get ony.'
'Well, I'll tell ye,' said the beadle,
'Whaur I got the stock of snuff.'

"'In the poupit
Low I stoopit,
An' the snuff and stour I soupit,
Then I brocht ye here a handfu',
For ye need it sair enough.'"

The old Scottish snuff-mill, which consisted of a small box-like receptacle into which fitted a conical-shaped projection with a short, strong handle was a more substantial affair than the rasp used by the French and English snuff-takers. (See page 232). Both, answered the purpose for which they were designed, the leaves of tobacco being "toasted before the fire," and then ground in the mill as it was called. The more modern snuff-mill is similar in shape, but is used to hold the snuff after being ground, rather than for reducing the leaves to a powder.

Boswell gives the following poem on snuff, in his "Shrubs of Parnassus:"

"Oh Snuff! our fashionable end and aim!
Strasburg, Rappee, Dutch, Scotch, what'eer thy name,

> Powder celestial! quintessence divine!
> New joys entrance my soul while thou art mine.
> Who takes—who takes thee not! where'er I range,
> I smell thy sweets from Pall Mall to the 'Change.
> By thee assisted, ladies kill the day,
> And breathe their scandal freely o'er their tea;
> Nor less they prize thy virtues when in bed,
> One pinch of thee revives the vapor'd head,
> Removes the spleen, removes the qualmish fit,
> And gives a brisker turn to female wit,
> Warms in the nose, refreshes like the breeze,
> Glows in the herd and tickles in the sneeze.
> Without it, Tinsel, what would be thy lot!
> What, but to strut neglected and forgot!
> What boots it for thee to have dipt thy hand
> In odors wafted from Arabian land?
> Ah! what avails thy scented solitaire,
> Thy careless swing and pertly tripping air,
> The crimson wash that glows upon thy face,
> Thy modish hat, and coat that flames with lace!
> In vain thy dress, in vain thy trimmings shine,
> If the Parisian snuff-box be not thine.
> Come to my nose, then, Snuff, nor come alone,
> Bring taste with thee, for taste is all thy own."

There seems to be as great a variety of design in snuff-boxes as among pipes and tobacco-stoppers. The Indians of both North and South America have their mills for grinding or pulverizing the leaves. In the East a great variety of snuff-boxes may be seen; they are made of wood and ivory, while many of them have a spoon attached to the box, which they use in taking the dust from the box to the back of the hand, whence it is taken by the forefinger and conveyed to the nose. In Europe we find greater variety of design in snuff-boxes than in the East. In Europe they are made of the most costly materials, and studded with the rarest gems.

In the East they are made of ivory, wood, bamboo, and other materials. Of late years boxes made of wood from Abbotsford or some other noted place have been used for the manufacture of snuff-boxes. Formerly when snuff-taking was in more general use by kings and courtiers than now—a magnificent snuff-box was considered by royalty as one of the most valuable and pleasing of "memorials." Many of these testimonials of friendship and regard were of gold and silver, and set with diamonds of the finest water.

Among the anecdotes of celebrated snuff-takers, the following from White's "Life of Swedenborg," will be new to many:

"Swedenborg took snuff profusely and carelessly, strewing it over his papers and the carpet. His manuscripts bear its traces to this day. His carpet set those sneezing who shook it. One Sunday he desired to have it taken up and beaten. Shearsmith objected, 'Better wait till to-morrow,' 'Dat be good! dat be good!' was his answer."

We copy the following article on the manufacture of snuff from a well-known English journal, "Cope's Tobacco Plant:"—

"Although snuff is still extensively consumed in this country (Great Britain), the mode of its manufacture is very little known to those who use it; and there are very few persons of even the most inquisitive turn of mind who can say they have ever penetrated into the mysterious precincts of a snuff-mill. Even those who have been privileged, and have had the courage to inspect the interior of such an establishment, have come away with very vague notions of what they saw. The hollow whirr of the revolving pestles, the hazy atmosphere closely resembling a London fog in November, a phenomenon which is produced by the innumerable particles of tobacco floating about, and causing the gas to flicker and sparkle in a mysterious way, and producing a lively irritation of the mucous membrane, all combine in placing the visitor in a state of amusing bewilderment, and he is compelled to make a speedy exit, having only had just a running peep at the interesting process of snuff-making. It is therefore our duty to give a description of a process which will be new to a large number of people, and will help to clear up some of the obscure theories that a great many more entertain of it.

"Those persons who have travelled on the Continent, and who have noticed on tobacconists' counters a small machine, somewhat like a coffee-mill, which a man works with one hand, while he holds a hard-pressed plug of tobacco about a pound weight against the revolving grater, and produces snuff while the snuff-taker waits for it, may imagine that snuff in England is produced on a somewhat similar small scale. But this, like many kindred theories, is quite a mistake. In this country there exist large snuff-mills worked by steam power, and in Scotland there is one water-mill which is driven by a water-power of the strength of thirty horses. The grinding of snuff is at present carried on much as it was one hundred years ago. The apparatus, although effective,

SNUFF-MILL A CENTURY AGO.

is very primitive, and would lead one to suppose that mechanical ingenuity had wholly neglected to trouble itself about improving that branch of machinery.

"All kinds of snuff are made from tobacco leaves, or tobacco stalks, either separate or mixed. This in the first instance goes through a kind of fermentation, and, like the basis of soup at the modern hotels, forms, as it were, the stock from which all the varieties in flavor and appearance are produced by special treatment and flavoring. Of course the strength and pungency of the snuff will depend a good deal upon the richness of the tobacco originally put aside for it. About one thousand pounds of tobacco would form an ordinary batch of snuff. The duty on this would amount to about £150, and this has to be paid before the tobacco is removed from the bonded warehouse. Having got his heap of material ready, the snuff-maker moistens it, then places it in a warm room and covers it over with warm cloths—coddles it, as it were, to make it comfortable, so that the cold air cannot get to it—and the heap is then left for three or four weeks, as the case may be, to ferment.

"In France, where, under the Imperial *régime*, snuff-making was a Government monopoly, the tobacco was allowed to ferment for twelve or eighteen months; and in the principal factory (that at Strasburg) might have been seen scores of

GRINDING THE LEAVES.

huge bins, as large as porter vats, all piled up with tobacco in various stages of fermentation. The tobacco, after being fermented, if intended for that light, powdery, brown-looking snuff called S. P., is dried a little; or if for Prince's Mixture, Macobau, or any other kind of Rappee, is at once thrown into what is called the mull. The mull is a kind of large iron mortar weighing about half a ton and lined with wood; and there is a heavy pestle which travels round it, forming, as it were, a large pestle and mortar.

These mulls are placed in rows and shut up in separate cupboards, to keep in the dust. The snuff-maker wanders from one to the other, and feeds them as they require.

"When the grinding of the snuff is completed it is then ready for flavouring, and in this consists the great art and secret of the trade. Receipts for peculiar flavors are handed down from father to son as most valuable heir-looms, and these receipts are in fact a valuable property in many instances, for so delicate is the nose of your snuff-taker that he can detect the slightest variation in the preparation of his favorite snuff. It is related of one old snuff-maker in London, who had acquired a handsome fortune and retired from business, that he made it a consideration with his successors that he should be allowed, so long as he lived, to attend one day in the week at the business and flavor all the snuff. Most people will also be familiar with some one of the numerous versions of the origin of the once famous Lundy Foote Snuff, better known as 'Irish Blackguard.'

"The excise are very rigid in their laws for regulating the manufacture of snuff; and with the exception of a little common salt, which is added to make the tobacco keep, and alkalies for bringing out the flavor, nothing is allowed to be used but a few essential oils. And here we must digress for a moment to correct a popular error, viz., that snuff contains ground glass, put there for titillating purposes. What appears to be ground glass is only the little crystals or small particles of alkali that have not been dissolved. So that fastidious snuff-takers may dismiss this bugbear at once and forever.

"The essential oils referred to form a very expensive item in the manufacture of snuff. The ladies would be much surprised to see a dusty snuff-maker drain off five pounds' worth of pure unadulterated otto-of-roses into a tin can, and as they (the ladies) would suppose, throw it away on a heap of what would appear to them rubbishy dust in one corner of the snuff-room. Of course the ladies would consider the proper place for it to be on the cambric handkerchief, but this idea would be about the last to occur to your matter-of-fact snuff-maker.

PERFUMING SNUFF.

"In addition to otto-of-roses, the scent-room contains great jars of essence of lemon, French geranium, verbena, oil of pimento, bergamotte, etc., all of which are used in the various flavoring combinations. There would most likely also be a few hundred-weight of fine Tonquin beans, and one of these beans is generally presented to any visitor who drops in, as souvenir to carry away in his waistcoat pocket. Snuff is very extensively used in the mills and factories of Lancashire. Those who toil long in heated and noisy mills seem to require, and doubtless do require, tobacco in some shape or other to keep them from flagging; and as chewing is not polite, and smoking in a mill not allowed, the only resource left to the operative is his snuff. A singular feature connected with this is, we believe, the fact that spinners in very few instances use snuff-boxes, they prefer having their supply of snuff screwed up in a piece of paper. One retail shop-keeper in a busy spinning town in Lancashire assured us that he retailed over four hundred weight of snuff a week in pennyworths.

"It is impossible to state the exact quantity of snuff used in this country; but, as far as we can arrive at it from statistics at hand, we should say it cannot be less than five hundred tons per annum. This seems an enormous quantity, considering the comparatively small number of persons who now use snuff; but the great bulk of snuff seems to be consumed by particular communities, such as the Lancashire operatives, and the consumption of it is therefore not generally observable; and further it should be remembered that those who do take snuff, individually use large quantities."

Snuff-manufacturing has in some cases been attended with considerable affluence. One instance is the London manufacturer already mentioned, whose profits accumulated to the extent of nearly a quarter of a million; another is the Lundy Foote business, and the third a Scotch manufacturer (Gillespie), who by the way, practised a bit of benevolence, in the shape of building an hospital, in return for the good things fortune had sent him. Of course an hospital, like many other things, may have a doubtful origin, as witness the famous Guy's, which stands as a lasting monument to the wonderful profits that used to be made out of the iniquitous advance note system. But we do not by any means wish to make comparisons which must be odious and although the profits of snuff-manufacturing are for a variety of reasons —amongst others the decreased consumption of the manufactured article—not nearly as large as they were fifty years ago; yet we are sure that the fortunes accumulated by some of the old snuff-makers were the result of honest, upright industry.

Of European tobacco used in the manufacture of snuff that of Holland and France (St. Omer) is considered to be equal to any grown in Europe. Of the varieties grown in America, Virginia leaf is used quite extensively for some grades of snuff and "good stout rich snuff leaf" commands excellent prices and meets with a ready sale.

A writer gives the following account of the love the Terra Del Fuegians have for tobacco.

"This morning we were up early, a large party going ashore for various scientific purposes, and the others taking the ship out in the channel to do a little dredging; both parties were very successful, and added much to our collection. As we on the shore were about ready to come off, we were visited by a party of Fuegians, five men, four women, and nine children, with three dogs. They came in an English-built boat, stolen or lost from some English ship. The men and dogs landed and came towards us with a great frankness of manner. They could talk neither English nor Spanish, except the few words, boat, fire, tobac, galleto, arco. But they understood the imperial manner of one of our officers, who said quietly but firmly, 'keep back those dogs,' and immediately drove back the barking curs with sticks and stones. They warmed themselves at our fire, and seemed disposed to be very civil and friendly. We gave them our remaining biscuit, and what little tobacco there was in our party to spare. One of them accepted a pinch of snuff and pretended to sneeze, crying 'Hatchee!' with mock solemnity.

An old man sat down on a stone and sang to us a low,

FUEGIAN SNUFF-TAKERS.

sweet recitation, or chant, in wild key, or mode, ending on a rising melody with each stanza.

They followed us to the ship, and we gave them some calico and beads, and tobacco, and also bought bows and arrows, and a sea-urchin, paying them in tobacco. They clung to the ship as we got under way, men and women, crying, 'Tobacco!' and frantic to catch any fragment of the precious weed thrown to them. But at length they let go, and we left the bay with the cry of tobacco ringing in our ears."

Having spoken of most of the modes of using snuff in both the Old and New World, we come now to a description of using snuff at the South, known as "dipping," and by some as "rubbing," both terms used to denote the same manner of use. The description of it as given by A. L. Adams is as follows:—

"In the South, and more especially in Virginia, where tobacco has been cultivated for more than two hundred and fifty years, and where a few pounds of it was the legitimate price for a wife, it is not surprising that it should be more highly prized and come into more general use than in any other section of our country. On the banks of the James River it was first successfully cultivated by the English colony, and this simple fact alone must forever throw a charm around it, which will foster the pride of the Virginian who has any respect for his ancestry, and hold him under sacred obligations to use, cherish, and defend the plant and its use—all of which he regards as no less a pleasure than a duty. Here too its many virtues were first discovered, and its soothing effects first felt and appreciated.

"To the old Virginian it is indeed a cherished weed, charming all manner of diseases, comforting in sorrow, soothing the ills of life, and preserving to a good old age and in a happy frame of mind all who use it. He believes in its superior virtues, and ascribes to it more good qualities than to any other known plant. He always carries it about with him, and if perchance he gets out he is truly miserable. He not only loves but worships it as a cure all. His wife and daughters know its virtues full well, and use it with equal grace and relish, believing it gives a lustre to the eye and a freshness to the cheek rarely surpassed. Among the variety of ways in which it is used none attracted my attention so much as the novel manner of snuff-taking in various parts of Virginia, West Virginia, the Carolinas and Georgia.

"In some localities the practice is unknown, while in many others it is very common. I first discovered young ladies putting snuff into their mouths as if eating it, when my curiosity was excited to an alarming extent, but on being invited to 'dip' with them I soon learned that they were not eating, but 'rubbing and chewing' it, as they called it, and in such a lively manner as to soon convince me that they appreciated it. I found the habit to be quite common even among the young of both sexes—all indulging in it as if it afforded real satisfaction to the appetite for tobacco in some form.

"The young ladies however seemed the more attached to the 'rubbing process,' as it has been appropriately styled, and defended it with equal logic and grace whenever it was assailed. The young gentlemen when in the society of the young ladies generally join them in this unique use of snuff, as they are always sure to be invited and urged if they decline, and to merit their favor of course they must appear social. I believe, in credit to their taste, however, that they really prefer a good cigar, and think it more in keeping with their ideas of manhood and neatness. I have seen young girls of ten 'rubbing and chewing,' as if they appreciated it as much as mother Eve did the apple in the garden of paradise.

"I have also seen old ladies with trembling limbs and few teeth 'rubbing and chewing,' as if it made them feel young again. I have frequently been ushered unexpectedly into the presence of young ladies, and found them puffing their cigarettes in a manner that convinced me that they knew how to smoke. There is nothing that will more surely and quickly bring a stranger into the fellowship and good graces of the ladies than to join them in their pet habit of snuff-rubbing. It seems to form a bond of friendship which they regard as sacred as the vows of wedlock.

"The older matrons 'rub' less and smoke more, which is in accordance with nature and philosophy: The older we grow the more we smoke. They find solid pleasure in sitting by the open grate after tea with fifteen inches of pipe's tail between their teeth, and slowly but gracefully puffing the perfumes of the exhilarating weed into the room, and watching with childish pleasure the hazy curling wreaths of smoke as they gently float around, changing in form and color until they finally disappear up the chimney, affording rich themes for meditation and profitable study, and perhaps suggestive of earlier days when grandmother, an innocent, blooming maid, was exchanged for the weed, the seed of which produced the plant she is now burning. Everywhere I marked only pleasant and soothing effects from the use of tobacco.

"The planter is never more indifferent to the ills of life and in sympathy with good feeling and pleasure, than when he sits down after dinner in his vine-thatched portico and lights his pipe, passing to his guests pipes, cigars, and tobacco in various forms, leaving them to choose their favorite mode of using it. Sambo is never more contented than when he burns the weed in a cob pipe, and draws the delicious smoke through an elder sprig or mullen stem. But the maid is happiest of all when with her lover she sits face to face, and they 'dip' together from the same magic plant—tobacco.

"In every walk of life throughout the sunny South tobacco in some form may be found, and its effects are always the same, whether drawn from the pocket of the beggar or taken with gloved fingers from the golden tobacco-box of the planter. For snuff the ladies have very nice round boxes with lids which, they always carry with them full of black snuff highly but pleasantly flavored. They also carry little brushes or sticks about three inches long with pliable ends; these they wet in the mouth, then dip into the snuff-box, and then place it in the mouth outside of the gums and rub earnestly for two or three minutes. 'Will you dip with me?'

SNUFF-DIPPING.

is the usual way of putting the invitation, when the box is drawn from the pocket and rapped slightly on the cover, sometimes by all present, who thus signify their readiness to 'dip,' then it is repassed open to all, and the 'dipping and rubbing' begins in earnest.

"The only advantage I ever discovered in this unnatural way of snuffing is in avoiding all unpleasant sneezing which snuffing is sure to produce, although it is claimed that it whitens and preserves the teeth and sweetens the mouth, and produces a beneficial effect on the lungs, all of which is true or not, just as you choose to believe. 'Will you dip and rub with me?' said one of the prettiest belles of Winchester, and in another city in another state the daughter of an ex-governor, handing me a silver-tipped brush and opening a rose-wood snuff-box richly inlaid with gold, politely asked me to 'dip' with her, expressing the belief that friendship would always follow. I have frequently been asked by ladies when travelling through the country and stopping at farm-houses, if I used tobacco—as a hint to offer them some, and it was a pleasure to comply, and receive the thankful smile of an appreciative heart."

In other parts of the country the habit of snuff-taking is confined principally to old ladies, who use any kind, either black or yellow, and who prefer themselves the cheaper kinds.

SNUFFERS.

But few varieties are used, and there seems to be but little taste manifested in the selection of the "dust." Foreign varieties are used only to a limited extent, being chiefly confined to those of transatlantic birth and tastes. The custom of chewing and smoking seems to be more popular with the male sex than snuff-taking, and one rarely finds a man addicted to the latter habit, unless it be one somewhat advanced in years.

Stewart in his admirable paper on snuff gives much useful information in regard to the universal custom of using it as well as its origin and distinguished uses of the great sternutatory.

"The luckless fate of inventors and originators has become proverbial, but the ingenious individual whose nostrils rejoiced in the first pinch of snuff stood in no need of the niggardly praise of contemporaries or the lavish gratitude of posterity. That first 'pinch' was its own priceless reward, far above present appreciation or future fame. What matters it, that his great name has not been reverently handed down to us: that posterity seeks in vain his honored tomb, on which to hang her grateful votive wreath; that zealous antiquaries have raised up innumerable pretenders to his unclaimed honors, and striven to rob him of his fame? Enough for that lucky inventor, wherever he may rest, that he enjoyed in his lifetime the reward for which ordinary benefactors of their kind are fain to look to the future.

"It is perfectly vain to attempt now to penetrate into the mystery which envelopes the name and nation of the first snuff-taker: long before rough, noble-hearted Drake cured his dyspepsia by the use of tobacco, or Raleigh transplanted some roots of that precious weed into English soil, there were European noses which had rejoiced at its pulverized leaves. Conjecture, lost in the mazy distance, gladly lays hold of something substantial in the shape of snuff's first royal patron. This was Catherine de Medicis, who, receiving some seeds of the tobacco plant from a Dutch colony, cherished them, and elevated the dried and pounded leaves into a royal medicine, with the proud title of 'Herbe à la Reine.' For in the beginning men took snuff, not as an everyday luxury, but as a medicament. Like tea—which a hundred years later was advertised as a cure for every ill—the new sneezing powder was hailed a universal specific; and so pleasant in its operation, that mankind, acting upon the wholesome aphorism that prevention is much better than cure, and eagerly anticipated the disease it was supposed to remedy."

"The use of 'the pungent grains of titillating dust' received a somewhat heavy and discouraging blow from an unexpected quarter. That ubiquitous power which hurled anathemas alike at the heresies of Luther and the length of clerical wigs, discountenanced its use, and at length fairly lost its temper in the contest with snuff. Whether from a prescience of the beneficial influence it was destined to exert upon mankind, or from a suspicion of its power of sharpening intellects, it is difficult to say; but Popes Urban VIII., and Innocent waged quite a miniature crusade against snuff, anathematizing those who should use it in any church, and positively threatening with excommunication all impious persons who should provoke a profane sneeze within the sacred precincts of St. Peter's pile; Louis XIV., that good son of the Church, filially complied with the paternal injunction, but his courtiers were less yielding; and the ante-chamber of Versailles frequently resounded with the effects of the pleasant stimulant.

"All persecution has a distinct tendency to establish the object of its hate, and so it was with the subject of our article—it only required to be loved; and I do not doubt that, had circumstances required them, snuff would have found its martyrs. Its use was not general in England until Charles II. introduced it, upon his return from exile, with other important fashions. It had been known and used before, as had the periwig, but it was not until his reign that it became common. When the Stuarts relieved the country of their presence for the second and last time, it had become firmly established; and, by the days of good Queen Anne, was such a necessary of life, that there were in the metropolis alone no less than seven thousand shops where the snuff-boxes of the Londoners could be replenished.

"At that time, indeed, gallants were as proud of their jewelled boxes of amber, porcelain, ebony and agate as they were of their flowing wigs and clouded canes, the handles of which were not unfrequently constructed to hold the cherished dust. We are told by courtly Dick Steel, that a handsome snuff-box was as much an essential of 'the fine gentleman' as his gilt chariot, diamond ring, and brocade sword-knot. We know them to have been manufactured of the costliest material, heavy with gold and brilliant with jewels, as they needed to be when their masters carried wigs 'high on the shoulder in a basket borne,' worth forty or fifty guineas, and wore enough Flanders lace upon their persons to have stocked a milliner's stall in New England.

"Unfortunately, but very naturally, this extravagance rendered snuff a butt for the wits (who all took it, by the way), to shoot at. Steele, whose weakness for dress and show were proverbial, levelled many of his blunt shafts at its use; and Pope, who himself tells us 'of his wig all powder and all snuff his band,' let fly one of his keener arrows at the beaux, whose wit lay in their snuff-boxes and tweezer cases. As the men laid by, in the Georgian era, much of the magnificence of their attire, so their snuff-boxes became plainer and decidedly uglier. Rushing into an opposite extreme, the most outrageous receptacles for the precious dust were devised. Boxes in the shape of bibles, boots, shoes, toads, and coffins outraged public taste. The strangest materials were used in their construction; the public taste leaning towards relics possessing historical interest. Thus the mulberry tree planted by Shakespeare, the hull of the Royal George, in which 'brave Kempenfelt went down, with twice four hundred men,' and the deck of the Victory, on which Nelson died 'for England, home, and beauty,' have alone been supposed to supply material for snuff-boxes to an extent which, if known, must considerably weaken the faith of their possessors in their genuineness.

FANCY SNUFF-BOXES.

"Nor has snuff itself been less liable to the rule of fashion than the boxes that held it. We will give a few familiar instances. In the naval engagement of Viga, in 1703, when a large Spanish fleet was taken or destroyed, a great quantity of musty snuff was made prize of, and patriotism ran high enough to cause the 'town' for some length of time to resist all that was not manufactured to imitate the flavor from which it took its well-known name of 'musty.' Nearer to our own time, a large tobacco warehouse having been destroyed by fire, in Dublin, a poor man purchased some of the scorched or damaged stock, and manufacturing it into coarse snuff, sold it to the poorer class of snuff-takers. Forthwith capricious fashion adopted it, endowing it with fabulous qualities, and Lundy Foot's Irish Blackguard (so it was termed) filled the most fashionable boxes.

"Again, during the Peninsular campaigns, in which the light division of the British army bore so memorable a part, the mixture used by and called after its gallant leader, General Sir. Amos Norcott, had a more extensive sale than any other. When Napoleon was at Elba, and folks began to tire of legitimacy, as they soon did, it became fashionable to use snuff scented with the spirit of violet, and significantly to allude to the perfume. Garrick, when he was manager of

Drury Lane Theatre, brought a mixture into fashion by using or alluding to it in one of his most famous parts. The tobacconist whom he thus favored was his under-treasurer, Hardham, whom no writer about snuff should omit to notice. He was a great favorite with Garrick, whom in his turn he almost revered. One of Hardham's most important duties was to number the house from a hole in the curtain above the stage; and it is amusing to fancy the little tobacconist, snuff-box in hand, calmly watching the pit fill, or from his elevated position admiring the histrionic talents of his gifted patron. His shop in Fleet street is also memorable. It was the general resort of theatrical men and tyros, who sought to reach the manager through his subordinates, and his little back parlor witnessed the *début* of many who afterwards gained applause from larger, though not more exacting audiences.

"Her Majesty Queen Charlotte has bequeathed her name to a once favorite mixture, and George the Fourth has some slight chance of being remembered by the famous 'Prince's Mixture,' which was so popular when it was the fashion to admire and imitate that gifted individual. It would be a grateful but almost an impossible task to enumerate the kings, soldiers, lawyers, poets and actors who had sought from and found in the snuff-box comfort and inspiration. Prominent among the rulers of the earth who have acknowledged the pleasing influence of snuff is Frederick the Great. His snuff-box was the pocket of the long waistcoats of that period, in which he kept large quantities loose—a dirty habit, which Napoleon, who was a great plagiarist, adopted. It would be easy to draw out a famous list of literary names attached to snuff, beginning with Dryden, who was particular enough to manufacture his own mixture, and selfish enough to preserve the secret of its excellence, with a view, probably, of enhancing the value of the pinch from his box, for which the beaux and wits at Will's intrigued.

"In the pulpit, at the bar, and on the stage, snuff has been equally valuable in adding to the persuasive eloquence and talent of its patrons. By the female portion of human-kind it was at one time pretty generally taken, nor was it uncommon for young and even pretty women to offer and accept a pinch in public. After the gentle sex had to a great extent given up the habit, some strong minded females were to be found who retained it. Mrs. Siddons, when she came off the stage after dying hard, as Desdemona, or harrowing the hearts of her audience by her representation of Jane Shore, could composedly ask those around for a pinch of the precious restorative. When we consider the beneficial influence which snuff has exerted over mankind generally, we cannot help regretting that its virtues were not sooner known.

"For we put forth the proposition seriously, that its effect upon the world has been to render it more humane and even-tempered, and that had the western hemisphere discovered the tobacco plant earlier, historians would have had more pleasant events to chronicle. For instance, it is not impossible—nay, most probable—that the fate of Rome, discussed by the Triumvirate over their snuff-boxes, would have been different. Is it likely that, under the humanizing influence of mutual pinches, Antony would have asked for, or Augustus resigned, the head of Cicero to his bloodthirsty colleague; or that the other details of the conscription which deluged the streets of Rome with the blood of her best citizens, would have been agreed to? Again, can any one imagine Charles the Ninth and his evil counsellors plotting the massacre of St. Bartholomew over pinches of the soothing dust? Is it probable that the High Court of Justiciary would have entitled its royal martyr to a special service in the Book of Common Prayer, if its deliberation had been inspired by the kindly snuff which since that time has so often softened the rigor of the law? My hypothesis may seem an absurd one, but history supports it.

"When Charles the Second introduced snuff into general use, men's hands had scarcely adapted themselves to more peaceable occupations than cutting their neighbors' throats, and the ashes of a long and bitter civil war needed little fanning to break into a blaze again; and yet, for forty years of misgovernment the nation kept its temper. How can this forbearance be accounted for? Was it that circumstances no longer called for as stern and as effectual remedies as before? No. Was the second Charles one whit more desirable than the first of that ilk? Was Clarendon more liked than Stafford? was Russell's head of less consequence than Prynne's ears? No. Again, wrongs as grievous as those which Hampden had died in resisting were to be avenged, but in a milder, better fashion; for mankind had in the meantime learned to take snuff. Much of the haste and irritation which had previously led to blows discharged itself in a good-natured sneeze. Snuff made men forbearing, even jocular over their wrongs. Who can doubt that the revolution which ended in placing William of Orange on his father-in-law's throne owed its bloodless character not a little to the influence of snuff. We read of difficulties in its course, which, fifty years previously, would inevitably have led to bloodshed, being easily, almost humorously surmounted. The plagued nation effected a revolution over its snuff-boxes in the happiest conceivable manner.

"Having ventured so far I am inclined to put forward a yet higher claim which snuff has upon our gratitude, and to hint that the great deeds of great men who were snuff-takers may be traced by a chain of reasoning—slight, yet conclusive —to this dearly prized luxury. The hackneyed saying that time is money, or money's worth, has more truth in it than most of the fallacies which are supposed to regulate our conduct. The most important events of our lives often hinge on moments. A moment to stifle passion, to summon reflection, to plunge into the past and bring up a buried memory, to consider results, is often of the utmost consequence, and this valued moment the pinch of snuff insures, when, without it, delay would be simply embarrassment. The pinch of snuff, taken at the right instant, secures an important reprieve, during which the unpleasant question may be evaded, the hasty reply reconsidered, or an angry *repartee* thought better of, while the same time gained serves to improve the diplomatist's *equivoque*, to point the orator's satire, and polish the wit's *mot*. In a word, its use on important occasions affords, to every one who needs them, better means of acting upon Talleyrand's mischievous yet clever aphorism—that language is useful rather to conceal than to express our thoughts. Moreover, the action necessary in conveying the tempting graces to their destination has not unfrequently been found useful. It employs the hasty hand that may itch to take illegal vengeance for fancied insults; it serves to hide the angry twitching mouth and passionately expanding nostrils, to give a natural expression to changes of the countenance which would otherwise indicate emotion, and to parry attention till reason has been summoned to supplant passion.

"It is denied (in a rather irritating way sometimes) that the subject of our article has any beneficial influence upon the intellects of its patrons. We are not about to claim for it any such exalted qualities, but we may be allowed to mention a fact or so which entitles it to some respect medicinally. As we have before stated, in its early days it was considered to possess powerful healing qualities, and even now is found of use in cases of headache and weak sight. It was also supposed valuable in cases of heaviness and obtuseness of intellect. Is it, therefore unreasonable to presume that it may have had some share in gaining for our brethren beyond the Tweed that shrewdness of national character which has become proverbial?

"The specimens which came in the reign of James I.,

CURING A HEADACHE.

southward, did not command much respect or admiration from our countrymen; indeed they were the bulls at which every satirist hurled his shafts, and blunt must have been that one which did not pierce some potent folly of language or manner. The town rang with anecdotes of their rags, beggary, and quarrels; ballad-singers made merry at their expense, and the stage resounded with uncomplimentary allusions. Indeed, in one of the most popular plays of that period, the king himself was not spared, and the actors (Ben Jonson among them) had very nearly lost their ears for their boldness. Nor was it at least for a hundred and fifty years after this period that the Scotch became noted for that enterprise and talent which now distinguish them.

"We do not deny that the union may have developed

their traits, but it is clear that within that time snuff had become a national stimulant. To the observer of men and manners there is something very characteristic in the various fashions in which the pinch of snuff is taken. 'The exercise of the snuff-box,' as it was once termed, was an acknowledged science, but few were the great proficients who could mutely express their feelings by its aid. We have not space to run through all its exercise, but we may mention the 'pinch military,' which Frederick, and after him, Napoleon practiced inhaling snuff copiously, and with much waste, as though it were human life they were throwing away; the 'pinch malicious,' of which Pope was perfect master; the 'pinch dictatorial,' which burly Jonson established; the 'pinch sublimely contemptuous,' such as Reynolds took when some travelling virtuoso hinted at excellence away from Leicester-square, and ruffled his complacent vanity; and, above all, the 'pinch polite,' which Talleyrand understood so well.

"From snuff to sneezing is but a step, which we purpose taking before we bring this cursory article to a close. The act of sneezing appears to have been variously regarded at various stages of the world's history, but from the earliest times of which we have any authentic record, it has been the customs of those around to give vent to a short benediction immediately upon its commission. The Robbins considering themselves bound to find a reason for this universal custom, and being hard pressed, gave the somewhat incomprehensible explanation that, previous to Jacob, man sneezed but once in his lifetime, and then immediately before death; so that those around, warned of his imminent journey, hastened to wish it a good termination. How it was that Jacob instituted a new order of things we are not told, but as a proof of the truth of their assertion they give the fact that in all nations of the earth a similar custom will be found existing.

"Strangely enough this assertion was corroborated by the first colonists of America, who found the habit to be in common use amongst the aboriginal tribes. The Greeks and Romans certainly had a similar habit, but far from attaching any ill-omen to the sneeze they regarded it as of good augury. Thus Catullus assures us that when Cupid upon a memorable occasion sneezed, all:

'The little loves that waited by
Bowed and blessed the augury.

DIFFERENT TASTES.

And in the 'Life of Themistocles,' Plutarch informs his readers that sneezing by the General on the eve of a battle was regarded as a certain sign of conquest. Strangely enough we find that in comparatively modern times, the custom of giving expression to good wishes when a friend sneezed was attributed to the fearful plague which periodically swept over Europe. Sneezing was one of its first and most dangerous symptoms, and those who were by, as they gathered their robes about them and fled from their doomed fellow-creature, would ejaculate a quick 'God bless you,' hurriedly invoking from a more merciful quarter the aid they feared to give. Violent sneezing was not only among the first, but was one of the last fatal signs of that fearful scourge, and was often too rapidly followed by death to give time for more than a short benediction. Anyhow, the custom still exists and one of the most pleasant reminiscences attached to the first pinch of snuff is the chorus of hearty good wishes of sympathizing friends which follows upon the inevitable sneeze."

The variety of taste in snuff is accounted for by the proverb, "So many men to so many noses." Highland gentlemen of every degree are mostly fond of Gillespie; while

HIGHLANDERS.

operatives from the Lowlands generally prefer plain Scotch. When two Highlanders meet, they usually exchange a pinch of snuff, mutually *preeing* the contents of their *mulls*, while their *colleys*, (dogs) after a fashion of their own, take a reciprocal *sniff* of each other. Cuba is the favorite of the gentlemen of the stock exchange; the tradesman's box usually contains rappee; high dried Irish is grateful to those who love to feel the taste of snuff in their throat. Sea-faring men seldom take snuff: a sailor with a snuff-box is as rarely to be met with as a sailor without a knife.

The history of the rise and progress of snuff-taking abounds in incidents and anecdotes, among the most curious of all that relate to the various modes of using the weed. Though once the most popular and fashionable manner of using tobacco it now falls far behind the other and more common and more popular forms of indulging in the herb. In France and Spain the introduction of tobacco ushered in this form of using it, and to inhale a few grains of the pungent dust was the delight of polished and favored courtiers who regardless of the forms royalty patronized and gave sanction to the custom. Thus its use in a short time became popular all over Europe and gave unlimited scope for the satirist and dramatist to ridicule the habit. In spite, however, of frown and ridicule this ancient custom though not now as popular or as fashionable, still claims many sincere votaries and doubtless will as long as the plant is cultivated or used in any form.

CHAPTER IX.

CIGARS.

"The poet may sing of the leaf of the rose,
And call it the purest and sweetest that blows;
But of all the leaves that ever were tried,
Give me the tobacco leaf rolled up and dried."

THE smoking of cigars is now considered the best as it is the most fashionable mode of using the weed. The word cigar is from the Spanish *cigarro*, and signifies a cylindrical roll of tobacco leaves, made of short pieces or shreds of the leaves divested of the stem and wound about with a binder, and enveloped in a portion of the leaf known by the name of wrapper—acute at one end and truncated at the other. In the East Indies a sort of cigar called *cheroot* is also made with both ends truncated. The smoking of tobacco in the form of cigars is doubtless the most general as well as the most ancient mode of its use. When Columbus landed in Hispaniola, the sailors saw the natives smoking the leaves of a plant, "the perfume of which was fragrant and grateful." But while cigars are of very ancient origin in the West Indies, they were not generally known in Europe until the beginning of the Nineteenth Century. In fact, of all the various works on gastronomy and the pleasures of the table, written and published from 1800 to 1815, not one speaks of this now indispensable adjunct of a good dinner. Even Britlat-Savarin, in his

Physiologie du Gout, entirely ignores tobacco and all its distractions and charms. Benzo gives the following account of the manufacture of a cigar in Hispaniola:—

"They take a leafe from the stalks of their great bastard corn (which we commonly called Turkie—wheat) together with one of these tobacco-leaves and fold them up together like a coffin of paper, such as grocers make to put spices in, or like a small organ-pipe. Then putting one end of the same coffin to the fire, and holding the other end in their mouths, they draw their breath to them. When the fire hath once taken at the pipe's end, they draw forth so much smoke that they have their mouth, nose, throat, and head full of it; and, as if they tooke a singular delight therein they never leave supping and drinking till they can sup no more, and thereby loose their breath and their feeling."

Sahagun, in his "History of New Spain," speaks of the natives as using the leaves of tobacco rolled into cigars, which they ignite and smoke in tubes of tortoise-shell or silver. The following article from the *New York Times* contains much valuable information in regard to cigars, especially Havanas:

"It is perfectly safe to say that there is more money spent every day in New York for cigars than for bread," (doubted.) "From the fine gentlemen, who buy their cigars at Del-

CIGARS.

monico's, or get them direct from the importers, down to the little barefoot boys in the streets, who buy theirs from the Chinamen at the corners or pick up the stumps that are thrown away, all smoke. In some countries pipes and cigarettes are made to do duty by the poorer classes, but in New York cigars seem to be almost invariably preferred. Now, while there is nothing better, in the way of something to smoke, than a first-class Havana cigar, there is nothing nastier than some of the cheap abominations made in that shape in New York. To the truth of this last proposition, anyone will readily testify who has ever been so unfortunate as to have had to ride from Harlem to New York in a late smoking-car, with half a dozen roughs smoking cheap cigars on board.

"The cigars sold in this market may be divided into three classes—the imported, those made of imported tobacco, and those made of domestic tobacco. These may be again classified under many different heads, as there are many kinds and grades of each. The cheapest cigars in New York are dispensed by dilapidated Chinamen, who have little stands about the streets and markets. These are certainly the vilest cigars made anywhere in the world, and are sold from one to five cents each. Next in order come the common domestic cigars. They are sold at five cents each, or six for twenty-five cents, and are of the kind kept at the cheap refreshment stalls, lager beer saloons, and low groggeries. After these are the more pretentious home-made cigars, manufactured of selected domestic tobacco, which are sold all over the city, and in the making of which Havana 'fillers' are supposed to be used. A filler, be it known, in technical parlance means that portion of the tobacco of which the inside of the cigar is made. Price, ten to fifteen cents. Then comes the best class of cigars in which domestic tobacco is used, those which are made with clear Havana fillers and Connecticut wrappers. Fifteen cents is the price, and many are palmed off on the unwise for the real imported article. Cigars made wholly of imported Cuban tobacco come next on the list. Some of them are excellent, and compare favorably with many of the imported. They bring from fifteen to fifty cents each at the cigar stores. Last in line, but best of all, is the genuine, imported Havana cigar. Few and rare are they, and great is the price of the higher grades thereof.

"There are some places in New York where an imported cigar of a reasonable size may be bought for fifteen cents, but they are few and far between. Twenty or twenty-five cents is the price usually charged, and from that to a dollar. All the cigars made in the United States are invariably put up in imitation Havana boxes, with imitation Havana labels and brands. It is doubtful, however, whether this transparent device deceives anybody, for in accordance with the United States Internal Revenue laws, all boxes of cigars manufactured in the United States must not only bear the manufacturer's label, giving his full name and place of business, and the number of his manufactory, but they must also bear the United States inspector's brand. Before the present law was in force, and the duties on tobacco were low, this scheme may have been profitable. But why the practice is still adhered to by the manufacturers is hard to imagine, for the boxes now used, being made of imported cedar, must be very costly, and must materially increase the price of cigars. Only those of the very poorest quality are packed in white wooden boxes.

"Some people seem to smoke not because they like it, but only to be in the fashion. Some days ago the writer of this article happened to be in a cigar-store, when two well-dressed young men came in and asked for some ten cent cigars. The clerk handed out the box, and after a critical inspection the purchaser asked: "Are these medium?" 'Yes, sir,' said the clerk. 'Then I'll take a dollar's worth.' After they had gone the writer asked the clerk what they meant by 'medium.' He said he didn't exactly know, but supposed they wanted to know whether the cigars were between strong and mild. 'I told them they were,' said he, 'because I thought they would buy if I said so, but they are all alike.' And in this connection it is very singular that although the Island of Cuba is so near to the United States and so many cigars are imported into this city, so little is known about the different sizes and brands of cigars, excepting, of course, by those in the business. It is a common thing here to see a man ask in a cigar store for a *Flor del Fumar*, a *Figaro*, or an *Espanola*. By this he means a cigar of a certain size, and does not seem to know that these are not the names which designate the

CIGAR-HOLDERS.

size, but are the names of the manufactories. In Havana, were a man to ask for a *Flor del Fumar*, the dealer would ask him what size he wanted.

HAVANA CIGARS.

"Every box of cigars packed in Havana has, at least, six distinctive works on it. First is the brand, which is burned in the upper side of the lid of the box, with an iron made for the purpose; second the label, this bears the name and address of the manufactory; third, the mark designating the size and shape of the cigars, this is usually put on with a stencil; there are not so very many regular sizes, or *vitolas*, made in Havana as might be imagined, a list of them may prove interesting. These are: Damos, Entre Actos, Opera, Concha, Regalia de Concha, Londres, Londres de Corte, Regalia de Londres, Regalia Britanica, Regalia del Rey, Regalia de la Reina, Reina Victoria, Panetelos, Trabucos, Embajadores, Especiales, Imperiales, Brevos, Prensados, Cilindrados, Millar Vegueros. The *Damos* (Dames) as their name indicates, are meant for the ladies, and are the smallest made. The *Cozadores* (huntsmen) are the longest, and the *Trabucos* (blunderbusses) the fattest. The *Prensados* (pressed) are flat, and *Cilindrados* (cylindrical) are so called because, when green, they are put in bundles of twenty-five, and tightly rolled in strong tissue paper, which is twisted at each end of the roll. When the cigars are dry the paper is taken off, and the bunch retains the cylindrical shape given it. The *Brevos* (figs) are also tied up while green, and and tightly pressed. This makes them stick together something like figs, hence their name. The *Vegueros* (plantation) take their name from the fact that they are supposed to be made like those made on the plantations, but they are not made in the same way.

" In the *Vegos* (plantations) the *veguero*, or planter, makes his cigar of a single leaf of tobacco, which he carries ready moistened for the purpose, by rolling it on his knee. Besides the above, some fancy sizes have been adopted of late years, but they are made by only a few of the larger manufacturers in Havana. Fourth is the color mark, which is also put on in stencil. Fifth, the class mark. All the round cigars made in Havana are separated into three classes: *Primera*, or first; *Segunda*, or second; and *Tercera*, or third. Some manufacturers never mark any of their cigars as of the third class, not because they do not make them, but because they think they sell better without the mark. They make the first class *Flor*, the second *Primera*, and the third *Segunda*. Others mark all their cigars as of the first class, and indicate the classes by the color of the labels, and in this way none but the wholesale purchaser knows the secret. Sixth, the last, is the mark denoting the number of cigars in the box. This is stenciled on the side in Arabic numerals.

"A theory has obtained that cigars made in Havana, by reason of some inexplicable climatic influence, are better than those made in New York, even should they be made of tobacco from the same plantation. This may be so, but it is doubtful whether this was ever fairly tested, or, indeed, whether it was ever tested at all. The truth is that all the best tobacco grown in the island of Cuba is bought up by the heavy manufacturers in Havana. The crops of the best plantations are contracted for in advance, and the old-established firms buy from the same *vegos* year after year. Hence it is why their cigars are so uniform in quality. All Cuban tobacco is not good, by any means. The tobacco from the Vuelta de Arriba is not so good as that from the Vuelta de Abajo, and yet there is but little difference in their geographical position. And in the Vuelta de Abajo, a short distance makes a difference in the quality of the tobacco. Some *vegos* are celebrated for their good crops, while others, perhaps not a hundred yards away, do not produce good crops at all. There are many poor cigars made in Cuba, as all who have ever been there know, and all over the island the Havana cigar is deemed the best. In Havana, and, indeed, in all parts of the island, green or freshly-made cigars are preferred, and the most esteemed cigar-cases are made of carefully prepared bladders, in which the cigars are rolled to prevent the evaporation of the moisture.

"When a Cuban gentleman gives a cigar to a friend, he does not, as we do, open his case, and offer it to him to choose from but he examines its contents carefully and critically, selects the one he thinks the best and offers it. And there is a great deal more in the choice of a cigar, by selecting it on account of its outside appearance, than one not accustomed to it would suppose. A wrapper which has that which the Cubans call *calidad* makes the cigar much stronger than one which does not possess it. That is to say, that the wrapper which has *calidad* contains more essential oil, is denoted by an abundance of small pustules on the surface of the leaf, and by a general rich, oily appearance. As a proof of the foregoing proposition, it is only necessary to know how cigars are made. A lot of tobacco is worked up into say 50,000. After they are all made, they are turned over to be assorted, according to color and class, and are packed and marked. The fillers are all alike, it is the wrappers that make the difference. To assort the colors a very, correct eye is required, and those who do this part of the work make better wages than those who make the cigars.

"The value of cigars does not increase in direct ratio with their size, for owing to the difficulty in getting good wrappers for the larger kinds, the expense of their manufacture is much increased. Upon one occasion, in Havana, a manufacturer received an order for a thousand cigars intended for the Queen of Spain's husband, Don Francisco de Asis, which he agreed to make for $1,000. They were delivered in due time, and packed in a richly-mounted cedar chest, were sent to the royal recipient. They were magnificent cigars, of the cazadores size, all of the same color, and so smoothly made as to look as if they had been turned out of hard wood instead of rolled tobacco. They were placed on exhibition for a few days before they were sent to Spain, and a gentleman who saw them, wishing to make a present to some dignitary, asked the manufacturer to make him a a like number at the same price. To his surprise, the order was refused. The manufacturer said he could not do it for the money. His explanation was that it was not the actual cost of the tobacco and labor of making them, but it was on account of the trouble and expense met with in selecting the wrappers. He said he had to pick over thousands of bales before he could secure a sufficient number of the proper length, color, and fineness.

"Some two years ago there was a story of a Cuban cigar-dealer in Broadway, who selected cigars for his more favored customers by ear. It was said that he put the cigar to his ear, and listened intently for a moment, and by the cracking of the tobacco was enabled to judge of its quality. This was a good advertising dodge, but in practice it was all nonsense. None but that wily Cuban ever heard of such a mode of trying a cigar. In the Island of Cuba that which we call a cigar is called a *tabaco* (a tobacco) and when it is required to discriminate between the manufactured and unmanufactured article it is called *tabaco torcido*, or rolled tobacco. This, however, is only necessary when used in the plural. In Mexico a cigar is called a *puro*, and in Peru* and some of the other Spanish American countries it is called a *cigarro puro*, in contradistinction to the *cigarro de papel*, or cigarette.

---

* Ballaert says that the consumption of cigars in Peru is enormous. "An old fisherman on being asked how he amused himself when not at his labors, replied, ' Why I smoke; and as I have consumed 40 paper cigars a day for the last 50 years, which cost me one rial each will you have the goodness to tell me how many I have smoked, and how much I have expended for tobacco?'"

Cigarettes in Cuba are called *cigarros*, and their consumption is enormous. Strange as it may appear, there are some confirmed smokers in Cuba who never use cigars at all, but confine themselves to cigarettes. To the New Yorker it looks curious to see a great, bearded man smoking a tiny cigarette; and, indeed were he to smoke his cigarette as the New Yorker would smoke his cigar, it would be labor lost, so far as getting any effect of the tobacco was concerned. But the cigarette smoker inhales the greater part of the smoke, it goes directly into his lungs, and into contact with a large surface of mucous membrane, and, indeed, with the blood itself. Were the New York cigar-makers to smoke a cigarette in the same way it would make him so giddy that he would be compelled to give it up long before it was consumed. That the smoke does go into the lungs is proved by the fact that a cigarette smoker can inhale the smoke and exhale it again after drinking a glass of water."

All tobacco grown upon the island of Cuba is not of the finest quality; the majority of it is far inferior to the best

LIFE IN MEXICO.

Mexican coast tobacco. The value of the tobacco lands of this last mentioned country has not been fully developed.

The variety of soil, exposure, climate, and atmospheric influences are greater than can possibly be in Cuba, and when the best is discovered, combining all the requisites, which undoubtedly will be the case with an increased culture of the plant, it will be found to be equal to the Vuelta Abogo of Cuba, and much more extensive. The subject of tobacco lands, evidently, is not well understood in Mexico, as it must be, from great experience, in Cuba. All of these varieties of lands and circumstances exist in Mexico, and it is safe to predict that, at some day, this country will stand pre-eminent over all others in this industry.

We extract the following from the *Tobacco Leaf* in regard to cigar-making in Cuba:—

"The rule is that a cigar-maker devotes all his ingenuity and diligence to one class of goods. For example, one workman makes only *Londres;* another only *Regalias;* another only *Milores Communes;* and so on. In the Cuban's factory the operatives are allowed to smoke as many cigars as they like when at work; and to take home with them, when they leave work in the evening, five cigars each. The immigration of Chinese laborers into Cuba has modified, and must further modify, the labor market there. In the cigarette factories at Havana, Chinese workmen are almost exclusively employed. Though objectionable for many of their moral habits, these workmen are nevertheless docile, ingenious, laborious, and contented."

A writer, alluding to the manufacture of cigars, says:—

"The colors or strengths are *Amarillo Claro*, bright yellow; *Amarillo Obscuro*, dark yellow; *Claro*, bright; *Colorado Claro*, bright red; *Colorado*, red; *Colorado Obscuro*, dark red; *Colorado Maduro*, red-ripe or mellow; *Maduro*, ripe or mellow; *Maduro Obscuro*, dark ripe or mellow; *Pajizo Claro*, bright straw-colored; *Pajizo*, straw-colored; *Pajizo Obscuro*, dark straw-colored; *Fuerte*, strong or heavy; *Entre Fuerte*, rather strong or heavy; *Flajo*, light. Then there are the indications of the qualities:—Superfine; *Firo*, not quite so fine; *Flor*, finest or firsts; *Superior*, next, or seconds; *Buenos*, next, or thirds. The cigar has a notable history. First has to be determined the part of the plant from which it is taken; then the part of the leaf from which it is taken, the tobacco being best which is furthest away from the root or middle of the leaf. One elaborate process follows another for the perfection of a work of art—for as such we must regard a cigar."

Hazard, in his admirable work on Cuba, devotes considerable space to cigars, their manufacture, varieties, and use, in which he speaks of the various brands as follows:—

"The brands known as '*Yara Mayau*,' and the '*Guisa*,' are perhaps the most celebrated made upon the Island. Of the '*Yara*,' which has some considerable reputation, particularly in the London market, I confess I cannot speak favorably. Cigars that I smoked made from this leaf, and which are much smoked in the vicinity of Santiago de Cuba, I found had a peculiar saline taste which was very unpleasant, as also a slight degree of bitterness; many smokers, however, become very fond of this flavor. When I state that in Havana alone there are over one hundred and twenty-five manufacturers of cigars, it will readily be understood there must be a great many inferior cigars made even in Cuba. Havana may be called the 'City of cigars,' from its reputation and the immense number of factories there are in it for the manufacture of cigars, from the smallest shop opening on the street, employing three or four hands to the immense *fabricos* erected expressly for this purpose, and employing five or six hundred.

"Let not any one imagine, then, that because he is in Havana he will get no poor cigars, for a greater mistake cannot be made, for just as vile trash can there be purchased as anywhere; and it appeared to me that in buying, from time to time in different *fabricos*, a few cigars it was rarely I found a really good one. It behooves, then, every lover of a good cigar to make himself familiar with the best makers and brands, and to purchase

CUBAN CIGAR SHOP.

those, and those only, that suit his taste. To the traveler in Havana, this is easy enough, as he can there buy sample boxes from any of the factories and of any of the brands. There are, in addition to these hundreds of other cigar factories, some of which, such as *Cabargos, Figaros, Luetanos, Victorias*, etc., are first-class, three or four at least in whose cigars every smoker may have perfect confidence, the brands of which are known all over the world. These are: *Cabaños, Uppmann* and *Partagas;* for whose brands, perhaps, one pays something more, but has always the satisfaction of finding them good. To the kindness of the gentlemen connected with some of these factories I am indebted for most of the information in this article, and particularly to Señor Don Avulmo G. del Valle, the present proprietor of the Cabaños Factory, who was good enough to show me through his establishment, carefully explaining to me its peculiarities. As the process of manufacture and description of grades and qualities are the same with all the best makers, I give here a detailed history of this factory and its products.

"The factory for Cabaños cigars has been established seventy-two years the founder of it being Don Francisco Cabaños, his son, Don de P. Cabaños, succeeding him, to whom has succeeded his son-in-law, Senor del Valle, the present proprietor and director of the factory. When it was founded, the cigars were sold to the public in bundles of twenty, only amounting to a total number per year, of four or five hundred thousand cigars, the sales of which kept constantly increasing until 1826, when there were sold two millions. At this period the demand for exportation commenced, increasing each year until 1848, when the number sold amounted to three and a half millions. At this time, the present director came in charge, and increased the sale to eight millions per year, until, in 1866, the total sales by this one house only, amounted to the enormous number of sixteen million cigars, which went to different parts of the world. The tobacco manipulated in this factory is, with some few exceptions, that grown upon plantations in the Vuelta Abojo, with the proprietors of which Señor del Valle has a special contract for their product. The most noted of these places are known as '*La Lena,*' '*San Juan aj Martin,*' '*Los Pilotos,*' '*Rio Hondo.*' The firm also own three *vegas*, as do also Partagas, Uppmann, and others, in a greater or less degree. The amount raised upon these *vegas* in connection with the Cabaños Factory, amounts to five thousand bales, of from first to eighth quality, leaving the most inferior qualities, which amount to about one thousand bales, for exportation, the factory not using such common grades. It is a custom of the manufacturers to keep a supply of the best qualities always on hand from year to year, in order that, should the tobacco crop, in any one year, be bad, the reputation of the house can be maintained by using the good tobacco in the store. The factory is a large stone building, opposite the Canipo de Moste, in which all the operations connected with cigar making are carried on (excepting the manufacture of boxes) by over five hundred operatives, all males. The following is the process of manufacture:

"Arrived at the factory, the tobacco bales, carefully packed and wrapped in palm leaves, are kept in a cool, dark, place on the first floor, being divided off into classes according to quality and value, which latter varies from twenty to four hundred dollars per bale of two hundred pounds. When wanted, the bales are opened, the *manojas* and *gabillos* are separated, and the latter carried in their dry state to the moistening room. Here are a number of men whose business it is to place the leaves, for the purpose of moistening and softening them, into large barrels in which is a solution of saltpetre in water; this done, the water is poured off, and other workmen spread out the leaves with their hands upon the edges of the barrels, ridding them as much as possible, of any surplus water; after which, the leaves, from being moistened, unfold very easily, and, with care, without tearing. The stem is then taken out, the process being known as *disbalillar*. These stems, with the refuse of other tobacco, are sometime used as filling for the commonest kind of cigars. The filling is known as *tripa*, the very best being selected, like the leaf, for the best cigars. Now comes the maker, and supplying himself with a handful of leaf (*copa*) for wrappers, and a lot of the *tripa* for filling or really making the body of the cigar itself he carries it to a little table, and spreading the wrapper upon the table, cuts with a short knife the different portions of the leaf, This is a very nice operation, requiring skill, knowledge, and experience; for it is in this operation that the different qualities of tobacco are separated, the outside of the leaf being generally the best; next that, another quality; and that portion adjoining the stem the worst.

"The general sorting of the tobacco is done by hands of great experience and judgment, who are the highest in consideration in the factories, some of them receiving large pay; thus for instance, the official *escojedor*, or chooser, gets from five to seven dollars (gold) per day, and the *torcedores*, or twisters, from two to four, the workmen being paid so much per thousand cigars, generally from two to four dollars. To show how very careful the maker must be in cutting out the leaf to make the most of it: Mr. del Valle was explaining to me the process of manufacture, and directed the maker to cut the leaf. This the man did drawing his knife in the manner denoted by the dotted lines in the engraving. This it appears was not making the most of the fine part of the leaf, for Mr. del Valle, annoyed, took the knife himself, and after rating the maker soundly for his carelessness, showed him how to cut it properly, as defined by the black line, the difference being, as far as I could judge, a slight inequality of color between the two parts.

TOBACCO LEAF.

The manufacture of the cigar is very simple. The cigar maker, being seated before a low work table, which has raised ledges on every side except that nearest him, takes a leaf of tobacco, spreads it out smoothly before him, and cuts it as in the drawing. He then lays a few fragments of tobacco (*tripa*) in the centre or a leaf strip and rolls the whole into the shape of a cigar, and taking then a wrapper, rolls it spirally around the cigar. If the workman is skillful, he makes it of just the right length and size, without any trimming of the knife. The cigars are assorted, counted, and done up in bundles of generally twenty-five each, and then packed in the boxes, ready for market, under their different names of *Londres*, *Regalias*, etc. These names are generally understood to have the same meaning throughout the trade, the '*Vegueros*,' for instance, being the plantation cigars, made at the *regas*, and much esteemed by smokers, though they are rarely to be met with for sale, or, if so, at an exhorbitant price. The '*Regalia Imperial*,' the finest and best, is nearly seven inches long, the price varying from one hundred and fifty to three hundred dollars per thousand (gold). The '*Regalia*' is not so large but fine, the '*Trabuco*,' short and thick; the '*Londres*,' the most convenient in shape, and most smoked in this country and England; the '*Dama*' the small sized one used by ladies(?) or by men between acts of the opera (*entr' operas*). There are also other names which each factory has for some particular kinds. Artificial flavors are given to cigars, when some particular taste is to be satisfied, by the use of flavoring extracts. Each of the above names has different qualities, as:

Londres '*superfine*' the very best of that size (delicious).
" '*fino*,' not quite so fine.
" '*flor*,' finest, or firsts.
" '*superior*,' next, or seconds.
" '*buenos*,' next, or thirds.

Again, these different qualities have different colors, known as: '*maduro*,' strongest; '*oscuro*,' strong (dark); '*colorado*,' medium; '*claro*,' mild; '*Brevors*,' means pressed. Thus, supposing one wanted a good cigar to suit his taste, he would perhaps order: 'Partagas' (maker), 'londres' (size), 'flor' (quality), 'Colorado' or 'oscuro' (strength), and he would get a good cigar, nice size, best quality, not too strong, or too mild.

"I must confess to a weakness for the Uppmann cigars, which I have found, without exception, to be good, and which have a fine reputation throughout the West Indies. A millionaire need not want a better cigar to smoke than their '*Londres superfine*,' at sixty dollars (gold) per thousand, in Havana, or their '*Cazadores*,' at fifty dollars. Partagas cigars of course, every one knows are good; and he keeps generally pretty well sold up, but fills orders as they come in. For a new experience, one of his '*Regalio Reyno flor*,' is something to try, even if they do cost out there eighty-five dollars, gold.

"In all the factories they make about the following rates: For every order of ten thousand, costing fifty dollars per thousand, five per cent. discount is allowed. Less than five thousand will pay five dollars extra. I should, perhaps, mention that no distinction is made to dealers, the only advantage they have over the private buyer is, that they are enabled to get the discount for large lots. The absurd notion so prevalent with us, that the Cubans only smoke their cigars green, is an error, since the leaf is entirely dried in the sun before being touched by the manufacturer. The Cubans are very particular indeed to preserve the aroma and fragrance of the cigars, by keeping them in wrappers of oiled and soft silks; it is, in fact, quite a sight to see with what ceremony some of these are produced at gentlemen's tables, with much unction, like the ushering in of old wine.

My chapter on cigars would be incomplete did I fail to note the beautiful and courteous way in which all Cubans no matter of what position, whether the exquisite at the club, or the *portero* at the door, ask you for a light. 'Do me the favor Señor?' and you present your cigar, the lighted end towards the speaker. He takes the cigar delicately between his thumb and fore-finger, lights his own, and then, with a quick, graceful motion, turns yours in his fingers, presenting you, with another wave, the mouth end, makes you a hand salute, utters his *gracios*, and leaves you studying out the 'motions' and thinking what a charming thing is national politeness."

In the selection of leaves for the manufacture of cigars in the factories only the large fine ones are used for *Regalias*, *Imperiales*, or *Medios Regalias*; and also for *Cazadores*, *Panetelos*, *Imperiales*, *Caballeros*, and so on; the smaller fine leaves for *Panetelos* and *Londres*; the dark inferior leaves for *Canones*. The commonest tobacco goes to form the *Milores Communes*; the worst is converted into cigars which are generally pressed flat, and known as *Prinsados*. For the smallest kind of *Londres* and for *Damos*, a proportionally small leaf is employed.

In Cuba and Luzon, one of the Philippine Islands, is found one of the largest factories for cigars in the world. In Manilla there are three factories where 7,000 families and 1,200 males are employed: one in Cavite, in which 5,000 operatives, mostly females, are engaged; and one in Malabar, which gives employment to about 2,000 more, also females. The tobacco is worked into both cigars and cheroots both of which have a variety of shapes. In both Manilla and Havana the custom of smoking is universal and one rarely meets with any of the male sex without a cigar between his lips.

A writer speaking of the universality of the custom says:

"In Havana, the custom of smoking is a universal one. There, young and old indulge freely in the use of the weed, dividing their attention pretty equally between the cigar and the cigarette. Even the ladies of the better class in many instances indulge; though not to so great an extent as is commonly reported."

"Smoking in Cuba" says an American writer, "is like the

habit of making shoes in Lynn, Massachusetts, everybody smokes!—in the house, and by the way; in the cars, and on horseback; everywhere, and at all times. You meet whole regiments of youngsters, from six to eight years of age, with black beaver hats, tail-coats, and canes, each with a cigar, nearly his own size, in his mouth. You feel like putting the miniature dandies into the water of the next fountain basin, which shallow as it is, would fully, suffice to drown the largest of them."

You have a right to accost any one smoking in the street, however much may be his superiority or inferiority to yourself, and to ask a light for your cigar; even negroes hatless and shirtless, thus address well-dickied gentlemen, and *vice versa*. Refuse to take a cigar with a Cuban, and you refuse his friendship. The negroes cannot work at all without their quota of cigars; "and looking out of the windows of a room in that magnificent hotel '*El Telegrafo*,' the writer remembers to have caught a glimpse more than once of the negro women at work in the laundry, every one of whom held a long cigar in her mouth, and puffed incessantly as the clothes were manipulated upon the washboards." In Havana, as throughout Cuba, there is a cigar etiquette, to infringe any of the rules of which is construed as an insult. It is, for instance considered a breach of etiquette when you are asked for a light to hand your cigar without first knocking off the ashes. A greater breach, however, is to pass the cigar handed for you to obtain a light from, to a third party for a similar purpose; the rule is to hand back the cigar with as graceful a wave as

WENCHES SMOKING.

you can command, and then if necessary, pass your own cigar to the third party.

The insult direct in cigar etiquette is for the party to whom you apply for a light, to pass on and leave you with the remains of his cigar, or to intimate to you, by word or action, that he has no further use for it, and that you can throw it away. In Cuba, where cigars are plentiful, the usual custom is, when you ask for a light, even if the party be a stranger, to pull out your case and offer him a cigar, by way of recognizing the civility in stopping to accommodate you. The Spaniards are naturally a polite people, and the stranger stepping into the Louvre and other public places of resort in Havana, is struck at once with the marked contrast in this respect to familiar gatherings elsewhere. In no place is a cigar more enjoyable than in Havana. Seated upon the roof of one of the large hotels in that city in a bright moonlight night, within hearing of the dreamy roll on the beach: the regular throb of the sea, lulling one into quiet-

A MOONLIGHT REVERIE IN HAVANA.

ness; the sigh of the summer breeze a lullaby to the senses; while a high-flavored prime cigar, as it wastes and floats away in air, is the fairy wand which opens the enchanted gates of Reverie and Imagination.

What need of a friend under such soothing circumstances? What need of the jolly *camarade* of former days to sigh back sigh for sigh, puff for puff, and wander in gentle reminiscences over the Lesbian labyrinth of the past, when Julia was most kind, or Cynthia, darling girl, delighted in the perfume of a capital havana? Here, in this quaint old city by the sea, is the place for dreams and reveries and the utter rendering of one's self up—to a good cigar. Is it not a place for reverie? Has not one with this most respectable weed, this prime *havana*, the concomitants of a thousand reveries? Will not one puff of that narcotic breath drowse deep all watching dragons, and make for him the sleeping beauties of his will? And, *presto*, there they are! and, oh! ye houris of the South, with what a smile and glance between the azure puffs! Well let me not forget myself. With a sterner morality he sees how the bending Bedouin fashions his pipe in the moistened ground; he sees the slender Indian reed with the flat bowls of Lahore and Oude, the pipe of the Anglo-eyed celestial, the red clay of Bengal, and the glittering gilded cups in which the dark-skinned races of Siam, the Malacca Isles, and the Phillippines, love to enshrine their dreamy opium-haunted spirits of the weed. He sees how in the squatter's hut the old squaw sits by her hunter lord, and puffs at the corn-cob sweetness, and how by lonely ways the traveler rests and thinks of home, and in the blue smoke greets once more the faces of the loved, perhaps forever gone. He sees how the Esquimaux, with his hollow Walrus-tooth, makes bearable the stifling squalor of his den; or, sterner and graver still, some item of historic lore mingles rudely with his dreams, and elbows sharply the airy spirits of his smoke-engendered thoughts. Softly tremble in the delicate blue mist and the azure spirals from his old Virginia clay—the domes of a sea-bathed city. Loftily pierce the tall white minarets into the quivering heavens, while the solemn cypress throws its shade below. Before him, silent-paced as in a dream, files the weird array of Arab camels, bowing their long necks tufted with crimson braids, and measuring the brown sands of the desert with ghost-like tread. 'Tis the moon of Egypt and the waters of the Nile; 'tis the palm-bough waves for him; and women, free-limbed, with flashing eyes, and antique water-vases on their heads, move past him from the low-rimmed shadowy wells. And he sees them there and smiles.

He sees on the beach by the sea the summer idler sitting beneath the jutting rock, gazing far out upon the sea, yet ignoring the white sails that pass up and down before him, as

BY THE SEA.

well as the open volume upon his knee, while his thoughts float outward and upward with the graceful wreaths of smoke that encircle his head; and if of a practical turn, he listlessly wonders why, if his own delightful land furnishes some twentieth of the whole Tobacco produce of the world, and does honor to her native weed by being its mightiest consumer, why, in the name of all disasters, the product is so dear—ay, doubly dear? And thus as his pipe burns low, a hundred other statistics; then, knocking out his whitened ashes on the floor, he reads sedately (his pipe being out) that the "Tobacco plant furnishes ashes to the amount of one-fourth of its bulk, being a much greater proportion than that of any other vegetable product," and, moreover, that "Tobacco exhausts the soil at the ratio of fourteen tons of wheat to one of Tobacco!" Oh, base insinuation! But, as he relights his pipe, and the graceful vapor circles in fresh buoyancy and grace before him, he only, in his contented mind, retains that one supreme expression—"*One ton of Tobacco!*" Ah,

"Think of it, picture it
Now, if you can!"

From "A Paper of Tobacco," *we extract the following humorous description of Yankee cigar smokers, which to a certain extent is true to life, but like most of the articles descriptive of American life by English Authors, who travel in America and write *a book* afterwards, it is exaggerated or overdrawn:

"The Americans, who pride themselves on being the fastest-going people on the 'versal globe'—who build steamers that can out-paddle the sea-serpent and breed horses that can trot faster than an ostrich can run—are, undoubtedly, entitled to take precedence of all nations as consumers of the weed. The sedentary Turk, who smokes from morn to night, does not, on an average, get through so much tobacco per annum, as a right slick, active, go-ahead Yankee, who thinks nothing, 'upon his own relation,' of felling a wagon-load of timber before breakfast, or of cutting down a couple of acres corn before dinner. The Americans, it is to be observed, generally smoke cigars; and tobacco in this form burns very fast away in the open air, more especially when the consumer is rapidly locomotive, whether upon his own legs, the back of a horse, the top of a coach, the deck of a steamboat, or in an open railway carriage. The habit of chewing tobacco is also

---

* London, 1839

prevalent in 'the States,' nor is it, as in Great Britain and Ireland, almost entirely confined to the poorer classes. Members of the House of Representatives and of the Senate, doc-

AN AMERICAN SMOKER.

tors, judges, barristers, and attorneys chew tobacco almost as generally as the laboring classes in the old country. Even in a court of justice, more especially in the Western States, it is no unusual thing to see judge, jury, and the gentlemen of the bar, all chewing and spitting as liberally as the crew of a homeward-bound West Indiaman. It must indeed be confessed that Brother Jonathan loves tobacco 'not wisely but too well,' and that the habits which are induced by his manner of using it are far from 'elegant.' The truth is, he neither smokes nor chews like a gentleman; he lives in a land of liberty, and takes his tobacco when and where he pleases. He spits as freely as he smokes and chews—upon the carpet or in the fire-place—for he is not particular as to where he squirts his copious saliva, and does not think with the late Dr. Samuel Parr, that a spitting-box is a necessary article of household furniture. The free-born citizen of the States laughs at the aristocratic restrictions imposed on smoking in England, where, on board of the numerous steamboats that ply on the Thames, conveying the pride of the city to Gravesend and Margate, no smoking is allowed abaft the funnel, and where, in public-houses ashore, no gentleman is permitted to smoke in the parlor before two o'clock in the afternoon, A pipe of tobacco, or a cigar, after a day's hard exercise, whether mental or bodily, and after the cravings of hunger and thirst are appeased, may be fairly ranked amongst the most delightful and most harmless of all earthly luxuries. It fills the mind with pleasing visions, and the heart with kindly feelings. A hard-working laborer, smoking by the side of his hearth at night, presents a perfect picture of quiet enjoyment. I see him now in my mind's eye. He is seated in an old high-backed, cushionless arm-chair, but an easy one, nevertheless, to him, who from dawn till sunset, has been engaged in ploughing, thrashing, ditching, or mowing. With one leg thrown over the other, he quietly reclines backward, and with an expression of perfect mental composure, he gazes on the smoke that ascends from his pipe. There is a sentiment-exciting power* in the smoke of tobacco when perceived by the eye, as well as a pleasing sedative effect when inhaled; and those smokers who have any doubt of the fact should take a pipe with their eyes closed. A person who smokes with his eyes shut cannot very well tell whether his cigar is lighted or not. How soothing is a pipe or a cigar to a wearied sportsman, on his return to his inn from the moors! As he sits quietly smoking, he thinks of the absent friends whom he will gratify with presents of grouse; and, in a state of perfect contentment with himself and all the world, he determines to give all his game away. Full of such kindly feelings, he retires to bed; but, alas, with day-light, when the effect of the tobacco has subsided, the old leaven of selfishness prevails, and his good intentions are abandoned. 'Mary,' said an old Cumberland farmer to his daughter, when she was once asking him to buy her a new beaver, 'why dost thou always tease me about such things when I'm quietly smoking my pipe?' 'Because ye are always best-tempered then, feyther,' was the reply. 'I believe, lass, thou's reet,' rejoined the farmer; 'for when I was a lad, I remember that my poor feyther was just the same; after he had smoked a pipe or twee he wad ha' gi'en his head away if it had been loose.'"

---

* The smoke ascending from the snuff of a candle could excite a sentimental feeling in the minds of Wordsworth and Sir George Beaumont, though it seems to have had no such effect on the mind of Crabbe.—*Lockhart's Life of Sir Walter Scott.*

ODE TO A CIGAR.

The following ode to a Cigar is no doubt familiar to many, yet will pay a re-perusal:

"And oft, mild friend, to me thou art
   A monitor, though still;
Thou speak'st a lesson to my heart
   Beyond the preacher's skill.

"Thou'rt like the man of worth, who gives
   To goodness every day,
The odor of whose virtues lives
   When he has passed away.

"When in the lonely evening hour,
   Attended but by thee,
O'er history's varied page I pore,
   Man's fate in thine I see.

"Oft, as thy snowy column grows,
   Then breaks and falls away,
I trace how mighty realms thus rose,
   Thus trembled to decay.

"Awhile, like thee, earth's masters burn,
   And smoke and fume around,
And then like thee to ashes turn,
   And mingle with the ground.

"Life's but a leaf adroitly rolled,
   And time's the wasting breath,
That, late or early, we behold
   Gives all to dusty death.

"From beggar's frieze to monarch's robe
   One common doom is passed;
Sweet nature's work, the swelling globe,
   Must all burn out at last.

"And what is he who smokes thee now?
   A little moving heap,
That soon, like thee, to fate must bow,
   With thee in dust must sleep.

"But though thy ashes downward go,
   Thy essence rolls on high;
Thus, when my body must lie low,
   My soul shall cleave the sky."

In Charles Butler's "Story of Count Bismarck's Life," a good anecdote is told of the Count and his last cigar:—

"'The value of a good cigar,' said Bismarck, as he proceeded to light an excellent Havana, 'is best understood when it is the last you possess, and there is no chance of getting another. At Königgrätz I had only one cigar left in my pocket, which I carefully guarded during the whole of the battle as a miser does his treasure. I did not feel justified in using it. I painted in glowing colors in my mind the happy hour when I should enjoy it after the victory. But I had miscalculated my chances.' 'And what was the cause of your miscalculation?' 'A poor dragoon. He lay helpless, with both arms crushed, murmuring for something to refresh him. I felt in my pockets and found I had only gold, and that would be of no use to him. But, stay, I had still my treasured cigar! I lighted this for him, and placed it between his teeth. You should have seen the poor fellow's grateful smile! I never enjoyed a cigar so much as that one which I did not smoke.'"

In European cities juveniles offer the smoker, at every street corner, a "pipe" or a "cigar light." The following description, entitled "Light, Sir," is from an English journal, and contains much interesting information on the various modes of lighting pipes and cigars.

"LIGHT, SIR."

"'Ere y'are, sir—pipe-light, cigar-light, on'y 'ap'ny a box—'ave a light, sir.' Every smoker of the larger cities knows the cry. Every tender-hearted smoker is familiar with the appeal, by day and by night, and remembers pangs of regret he has felt when the want of ha' pence or the repletion of his match-box has prevented his much-besought response. There is no need now to enlarge upon the sufferings, the adventures, the dangers of these peripatetic juvenile trades folk, sparse of clothes and food, and full of the material which may make or mar a nation; for all this was done, and even overdone, by the graphic sensationalists of the London penny dailies when Chancellor Lowe proposed a tax on matches. We may, upon occasion, feel for the manufacturers and venders of 'lights,' but more generally we find ourselves constrained to sympathize with the purchasers of such contrivances for the ignition of pipes and cigars. The smoking of tobacco is an art; an art which, in its proper exercise, requires much care, much prudence, and not a little skill. This is a proposition which must, from its very nature, be startling to non-smokers, and surprising to many smokers. The tobacco hater (invariably an illogical creature, who hates that which he knows not) will hold up hands in amazement, and sniff with the nose in contempt, to whom reply would be superfluous.

"With the smoker the case is otherwise. A German writer recently said that the English were better smokers than the Germans; because, whereas the German smoked incessantly, without rule, system, or moderation, the English smoked with care, with slow and appreciative lovingness, and the determination not to overstep the bounds of rational enjoyment. Had he known more of English smokers, he would not have made so wild a statement; and had he known English women better, he would never have attributed to their sweet influence the fancied superiority he describes in English as compared with German smoking. In truth, the art of tobacco using is nowhere more ignored, nowhere more contemptuously neglected than in these 'favored isles.' For one man who smokes with a reason, for a purpose, or by system, you shall find a thousand who smoke without either; and the result is that those who smoke have little defense, in the general way, for their practice, while those who condemn the habit have far better grounds for their opposition than they have ever yet been able to explain. To those who do know why they use tobacco, it is well-nigh incredible that so many of their fellow-smokers should be ignorant of the properties, the uses, the abuses, of the weed they burn and the fumes in which they delight. Yet, even this is not so surprising as the fact that so few of those who smoke—smoke much, often and constantly—should be ignorant of, or indifferent to, the conditions which are necessary to their own adequate enjoyment of the weed.

"You will see a man light a cigar so carelessly that one side of the roll will burn rapidly, with prodigious fumigation and giving out a dark and offensive cloud, while the other side remains untouched by the fire, only to wither and crackle and twist into uncouth shapes, until the smoker flings the cigar away, with an accompaniment of expletives which attach rather to his own stupidity than to the piece of tobacco he has so abominably abused. You will see another with a good pipe, laden with good tobacco, well lit, blowing incessantly down the mouth-piece and the stem until the moisture introduced with his breath into the bowl of his pipe effectually prevents the tobacco from burning, and puts out the fire; and then you will hear him lament that he should have paid so good a price for a pipe so bad that it 'fouls' before he has smoked a single hour. You will see another who, while he talks to his friends, allows his tobacco to go out every three or four minutes, so that at length his mouth is sore and his palate nauseated with the combined fumes of lucifer matches, burnt paper and exhausted tobacco dust; and he inveighs against the 'cabbage-leaf which that rascally tobacconist sold him for good Shag or Cavendish.' Another knows so little of the art of smoking that he never 'stops' his pipe, and so allows the light dust of the burnt weed to fly about him in flakes and minute particles, to the permanent damage of his own and his neighbors' clothes. But in nothing is the inartistic character of English smoking so conspicuously exemplified as in the use of 'lights.' Those who form the great majority of smokers amongst the English-speaking races seem to consider that, so long as their pipes are set alight, it matters not how or from what source the light is obtained. Thus, one will place his pipe-bowl in a flame of gas, and pull away at the stem till his tobacco is on fire; another will thrust the bowl into the midst of a coal fire, and when he sees a glow in the bowl withdraw it, and contentedly puff away; another stops an obliging policeman or railway guard, and ignites his tobacco by hard pulling at the flame of an oil-lamp; another will stick the end of a choice cigar into the bowl of a pipe filled with coarsest Shag, thus ruining the flavor of his 'prime Havana' forever; while yet another will light lucifer matches, and apply the blazing brimstone to his pipe or cigar, thus saturating the whole mass with sulphurous and phosphoretic fumes, to the ruin of the weed and the injury of his own health.

"How much wiser the West Indian negro, who takes a burning stick from the wood fire, and tenderly lights his weed therewith, or joyfully brings a handful of the white-hot ashes in his thick-skinned palm, that 'massa' may fire his cigar! Or the travelling peddler or tinker, who, as he sits by the way-side, patiently wooes the sun with a 'burning-glass' till his tobacco ignites, or uses with equal prudence and skill the ancient but inimitable tinder-box.

"But this is the age of Fusees. What a name! When, in our youth, those longitudinal strips of tinder, semi-divided

BRINGING A LIGHT.

into innumerable transverse slips, all tipped with harmless, ignitable matter, first assumed the title, we had little notion of the atrocities which would come to be dignified by their name. This was soon after the world had been delighted by the Congreves, which drove Lucifer to the wall, and before English and German ingenuity had taught us to find 'death' in the box, as well as 'the pot.' The innocent old fusee had his faults, certainly. He would not always light; he had a bad habit of turning back on your finger-nail and burning its quick when you struck him; and he would occasionally light up, all by himself, and set fire to fifty of his fellows in your waist-coast pocket, or the tail of your best dress-coat. (Those were the days when waist-coats were gorgeous and tail-coats immense.) But what were these peccadilloes compared with the sins of the modern 'cigar-light?' 'Fusees,' forsooth! More like bomb-shells, military mines, torpedoes, and nitroglycerine trains. Who has not had them explode in his eye, on his cheek, down his neck, scarring his skin, burning holes in his coats and trousers, frightening passers-by, and doing all manner of deep-dyed devilment? Nor is this the worst. Those who will trust their skins, and their eyes, and their clothes to 'Vesuvians,' 'Flamers,' and the like, are not to be pitied; for they are more cruel to their tobacco than the fusees are to them. Our grievance is that so many engines of destructiveness and offensiveness should be so largely patronized by smokers, to their own discomfort, the ruination of their tobacco, the scandalization of gentle and simple, and the encouragement of vicious manufactures. Now, we are not going to particularize too closely, for fear of consequences. In these days, when a man may bring an action for libel because it has been said of him that he sells bad soup at a railway station, prudence is the better part of valor. But, just examine this heterogeneous pile of 'cigar-lights,' which rears its audacious head upon the table. Here are Palmers, Barbers, Farmers, Lord Lornes, Tichbornes, Bryants and Moys, Bells and Blacks, Alexandres, Bismarcks, King Williams, Napoleons, and scores of other varieties. Some light 'only on the box,' some light anywhere, some everywhere, and some nowhere. Some are on wood, some on porcelain, some on glass, some on dire deeds intent. There are vestas, safety-matches, patent flint-and-steel contrivances, with silver tubes and marvellous screws wherewith to put them out when they have served your turn. Some are excellent, many passable, still more intolerable. One of these times it may be worth while to speak of the good ones, but at present we care only to treat of those that are bad, and that briefly.

"Here's a 'Flamer'—we name no names—everybody seems to make flamers; and this one deserves his title. We want to light a peaceful pipe, and he bursts out in a fury like unto nothing on earth so much as Etna in convulsion, or the Tuilleries in petroleum blaze. But, if you have any respect for your tobacco, your lips, your nostrils, or your lungs, you will let him get rid of his flames before you apply him to your cigar; and, when you do venture so far, he drops off the stick and burns a hole in the carpet. Or, if you be daring enough to take a light from the flamer while he flames, you spoil your tobacco, foul your mouth, and get a taste of sulphur-suffocation such as Asmodeus might have were he to take a whiff of a smoke-and-fire belching chimney in the Black Country as he flies across that district by night.

Haven't got a light? Glad of it. Try a Vesuvian-round, black and tipped with blue. There's a pyrotechnic display for you! Now, in with it, after the approved style illustrated by the two human hands engaged in lighting a cigar on the illuminated cover of the box. 'Ugh!' you say. Just so; you've got a mouthful of choice abominations, which will cost you much waste of saliva, several shivers, and the whole piece of tobacco you were about to enjoy. Here, put that away; take another, light it quietly with this wax-vesta, or this wooden 'spill,' or this screw of paper; smoke gently, don't let the fire out, and you'll be all right. In future, you may be wise enough to avoid cheap cigar-lights and pipe-lights, even for use in the streets. Our word upon it—they are far dearer than those which cost more."

The following description of "Home Made Cigars" is from *All the Year Round*, and will doubtless be read with interest by many growers of the weed who may recall similar scenes:

" 'Apropos of cigars,' said Wilkins, lighting a second fragrant Havana with the stump of the first, 'let's go and see the farmer's establishment for making them. You see that field of tobacco over yonder? Old Standish raises his own weed, dries it in the big open sheds behind the barn, cures it—I don't quite know the whole process—and then has it made into sixes and short fives, Conchas and Cabanas, like a Cuban señor. I went over the establishment about a year ago, and it is worth seeing.'

" We strolled first over to the tobacco field. The weed was then just at its full ripeness, and the long, flappy, delicately-furred green leaves bent gracefully over toward the ground, growing smaller and smaller the higher they were on the stout stalk. Few foreigners know that even as far north as New England, in the sunny valleys of Connecticut, sheltered as they are from the bleak east winds of the Atlantic and accustomed to a long and steady summer heat, tobacco is grown in large quantities, flourishes exuberantly, and is one of the chief sources of profit to the farmers. It needs a rich warm soil and careful tending; but it gives in its growth, a sentimental reward to the cultivator; for it comes up gracefully, rapidly, and beautifully, and is with some care, one of the most satisfactory crops to 'handle." Having gazed at and tasted the thick leaves, we sauntered behind the barn, and there saw the long open shed, with beams running parallel from end to end, where the gathered tobacco leaves were hung to be thoroughly dried by the sun.

"Then Wilkins conducted us for some distance along the river bank; we jumped into a boat and rowed perhaps half a mile, landing by the side of a little shop-like building, where we heard the hum of voices and the commotion of many busy persons. We entered and found ourselves in a long, low room, having wide tables ranged along the walls; here, working rapidly, were rows of chatty country

MAKING CIGARS.

girls, who, as they worked, laughed and talked, and now and then hummed a verse of some familiar ballad. Neatly packed piles of the dried and cured leaf lay upon the table before them.

"Each was armed with knives and cutters, and we watched the quick transformation of the flat leaves into the smooth and compact cigars. The tobacco grown upon the farm was, we discovered, only used as wrappers for the cigars. The good farmer imported, for the interior filling, a fine tobacco from Havana. Strips and little pieces of this the girls placed in the centre of the cigar, wrapping the Connecticut tobacco in wide strips tightly about it, then pasting each of the last with some paste in a pot by their side. It seemed to be done almost in an instant; the Havana slips were laid down, cut and trimmed, and pressed into shape in a twinkling; the wrappers were cut as quickly; and, more rapidly than I can describe it, the cigar was made. These girls were mostly daughters of neighboring farmers, who received so much per hundred cigars made; intelligent, bright-eyed and witty; many of them comely, with rosy cheeks and ruddy health; educated at the common schools, and able, their day's work over, to sit down at the piano and rattle away *ad infinitum*.

"His stock of cigars thus made up, from the first sowing to the last finishing touch, the good squire (being Yankee-like, a sort of Jack-of-all-trades,) would have them put up in gorgeously labeled boxes, carry them to town, and sell them to retail dealers; not disdaining himself, twice or thrice a year, to go through the neighboring States with samples, and acting as his own commercial traveler."

This description, however, may not convey a correct idea of the exact mode of manufacture to many growers of tobacco in the Connecticut Valley inasmuch as many planters of the "weed" make the entire cigar (more particularly for their own use) wrapper, binder and filler wholly of seed-leaf tobacco, such cigars do not readily sell to the trade except at inferior prices which admit of but a small profit to the manufacturer. The following spicy article from the "London *Figaro*" may be interesting to all smokers as well as guide them in the selection of a good cigar.

"I am an imaginative person, and 'society' has treated me shamefully of late—its tangible delights are absent from me. Allow me, then, to console myself by the 'creations of smoke,' as Lord Lytton puts it. I am scouted by society because I am in love. I am told I look:

> "As hyenas in love are supposed to look, or
> A something between Abelard and old Blücher."

And, moreover, I am an ugly man, but there was only a fortnight's difference in gaining a woman's love between John Wilkes and the handsomest man in England, courage, Jehu! I like idleness, because it shows that one can afford it; so I am puffing idly—ah! the balmy fragrance of this mild Havana! 'Oh! the effect of that first note from the woman one loves!' says one; 'Oh! the kiss on the dimpled cheek, the sound of the silver voice!' says another; but what can compare to the dreamy exquisite luxury of a good cigar? But, heavens, what am I saying? I am in love, and Julia reads the "*Figaro*!" The paleness of Flaxman's illustrations spreads over me—please, reader, look upon the sentiment as sarcastic. I am in a fog of smoke, and am quaffing claret from the silvered pewter. There's plenty of it; and no soul can say:

> "That in drinking from *that* beaker
> I am sipping like a fly.'

How changed from the long, long days ago, when I was a connoisseur in Parparillo cigars, brown-paper cigarettes, and cane cheroots! Then I fondly adored Sir Walter Raleigh as my earthly idol, for giving me tobacco—when I had the halfpence to buy it—and delighted in the story, told by queer Oldys, of Sir Walter's servant extinguishing the Virginny smoke that issued from his master's lips, by drenching him with ale. Alas! my idol is shattered by Hawkins. The Spaniards say, 'The lie that lasts for half an hour is worth telling.' History has lied for longer, by a considerable period. Fond even as I was of my brown-papered cigarettes when baccy failed, I must confess I never reached the stage attained by Sir Christopher Haydon's chaplain, William Breedon, parson of Thornton, in Bucks, who was so given to

> "October store and best Virginia,"

that when he had no tobacco (and too much drink) he used to cut the *bell-ropes* and smoke them!

> "The Polyglot—three parts—my text;
> Howbeit—likewise—now to my next."

" On Smoke.—It is a vulgar, ludicrous, and foolish custom to bite off the nose of a cigar. Don't be a Vandal—you are not a Sandwich Islander, about to chew your *Kava*. A cigar should be handled daintily; it is a fragile, graceful creature —don't mar its beauty, Tear off the twist, and the pleasure of smoking is at an end! The outer leaf becomes untwirled. Ere it is half finished, you have a ragged end between your lips—nasty, foul, and unsightly—through which the smoke comes in huge clouds to your mouth, instead of slender streams on the palate. 'How, then,' say you; 'prick it, or cut it, or what?' Tear it not, cut it not; nor yet puncture it. Don't be frightened of the cigar—thrusting a half-inch alone into the mouth; but, when you begin, take a good half of it in the mouth; pull at it lustily for a few seconds, to open its pores; then draw it out, allowing but an inch to be held within the lips—believe me, you will enjoy it a hundred-fold more; and there are but few cigars that will not allow of their virtue being drawn though their leaves. Never bite the end off, and never use your cigar cruelly, by squeezing it, biting it, or re-lighting it. Cigar-holders, tubes, quills, and such like inventions, we despise. If you cannot bear the cigar in your mouth—aye, and enjoy it—you have no business with it: go back to your brown paper and cane!

"What is the best beverage to imbibe whilst inhaling the precious weed? Momentous question! Coffee, or claret, says Jehu. I do not believe in bitter, as an accompanying liquid to a cigar. The Corporation of Christ-church, years ago, smoked cigars, and drank with them that then famous concoction known as 'Ringwood Beer.' What was the result? The first toast at every civic banquet held for years in that borough was gravely given out, and bumpered with due solemnity, as follows:—

'Prosperation to this Corporation.'

Brandy is a perfect antidote to inebriation from beer, so we are told. The Corporation should have known this, and been awakened from their long and pleasant dream of *prosperation*. Brandy I should hardly reckon amongst the drinks that ought to be with cigars, notwithstanding that Tennyson has asked:—

'For what delights can equal those
Which stir, with spirits, inner depths? &c.'

Brandy-and-water, gin, whisky, and the likes are only fit for those who nocturnally lay the foundation for matutinal 'hot coppers,' with the vilest shag in the most odorous of yards of clay. 'Smoking leads to drinking,' has been a favorite old woman's saying for time out of mind. How I hate old women's sayings! A grain—requiring to be picked out with a pin and microscope—of truth, with a bushel of bunkum or cant. How is it, that ever since the days of James I, of 'hateful to the nose, harmful to the brain' memory, there have always been carpers on the injurious effects of smoking? 'Nicotine!' they say, with a would-be-taken-for-know-all-about-it-air. Quite so; but, as recent investigations have proved that, so far as the actual 'poisoning' is concerned, it would take upwards of a thousand years to kill the most inveterate of healthy smokers, we have still time to breathe— and 'it please the pigs.' *Mem.* for pipers—French tobacco contains the greatest, Turkish the least, per-centage of nicotine. Havana, two and one-half per cent.

"But an unique old woman of Jehu's acquaintance goes further still; boldly asserting that 'smoking is well for making good soldiers, well for making good sailors, well for making sometimes good lawyers; not so well for making good Christians.' Oh! ashes of Hawkins and Raleigh, shudder for the results of 'baccy on degraded human nature.' There must be a rarity of good Christians, then amongst the parsons; they are all fond of it. Dean Aldrich was, perhaps, the greatest smoker of his day. His excessive attachment to this habit was the cause of many wagers. Here's one:—At breakfast, one morning, at the 'Varsity, an undergraduate laid his companion long odds that the Dean was smoking at that instant. Away they hastened; and, being admitted to the Dean's study, stated the occasion of their visit. The Dean replied, in perfect good humor, to the layer of the bet, 'You see, sir, you have lost your wager; for I am not smoking, but filling my pipe.' But—my cigar has reached its last dying speech, and there is but a drop left in the beaker.

> 'I'll not leave thee, thou lone drop!
> 'Twould be mighty unkind,
> Since the rest I have swallow'd,
> To leave thee behind.'

"Final exhortation. Choose the small, sound, tolerably firm, and elastic cigar: the dwarf contains stuff within which the giant hath not. Don't flatter yourself you're smoking cabbage, if not tobacco—its any odds on rhubarb!

> 'For me there's nothing new or rare,
> Till wine deceives my brain;
> And that, I think, 's a reason fair
> To fill my pipe again.'"

Charles Lamb, "the gentle Elia" was during a portion of his lifetime a famous smoker. In a letter to Hazlitt he writes, "I am so smoky with last night's ten pipes, that I must leave off." It is said that he smoked only the coarsest and strongest he could procure. Dr. Parr inquired of him how he acquired his "prodigious smoking powers." "I toiled after it, sir," was the reply, "as some men toil after virtue!" Lamb was constant in his use of tobacco, and among all the great luminaries of English literature we know of none more addicted to the use of the pipe. Lamb might often be seen in his chambers in Mitre Court Building, puffing the coarsest weed from a long clay pipe, in company with

Parr who used the finest kind of tobacco in a pipe half filled with salt. It was no easy task to relinquish the use of tobacco and it cost him many a struggle and much determined effort. In writing to Wordsworth he says:—" I wish you may think this a handsome farewell to my 'Friendly Traitress.' Tobacco has been my evening comfort and my morning curse for these five years. I have had it in my head to do it (Farewell to Tobacco) these two years; but tobacco stood in its own light when it gave me headaches that prevented my singing its praises."

Lamb's poem is without doubt one of the finest pieces of verse ever written on tobacco, and seemingly contains both words of praise and dispraise—the latter however in some sense are insincere.

"May the Babylonish curse
  Straight confound my stammering verse
  If I can a passage see
  In this word-perplexity,
  Or a fit expression find,
  Or a language to my mind,
  (Still the phrase is wide or scant,)
  To take leave of thee, GREAT PLANT!
  Or in my terms relate
  Half my love, or half my hate;
  For I hate, yet love thee so,
  That whichever thing I show,
  The plain truth will seem to be
  A constrain'd hyperbole,
  And the passion to proceed
  More from a mistress than a weed.
  Sooty retainer to the vine,
  Bacchus' black servant, negro fine;
  Sorcerer, thou mak'st us dote upon
  Thy begrimed complexion,
  And for thy pernicious sake,
  More and greater oaths to break
  Than reclaimed lovers take
  'Gainst women: thou thy siege do'st lay
  Much too in the female way,
  While thou suck'st the lab'ring breath
  Faster than kisses or than death.
  Thou in such a cloud do'st bind us,
  That our worst foes cannot find us.
  And ill fortune that would thwart us,

Shoots at rovers shooting at us;
While each man through thy height'ning steam
 Does like a smoking Ætna seem,
 And all about us does express
(Fancy and wit in richest dress)
A Sicilian fruitfulness.
Thou though such a mist dost show us
That our best friends do not know us,
And for those allowed features
Due to reasonable creatures,
Liken'st us to feel Chimeras
Monsters that, who see us, fear us;
Worse than Cerberus or Geryon,
Or, who first loved a cloud, Ixion.
Bacchus we know, and we allow,
His tipsy rites, but what art thou,
That but by reflex canst show
What his deity can do,
As the false Egyptian spell
Aped the true Hebrew miracle?
Some few vapors thou may'st raise,
The weak brain may serve to amaze,
But to the reins and nobler heart
Canst nor life nor heat impart.
Brother of Bacchus, later born,
The old world was sure forlorn,
Wanting thee, that aidest more,
The gods' victories than before
All his panthers, and the brawls,
Of his piping Bacchanals.
These, as stole, we disallow
Or judge of thee meant: only thou
His true Indian conquest art;
And, for ivy round his dart,
The reformed god now weaves
A finer thyrsus of thy leaves.
Scent to match thy rich perfume—
Chemic art did ne'er presume,
Through her quaint alembic strain,
None so sov'reign to the brain.
Nature, that did in thee excel,
Framed again no second smell.
Roses, Violets but toys
For the smaller sort of boys;
Or for greener damsels meant;
Thou art the only manly scent.

Stinking'st of the stinking kind,
Filth of the mouth and fog of the mind,
Africa, that brags her fois on
Breeds no such prodigious poison,
Henbane, nightshade, both together,
Hemlock, aconite——
           Nay, rather,
Plant divine of rarest virtue:
Blisters on the tongue would hurt you.
'Twas but in a sort I blamed thee;
None e'er prospered who defamed thee;
Irony all, and feigned abuse,
Such as perplex'd lovers use,
At a need, when in despair,
To paint forth their fairest fair,
Or in part but to express
That exceeding comeliness
Which their fancies doth so strike,
They borrow language of dislike;
And instead of Dearest Miss,
Jewel, Honey, Sweetheart, Bliss,
And those forms of old admiring,
Call her Cockatrice and Siren,
Basilisk, and all that's evil,
Witch, Hyena, Mermaid, devil,
Ethiop, Wench, and Blackamoor,
Monkey, Ape, and twenty more;
Friendly traitress, loving foe,
Not that she is truly so,
But no other may they know,
A contentment to express,
Borders so upon excess,
That they do not rightly wot,
Whether it be pain or not;
Or, as men constrained to part
With what's nearest to their heart,
While their sorrow's at the height
Lose discrimination quite,
And their hasty wrath let fall,
To oppose their frantic gall,
On the darling thing whatever
Whence they feel it death to sever,
Though it be, as they, perforce,
Guiltless of the sad divorce.
For I must (nor let it grieve thee,
Friendliest of plants,

That I must) leave thee.
For thy sake, TOBACCO, I
Would do anything but die,
And but seek to extend my days
Long enough to sing thy praise.
But as she who once hath been,
A king's consort, is a queen
Ever after, nor will bate
Any title of her state,
Though a widow, or divorced,
So I, from thy converse forced,
The old name and style retain,
A right Katherine of Spain,
And a seat, too, 'mongst the joys
Of the blest Tobacco Boys;
Where, though I, by sour physician,
Am debarred the full fruition
Of thy favors, I may catch,
Some collateral sweets, and snatch,
Sidelong odors, that give life
Like glances from a neighbor's wife;
And still live in the by-places,
And the suburbs of thy graces;
And in thy borders take delight,
An unconquered Canaanite."

Thomas Jones, in the following neat little tribute to tobacco, pays a deserved compliment, not only to the plant, but to the great English smoker, " ye renowned Sir Walter Raleigh."

"Let poets rhyme of what they will,
Youth, Beauty, Love or Glory, still
My theme shall be Tobacco!
Hail, weed, eclipsing every flow'r,
Of thee I fain would make my bow'r
When fortune frowns, or tempests low'r,
Mild comforter of woe!

"They say in truth an angel's foot
First brought to life thy precious root,
The source of every pleasure!
Descending from the skies he press'd
With hallow'd touch Earth's yielding breast,
Forth sprang the plant, and then was bless'd,
As man's chief treasure!

"Throughout the world who knows thee not?
  Of palace and of lowly cot
    The universal guest;
The friend of Gentile, Turk and Jew,
To all a stay—to none untrue,
The balm that can our ills subdue,
    And soothe us into rest.

"With thee the poor man can abide
Oppression, want, the scorn of pride,
    The curse of penury,
Companion of his lonely state,
He is no longer desolate,
And still can brave an adverse fate,
    With honest worth and thee!

"All honor to the patriot bold,
Who brought instead of promised gold,
    Thy leaf to Britain's shore;
It cost him life; but thou shalt raise
A cloud of fragrance to his praise,
And bards shall hail in deathless lays
    The valiant knight of yore.

"Ay, Raleigh! thou wilt live till Time
Shall ring his last oblivious chime,
    The fruitful theme of story;
And man in ages hence shall tell,
How greatness, virtue, wisdom fell,
When England sounded out thy knell,
    And dimmed her ancient glory.

"And thou, O Plant! shall keep his name
Unwither'd in the scroll of fame,
    And teach us to remember;
He gave with thee content and peace,
Bestow'd on life a longer lease,
And bidding ev'ry trouble cease,
    Made Summer of December!"

The smoker of cigarettes is passionately attached to his "little roll" and regards this mode of obtaining the flavor of tobacco the best. The finest are made in Havana and, vast quantities are used by the Cubans and Spaniards. A writer in "The Tobacco Plant" gives this pleasing effusion in regard to them:—

"Your cigarette is a sort of hybrid—half-pipe and half-cigar; neither the one nor the other; neither the delight of the epicure nor the solace of the true tobacco-lover. Far be it from us to deny, or even to question, its value, its utility, or its charm. We have smoked too many to dream of treating them with scorn—cigarettes of Virginia shag, strong, pungent, luscious; of light and fragrant Persian, innocuous and soothing; cigarettes rolled by ladies' dainty fingers, compressed by elegant French machines of silk and silver, cut, stamped, and gummed by prosy, matter-of-fact, and even vulgar Titanic engines in great tobacco-factories. But the thorough-paced smoker renders to his cigarette only a secondary and diluted adoration: it is nice, it is delicate, it is pretty—a thing to be toyed with, to be fondled, even to burn one's fingers (or, perchance, one's lips) withal; but by no means an object to call forth a passion.

"But just as the world would be a tame and an insipid institution were all men's tastes alike, so the world of smokers would lose much of its romance were all the lovers of the weed of temperament too robust to love a cigarette. Brevity and sweetness are proverbially held to constitute claims upon the respect and admiration of the voluptuous, and to the cigarette these cannot be denied. There is something touching in the self-abnegation of a tobaccoite who will devote five mortal minutes and the sweat of his refined intelligence, with the skill of his delicate fingers, to the preparation of a tiny capsule of the weed, which burns itself to ashes in five minutes more. There is a butterfly-beauty about the cigarette to which the cigar and the pipe can lay no claim—a summer charm to stir the dreamy rapture of a poet, and to excite the Lotus-eating philosopher even to analogy. Just as the suns, and flowers, and balmy zephyrs of a century have gone to form the gauzy, multi-colored insect that flits across your path throughout a single summer's day, and then returns to dust and vapor, so the harvest of West-Indian and East-Asian fields, the long voyage of the mariner, the merchant's hours of soil, the steam-power and manual labor of the factory, the thoughtful calculations of the trader, the skill of the tissue-paper maker, all have gone, and more than these, to the creation of a fairy-cylinder of Tobacco, which glows, delights, expires, and meets its end in ten or fifteen fleeting minutes."

Although the cigarette is not a favorite with us, still we admire its use as a sort of appendage to a good dinner, and as preparatory work for a "good smoke." The Spaniards have always been great lovers of their minute rolls, and with them, no other method of burning tobacco appears so delicate or refined. Especially is this true among the ladies, who prefer "Seville cigarettes" to all others. Many smokers make their own cigarettes, sometimes using Havana tobacco, and sometimes making them of two or more kinds. An excellent cigar is made by using equal parts of Virginia and Perique tobacco, or equal parts of Havana and Perique. A fine flavored cigarette is also made from Yara and Havana tobacco, equal parts of each being used. Thos. Hood has signalized his attachment to cigar in the following pleasing little poem:—

THE CIGAR.

"Some sigh for this and that,
   My wishes don't go far;
The world may wag at will,
   So I have my cigar.

"Some fret themselves to death
   With Whig and Tory jar;
I don't care which is in,
   So I have my cigar.

"Sir John requests my vote,
   And so does Mr. Marr;
I don't care how it does,
   So I have my cigar.

"Some want a German row,
   Some wish a Russian war;
I care not. I'm at peace,
   So I have my cigar.

"I never see the Post,
   I seldom read the Star;
The Globe I scarcely heed,
   So I have my cigar.

"Honors have come to men
   My juniors at the Bar;
No matter—I can wait,
   So I have my cigar.

"Ambition frets me not;
   A cab or glory's car
Are just the same to me,
   So I have my cigar.

"I worship no vain gods,
 But serve the household Lar;
I'm sure to be at home,
 So I have my cigar.

"I do not seek for fame,
 A General with a scar;
A private let me be,
 So I have my cigar.

"To have my choice among
 The toys of life's bazar,
The deuce may take them all,
 So I have my cigar.

"Some minds are often tost
 By tempests like a tar;
I always seem in port,
 So I have my cigar.

"The ardent flame of love
 My bosom cannot char,
I smoke but do not burn,
 So I have my cigar.

"They tell me Nancy Low
 Has married Mr. R.;
The jilt! but I can live,
 So I have my cigar."

Lord Byron, a "good smoker" as well as a great poet, has immortalized his love of the cigar in the following graceful lines:—

"Sublime Tobacco! which from east to west,
Cheers the tars labors, and the Turkman's rest—
Which on the Moslem's ottoman divides
His hours, and rivals opium and his brides;
Magnificent in Stamboul, but less grand,
Though not less loved in Wapping or the Strand;
Divine in hookhas, glorious in a pipe,
When tipped with amber, mellow, rich, and ripe,
Like other charms, wooing the caress
More dazzlingly when dawning in full dress.
Yet thy true lovers more admire by far
Thy naked beauties—Give me a Cigar!"

Having given a general description of the cigar and its mode of manufacture, we come now to a more particular account of the various kinds known as the best and of world-wide reputation. Standing at the head of the various kinds of cigars, either of the Old or New World, are those known to all smokers as:

HAVANA CIGARS.

These are, by common consent, the finest in the world. They possess every quality desirable in a cigar, and seemingly to its greatest extent. Grown in the richest portion of the tropical world, the leaf has a rich, oily appearance, and, when made into cigars, possesses a flavor as rich as it is rare. Unlike most tobaccos suitable for cigars, every taste can be met in the Havana cigars, its many varieties of flavor and strength suiting it alike to both sexes, and to the making of the delicate cigarette or the largest Cabanas. These cigars are made up of all the various colors and parts of the leaf,

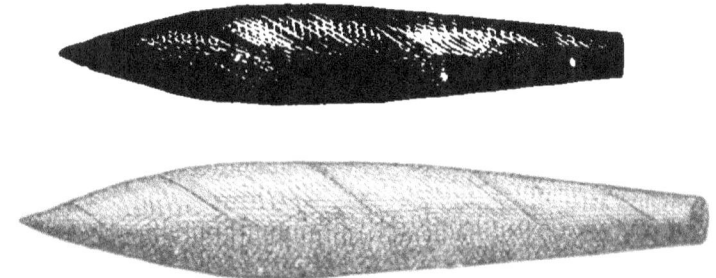

HAVANAS.

and also of all sizes common to the trade. In shape they are usually round, though sometimes pressed (flat), and in color are (according to our description) light and dark brown, light and dark red, straw colored and dark straw colored, and some other shades or strengths. It is necessary to have all the various shades of color in order to meet the demand for the various flavors desired. Without doubt a greater variety of flavors can be found among Havana cigars than in any other kind, owing to the many shades of color, which determines the strength and flavor of the cigar. The Havana cigar is made of a leaf tobacco well known for its good burning qualities, when properly cured and sweated,—burning with a clear, steady light, leaving a fine white or pearl-colored ash, according to the color chosen. These cigars rarely "char" in burning; certainly not, if made of good quality of tobacco and thoroughly sweat. If a full-flavored cigar is desired, choose the dark colors, and the lighter if a mild cigar is preferable. The lighter the color of the tobacco the lighter the ash and the milder the flavor of the cigar. Light-colored cigars usually burn freer and more evenly than dark ones. In selecting a cigar for its good burning qualities, choose those (if such are to be had) covered with white specks, or white rust; such cigars burn well, as white rust is found only on well-ripened leaves. Select a firm, well-made cigar—one that contains a good quantity of fillers—avoiding, however, in Havana cigars, one made *too* nicely, as it is sometimes the case that superior external appearance is made to cover defects in the more important qualities.

Such a selection will insure a cigar of good quality; one that will hold fire and last the length of time appropriate to its size. A cigar should not be chosen simply because it is made well, and neither because its outside appearance (wrapper) is fine, both in color and quality of leaf; rather depend upon the manufacture of the brand. Havana cigars have as many distinct flavors as there are colors of the leaf, ranging from very mild to very strong.

The first great requisite of a cigar is its burning quality, and the second its flavor; without the first the latter is of little value. A cigar made from leaf that does not burn freely will not possess any desirable flavor, but will char and emit rank-smelling smoke, without any desirable feature whatever. When both of these qualities are in a measure perfect the cigar will prove to be good. There are two varieties, at least, known as non-burning tobacco, of which we shall speak hereafter. The flavor and burning quality of a cigar always determine its character, and are found in perfection in those made of fine even-colored leaf. Dark cigars have a thicker leaf or more body, and consequently are stronger than light-colored cigars. When the cigar is made of fine, well-sweat tobacco, and contains the full quantity of fillers, the pellet of ashes will be firm and strong, and should possess the same color all through, if the filler, binder and wrapper are of the same shade of color. The finest-flavored cigars are those of a medium shade, between a light and a dark brown,—not so dark as to be of strong, rank taste, or so mild as to be deficient in a decided tobacco flavor, but simply possessing sufficient strength to give character to the cigar.

YARA CIGARS.

This variety of cigars is made from tobacco grown on the Island of Cuba, bearing the same name as the cigars. They are highly esteemed by those who smoke only this kind, but are not liked by most smokers of Havana cigars. Most of them are exported to Europe, very few of them finding their way to this country. It is somewhat difficult to compare them with Havana cigars, as the flavor is essentially different.

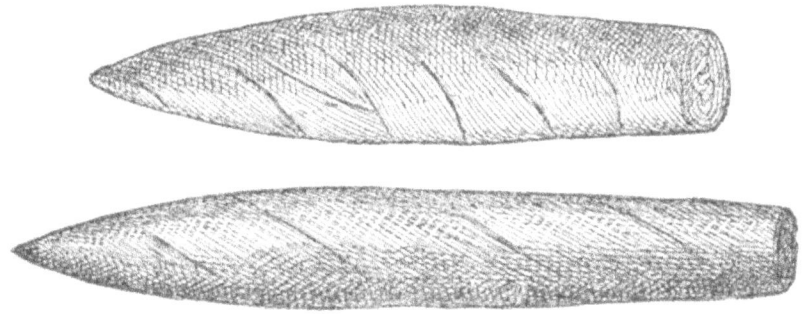

YARA CIGARS.

In comparison with other brands made upon the Island, the Yara holds an unimportant place, yet, in some parts of Cuba, it is preferred to any other kind. In London the Yara is a favorite with many old smokers, who use no others. Old smokers describe the Yara cigar as having a "sweet" flavor, but one unaccustomed to them, like Hazard and others, pronounce them bitter, and having a "peculiar saline taste." It can, doubtless, be said with truth concerning the Yara cigar, that unlike other varieties, such as Havana, Manilla, Paraguayan, Swiss and Brazil, the taste for them is not natural, but, when once formed, becomes very decided. As a general rule smokers of Yara cigars think other kinds are deficient in flavor, and are wanting in quality, because they lack the peculiar flavor belonging only to Yara cigars. Be this as it may, we hardly think the Yara cigar suited to the cigarist's taste at the present time. Its aromatic flavor is not adapted to the general taste, and some little time is required to develop a decided love for it. We prefer the "Cubas," made from a good quality of leaf grown near Trinidad, Puerto-Principe, and other cities east of Havana. The peculiar flavor of Yara cigars is owing to the character of the soil, rather than to any artificial process employed in manufacturing. In moistening Havana leaf Catalan wine is used, and other flavoring extracts. This may (and does) change the condition and quality of the tobacco, but even *with* this treatment, the flavor of Yara tobacco would be unlike that of Havana leaf.

MANILLA CIGARS.

This well-known variety of cigars is manufactured from Manilla tobacco grown in Luzerne, one of the Phillippine Islands, which is known as superior leaf for cigar purposes. Manilla cigars have an extensive reputation, but principally in the East and in Europe. These cigars are made in various

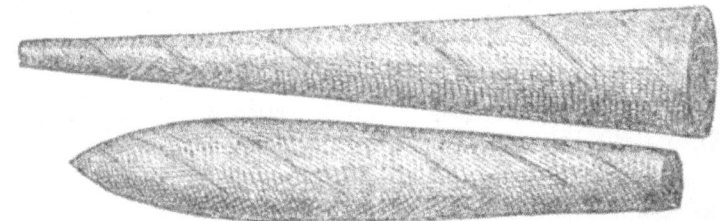

MANILLA CIGAR AND CHEROOT.

forms and shapes, some of them are called cheroots (the term used in the East for cigars) and are principally known for their aromatic flavor, entirely distinct from that of Havana cigars.

Some smokers think that they have the same effect as varieties of tobacco that have been moistened with the juice of the poppy, giving the cigar a flavor like that of opium, and as a natural result, securing a light-colored ash. There are not as many colors of Manilla cigars as there are of Havana, and they are not as closely assorted. Some of them are a high-cinnamon color, and are far from being a strong cigar. Their flavor is not always uniform, and is not denoted by the color as in other varieties. The flavor is not unpleasant, but is better suited to those who prefer a mild rather than a full flavored cigar. The aroma is pleasant and mild, and to those but little acquainted with them, agreeable. Manilla tobacco usually burns well, if the leaf is of good quality and well sweated, still it is known as a non-burning tobacco. As the tobacco is of good body, the cigars do not usually burn as well as other kinds. Select a light-colored rather than a dark cigar if one of good quality is desired. Both the cigars and cheroots are made of the same quality of leaf, and are of about the same size—differing, however, in shape. There are but few grades of Manilla cigars, and most of them are solid and well wrapped. They are flat rather than round, and draw well but do not hold fire like some other cigars. The leaf makes a very good wrapper for a tobacco of its thickness and strength.

SWISS CIGARS.

These well-known cigars have but little reputation in this country, owing to the fact of their being but little known. In Europe the cigars of Luzerne have no insignificant reputation, and are generally liked by smokers who prefer a mild and agreeable cigar. These cigars are usually dark-colored, but not strong, and have but little variety of flavor. Travelers and tourists through Switzerland speak of Swiss cigars as being of agreeable flavor, and unlike any other found in Europe. With American tobacco, those of a dark color are usually strong, but with European tobaccos this is not always the case—they possess much less strength, and can be used more freely than the tobacco of America. These cigars are usually pressed, and burn well, leaving a dark-colored ash, and emitting a fragrant odor. Most of those used in this country may be more properly termed cheroots, both ends

SWISS CIGARS.

being cut, allowing a free passage of air, which is usually the case with all kinds of cheroots, or Eastern and European cigars. There is not that freshness of flavor to Swiss cigars peculiar to Havana's, and they lack that essential quality which renders the latter so delicious and enjoyable. The Swiss cigar is in perfection when just made or rolled, and such should be chosen instead of those that have been made for some time and closely packed and dried.

PARAGUAY CIGARS.

These cigars are made of one of the finest varieties of leaf tobacco known to commerce. Although unknown to this country—both the cigars and the leaf tobacco have a deserved reputation in Europe, and it is beyond all question one of the finest tobaccos in the world for cigars. These cigars have a delicacy of flavor unapproachable in any other variety, and may justly be termed the finest at least of all South American cigars. It is one of the finest burning tobaccos in the

PARAGUAY CIGARS.

world, and does not fail to suit the taste of the most fastidious of smokers. The finest are of dark color and wholly free from any rank or unpleasant taste. These cigars are uniformly mild and have but little variety of flavor, the ash is dark-colored, firm and strong, clinging with tenacity to the cigar, which is the best evidence of the quality of the leaf. In Paraguay they are considered superior to all other kinds and are smoked continuously without any seemingly ill effect. Page alludes to the custom of smoking as being universal, "Men, women, and children—delicate, refined girls, and youngsters who would not with us be promoted to the dignity of pantaloons—smoke with a gravity and gusto that is irresistibly ludicrous to a foreigner." The Paraguayans consider excessive smoking of other tobacco as injurious but not of the delicate flavored leaf of Paraguay. These cigars are rolled firm and strong usually small and hold fire until the entire cigar has been consumed.

GUATEMALA CIGARS.

This variety of cigars, although of excellent flavor, is hardly known outside of Central America. They are made from Guatemala tobacco—one of the few varieties of tobacco bearing white blossoms, and possessed of a similar flavor to Mexican tobacco. Although Guatemala tobacco has not been thoroughly tested by the great manufacturers of cigars either in Europe or America, it doubtless is well suited for cigars. It is a distinct variety from those kinds bearing pink and yellow blossoms, and its growth and quality would seem to suggest some doubt as to its quality and adaptability for cigars. Stephens and other travelers seem to regard it as tobacco of excellent quality, and allude to its constant use by the ladies who smoke *puros*, a cigar made of a single leaf, or formed entirely of tobacco. They also use the *papelotes* wrapped in paper and sometimes in the dried leaf of maize. It would seem probable from the climate of Central America, that Guatemala tobacco would be exactly suited for the manufacture of cigars, but so little is known concerning it, and its cultivation is so limited, that at present it is simply a matter of conjecture.

BRAZILIAN CIGARS.

The cigars of Brazil, like those made of South American

tobacco, are noted for their superior flavor. They are made of "Brazilian Aromatic" one of the finest tobaccos of Brazil. Although but little known in this country, both the tobacco and the cigars are highly esteemed in Europe, where most of the leaf is sent. Both Brazilian cigars and the celebrated "Tauri Cigarettes" possess a delicacy of flavor, described by travelers as unapproachable by any other variety of cigars and cigarettes. A late traveler says concerning them:—" Accustomed to smoke only Havana cigars, I was unprepared to recognize any others as being worthy even of the name of cigars. I was presented with a box of Brazilian cigars of commendable size and finish, of a dark color and of a good flavor, before trying them, I ignited one, merely to test their quality and not from any impression that they were worth even the value of the cheapest Havanas. Great was my surprise to find them of an agreeable flavor and very pleasant to the taste."

The leaf is very thin, and without doubt, well suited for a cigar wrapper. The flavor of all cigars made from South American tobacco is similar, especially those made from tobacco grown east of the Andes. A writer, alluding to their mode of manufacturing cigars for their own use says:

"They take the leaf after it is cured and ready for manufacture into cigars, and dampen it, not with pure water but with water containing the juice of the poppy so as to produce the effect of opium. When prepared in this manner they are much esteemed by the Brazilians and especially by the herders."

AMERICAN CIGARS.

This was the name given to cigars made some forty or fifty years ago composed of Connecticut seed-leaf, or as it was then called, American tobacco. The fillers were selected from various kinds of tobacco, including Virginia, Kentucky, and Spanish, using for a wrapper Spanish, American or Maryland leaf. At this time the tobacco was not sorted as now, and was made up into cigars after being stripped, but the cigars after being manufactured were kept for some time before they were sold. At this time but little pains comparatively was taken in their manufacture: they were not assorted or shaded according to the present standard, and were packed in chestnut instead of cedar boxes containing from one to five hundred cigars each. A manufacturer of cigars nearly fifty years ago gives the following account of his method: "We selected for wrappers those leaves having white specks (white rust) upon them, which greatly increased the sale of the cigars, and which were considered by smokers to be much better than those not wound with fancy wrappers. After the cigars were packed in the boxes a little Spanish bean was grated upon the cigars, or a single bean was placed between the cigars in the box." At this time some little taste was evinced for colors, and cigars of a "bright cinnamon red," and afterwards, of a dark brown, were considered the finest, while leaf that was black was considered worthless for wrappers. A kind of cigar which is distinctly American and which is made to a considerable extent, is called a seed cigar, and is made from tobacco grown in Connecticut, New York, Pennsylvania, or Ohio. These cigars have but little reputation, and are of inferior quality and manufacture. A very good cigar, call a "sprig cigar," is made from Havana and Connecticut seed-leaf filler wound with a seed wrapper which gives a good flavor similar to clear Havana.

A full flavored cigar like a sip of rare old wine is inspiring to a lover of the "royal plant" and amid the sublime and companionable thoughts that its fragrance engenders, one is led oftentimes to reflect on its rare virtues and the benign effects it produces wherever known. Thus it lightens the toil of the weary laborer plodding along the highway of life. The student poring over musty tomes sees with a clearer perception as its fragrance accompanies him along the pathway of science and of history. The poet "as those wreathes up go" sees Helicon's fresh founts flowing clearer and purer. The musician "lord of sounds," evokes tones from his instrument never before heard by mortal ear. The warrior, "fresh from glory's field" is charmed by its fragrance as he dreams of shattered battalions and sleeping hosts. The farmer nurtured amid the odors of the "balmy plant" honors the "useless weed" as a promoter of happiness and an increaser of gains. While:

> " Kings smoke when they ruminate
> Over grave affairs of state."

The exile too, far from home and kindred smokes on as he muses of happier hours gone never to return. And thus amid all the varied ranks and walks of life this solace of the mind and comfort of life exhales its fragrance and breathes its benedictions over all.

CHAPTER X.

TOBACCO PLANTERS AND PLANTATIONS.

THE grounds selected for the cultivation of tobacco are called by various names even in the same countries. Thus in the Connecticut Valley, such lands are called tobacco fields, at the South they are known as tobacco plantations, while in Cuba they are called Vegas or tobacco farms. In Cuba almost the entire tobacco farm is planted to tobacco while at the South and in New England this is rarely the case unless the plantations or tobacco farms are small and contain but a few acres. In the Connecticut Valley and more especially along the banks of the Connecticut River, where the farms are frequently small, this is sometimes the case but farther removed from the river, where the farms are much larger but a few acres of the best land is used for this purpose.

In the Connecticut Valley the tobacco fields average from one to forty acres, rarely exceeding the latter and indeed seldom including as large an area. The average size of tobacco fields is about five acres—sometimes all in one lot but oftener divided into several small pieces on various parts of the farm.

The Connecticut planter is deeply interested in the plant and gives it his undivided attention from seed-sowing until it is sold to the speculator or manufacturer. All other crops in his opinion are of but little importance compared with the great New England product, one crop is frequently not off his hands before he is preparing for another. The Connecticut planter stands first in the rank of tobacco growers; he is thoroughly acquainted with the nature of the plant and knows just what land to select and what kind of fertilizers to apply. He has reduced the cultivation to almost an exact science and can obtain (the season being favorable) the color most desirable. He has thoroughly tested all kinds of fertilizers, and knows just what kinds will produce the various shades of color as well as the desired texture and size of leaf. No other tobacco planter so thoroughly understands the methods of curing, sweating and doing up the crop, and he takes no little pride in showing his crop to the buyer.

It is his aim to obtain not only the best leaf for a cigar-wrapper but also a tobacco of the finest possible flavor; hence he tries the principal varieties grown in Cuba, Brazil and other countries in order to judge of their quality and whether they can be cultivated with profit on his lands. He has the best constructed sheds for hanging and curing and the latest and most improved agricultural implements for the cultivation of the plants. The greatest pains are taken in securing the crop and harvesting and handling the plants without injur-

CONNECTICUT TOBACCO FIELD.

ing the leaves. The tobacco fields are kept in the best possible condition, no weeds or grass is allowed to grow and the entire surface is as free from stones as a lawn. He usually, if his farm is small, plants the same field year after year, securing a much finer leaf and by yearly manuring keeping the ground fertile and in good condition. When the tobacco is stripped the utmost care is taken to assort the leaves and he frequently shades or assorts the colors, obtaining fancy prices for such "selections."

The Connecticut grower is well acquainted with the different soils, and is able to judge with considerable accuracy in regard to selecting the right fields for tobacco. The warmest land is chosen—mellow and free from stones or shaded by trees and prepared as if for a garden. All of the improved methods of obtaining early plants as well as transplanting, he adopts, and in spite of early freezing, is generally able to outwit Jack Frost, and secure the plants before this great foe of the weed ravages the fields. It may safely be said of the Connecticut planter that he secures more even crops than any other grower of the plant, and obtains the finest colored leaf for cigar wrappers.

The growers are thoroughly informed as regards the prices, and although the buyers may steal suddenly upon them, are generally prepared to "set" a price upon their crops. Some refuse to sell on the poles, or even after it is stripped, preferring to pack their tobacco until it has passed through the sweat, when larger prices are obtained. Many growers not only pack their own crop, but buy up that of others, thus acting as both producer and buyer. During the growing of the crop, and particularly after it has been cured and stripped, the growers congregate together, and talk over the condition of the crop and the prices likely to be realized. Sometimes they form an association or club, agreeing to "hold" the tobacco for satisfactory prices, and frequently employing an agent to sell the crop. Many of the tobacco fields or farms in the Connecticut valley are very valuable, especially those near large cities and means of transportation; such lands often selling for one thousand dollars per acre.

The finest tobacco lands in the Connecticut valley are

located in the vicinity of Hartford about fifty miles from Long Island Sound. These lands are near enough to the sound to get the salts in the atmosphere from the south winds that blow up the valley in the precise amount which the plant needs. Not much farther north does the atmosphere possess this peculiar quality, while lower down the river the salt air is too strong for the plant, and the leaves in consequence are thick and harsh. Fine tobacco leaves can be manufactured as well as fine broadcloth or costly silks. These results depend in a great manner upon the proper soil and the fertilizers, applied together with the most thorough cultivation of the plants. The soil of our best Connecticut tobacco fields is alluvial, varying in composition from a heavy sandy loam to a light one containing very little clay.

For the past few years light soil has been preferred for the tobacco field, on account of the demand for light colored leaf. The soil can hardly be too light when leaf of a light cinnamon color is desired; as the color of all kinds of tobacco depends upon the soil and the fertilizers used.

A quarter of a century since Havana tobacco commanded very high prices, both in this country and in Europe. It burnt freely and purely. The Cuban planters, although getting rich on the ordinary crops, were not satisfied with their gains, and attempted to increase their crops by the use of guano and artificial fertilizers. They secured heavier crops, but the quality became poorer. The prices fell off and the planters did not realize as much for their crops as formerly, although the growth was larger. About this time Connecticut seed leaf became known as a cigar wrapper, and in a short time took the lead for this purpose, as it still continues to. It cured finely, burnt white and free, and in a short time brought high prices. The profit realized from its growth led some Connecticut growers into the same mistake as it did the Cuban planters, when they, by misguided culture, nearly ruined their crops and injured the reputation of Cuban tobacco.

Artificial fertilizers and strong manure produce a leaf

larger and heavier, but their effect on the character of the leaf is injurious, the salts destroying its fine qualities, so that it sweats and cures poorly, and compared with the finest leaf burns dark and emits a rank and unpleasant odor.

The Connecticut tobacco grower requires considerable capital when engaged extensively in the business, as ordinarily he buys large quantities of fertilizers and requires many hands to cultivate the crop. On the largest tobacco farms the sheds or "hanging houses" are built near or in the field, and are sometimes very large, say two or three hundred feet in length, and capable of holding the crop of from five to ten acres.

His broad fields of the weed can usually be seen from his house and he loves to show to visitors the plants growing in

HOME OF THE CONNECTICUT PLANTER.

all their luxuriance, or to sit on his piazza and call attention to their waving leaves and graceful showy tops. Few tobacco-growers can discuss the relative merits of the numerous varieties like the Connecticut planter; and he is well acquainted not only with the various kinds grown in his own country but also with those of others. Indeed you may often see growing in his garden specimens of Cuban, Brazil or Latakia tobacco; such is his love for all that pertains to this great tropical plant. He considers it one of the greatest of all the vegetable products and never tires of lauding the plant and its use. He sincerely hates all anti-tobaccoites and has a supreme disgust for the memory of King James I. and all royal foes of the plant. He is, however, a man of large and liberal views and bestows his favors with a princely hand. If fortune frowns he may lessen his crop but never his attachment for the plant. Amid all the cares and perplexities incident to life, he puffs away and as the ashes drop from his cigar meditates upon the probable future of tobacco growers and all users of the weed.

The Connecticut tobacco grower is in all respects a man of genuine refinement and nobility of soul. He is always ready to give information on his particular system of culture, and how he obtains such large and fine crops. He is a good judge of leaf tobacco, and can tell in a moment the quality of his famous variety. He is thoroughly awake to modern improvements, and always willing to try new implements, such as tobacco hangers or transplanters in his sheds or fields. He is just the person one likes to meet, jovial and good-natured; he naturally loves the plant he cultivates and uses it freely; lighting his after-dinner cigar or evening pipe with a gusto that is peculiar to the grower of tobacco everywhere. Indeed he is hardly in a proper frame of mind to converse about tobacco until he lights a cigar.

No other cultivator of the soil gains as many friends as the tobacco-grower. His table is well supplied from the choicest his larder affords and he cheerfully welcomes all to its side. He is the friend of the poor and the companion of the rich. No meanness or low chicanery is his. His attachment for home, friends, and country is as firm and strong as for the plant he cultivates.

Olmsted in his work "The Seaboard Slave States"

gives the following description of a Virginia plantation:

"Half an hour after this I arrived at the negro quarters—a little hamlet of ten or twelve small and dilapidated cabins. Just beyond them was a plain farm gate at which several negroes were standing; one of them, a well-made man, with

NEGRO QUARTERS.

an intelligent countenance and prompt manner, directed me how to find my way to his owner's house. It was still nearly a mile distant; and yet, until I arrived in its immediate vicinity, I saw no cultivated field, and but one clearing.

"In the edge of this clearing, a number of negroes, male and female, lay stretched out upon the ground near a small smoking charcoal pit. Their master afterwards informed me that they were burning charcoal for the plantation blacksmith, using the time allowed them for holidays—from Christmas to New Years—to earn a little money for themselves in this way. He paid them by the bushel for it. When I said that I supposed he allowed them to take what wood they chose for this purpose, he replied that he had five hundred acres covered with wood, which he would be very glad to have any one burn, or clear off in any way. Cannot some Yankee contrive a method of concentrating some of the valuable properties of this old field pine, so that they may be profitably brought into use in more cultivated regions? Charcoal is now brought from Virginia; but when made from pine it is not very valuable, and will only bear transportation from the banks of the navigable rivers whence it can be shipped, at one movement to New York. Turpentine does not flow in sufficient quantity from this variety of the pine to be profitably collected, and for lumber it is of very small value.

"Mr. W.'s house was an old family mansion, which he had himself remodeled in the Grecian style, and furnished with a large wooden portico. An oak forest had originally occupied the ground where it stood; but this having been cleared and the soil worn out in cultivation by the previous proprietors, pine woods now surrounded it in every direction; a square of a few acres only being kept clear immediately about it. A number of the old oaks still stood in the rear

THE PLANTER'S HOME.

of the house, and, until Mr. W. commenced his improvements, there had been some in its front. These, however, he had cut away, as interfering with the symmetry of his grounds, and in place of them had ailanthus trees in parallel rows.

"On three sides of the outer part of the cleared square there was a row of large and comfortable-looking negro quarters, stables, tobacco-houses, and other offices, built of logs. Mr. W. was one of the few large planters, of his vicinity, who still made the culture of tobacco their principal business. He said there was a general prejudice against tobacco, in all the tide water regions of the State, because it was through the culture of tobacco that the once fertile soil had been impoverished; but he did not believe that, at the present value of negroes, their labor could be applied to the culture of grain with any profit, except under peculiarly favorable circumstances. Possibly the use of guano might make wheat a paying crop, but he still doubted. He had not used it, himself. Tobacco required fresh land, and was rapidly exhausting, but it returned more money, for the labor used upon it, than anything else; enough more, in his opinion to pay for the wearing out of the land. If he was well paid for it, he did not know why he should not wear out his land. His tobacco-fields were nearly all in a distant and lower part of his plantation; land which had been neglected before his time, in a great measure, because it had been sometimes flooded, and was, much of the year, too wet for cultivation. He was draining and clearing it, and it now brought good crops. He had had an Irish gang draining for him, by contract. He thought a negro could do twice as much work in a day as an Irishman. He had not stood over them and seen them at work, but judged entirely from the amount they accomplished: he thought a good gang of negroes would have got on twice as fast. He was sure they must have 'trifled' a great deal, or they would have accomplished more than they had. He complained much of their sprees and quarrels. I asked why he should employ Irishmen, in preference to doing the work with his own hands. 'It's dangerous work, (unhealthy!) and a negro's life is too valuable to be risked at it. If a negro dies it's a considerable loss, you know.' He afterwards said that his negroes never worked so hard as to tire themselves—always were lively, and ready to go off on a frolic at night. He did not think they ever did half a fair day's work. They could not be made to work hard: they never would lay out their strength freely, and it was impossible to make them do it. This is just what I have thought when I have seen slaves at work—they seem to go through the motions of labor without putting strength into them. They keep their powers in reserve for their own use at night, perhaps.

"Mr. W. also said that he cultivated only the coarser and lower-priced sorts of tobacco, because the finer sorts required more pains-taking and discretion than it was possible to make a large gang of negroes use. 'You can make a nigger work,' he said, 'but you cannot make him think.'"

In speaking of the early tobacco culture of Virginia, he says:—

The light, rich mould resting on the sandy soil of Eastern Virginia was exactly suited to the cultivation of tobacco, and no better climate for this plant was to be found on the globe. This had just been sufficiently proved, and a suitable method of culture learned experimentally, when the land was offered to individual proprietors by the king, (James I.) Very little else was to be obtained from the soil which would be of value to send to Europe, without an application to it of a higher degree of art than the slaves, or stupid, careless servants of the proprietors could readily be forced to use. Although tobacco had been introduced into England but a few years, an enormous number of persons had initiated themselves in the appreciation of its mysterious value.

"The king, having taken a violent prejudice against it, though he saw no harm in the distillation of grain, had forbidden that it should be cultivated in England. Virginia, therefore, had every advantage to supply the demand. Merchants and the super-cargoes of ships, arriving with slaves from Africa, or manufactured goods, spirits, or other luxuries from England, very gladly bartered them with the planters for tobacco, but for nothing else. Tobacco, therefore, stood for money, and the passion for raising it, to the exclusion of everything else, became a mania, like the 'California fever' of 1849.

"The culture being once established, there were many reasons growing out of the social structure of the colony, which, for more than a century, kept the industry of the Virginians confined to this one staple. These reasons were chiefly the difficulty of breaking the slaves, or training the bond-servants to new methods of labor, the want of enterprise or ingenuity of the proprietors to contrive other profitable occupations for them, and the difficulty or expense of distributing the guard or oversight, without which it was impossible to get any work done at all, if the laborers were separated, or worked in any other way than side by side, in gangs, as in the tobacco-fields.

"Owing to these causes the planters kept on raising tobacco with hardly sufficient intermission to provide sustenance, though often, by reason of the excessive quantity raised, scarcely anything could be got for it. Tobacco is not now considered peculiarly and excessively exhaustive; in a judicious rotation, especially as a preparation for wheat, it is an admirable fallow crop, and, under a scientific system of agriculture, it is grown with no continued detriment to the soil. But in Virginia it was grown without interruption or alternation, and the plantations rapidly deteriorated in fertility. As they did so, the crops grew smaller in proportion to the labor expended upon them; yet, from the continued importation of laborers, the total crops of the colony increased annually, and the market value fell proportionately to the better supply.

"With smaller return for labor and lower prices, the planters soon found themselves bankrupt, instead of nabobs. How could they help themselves? Only by forcing the merchants to pay them higher prices. But how to do that, when every planter had his crop pledged in advance, and was obliged to hurry it off at any price he could get for it, in order to pay for his food, and drink, and clothing, and to keep his head above water at credit for the following year. The crop supplied more tobacco than was needed, but no one man would cease to plant it, or lessen his crop for the general good. Then it was agreed all men must be made to do so, and the colonial legislature was called upon to make them.

"Acts were accordingly passed to prevent any planter from cultivating more than a certain number of plants to each hand he employed in labor, and prescribing the number of leaves which might be permitted to ripen upon each plant permitted to be grown. An inspection of all tobacco, after it had been prepared for market, was decreed, and the inspectors were bound by oath, after having rejected all of inferior quality, to divide the good into two equal parts, and then to burn and destroy one of them. Thus, it was expected the quantity of tobacco offered for sale would be so small that merchants would be glad to pay better prices for it, and the planters would be relieved of their embarrassment."

Mrs. M. P. Handy gives the following interesting sketch, entitled "On the Tobacco Plantation":—

"Riding through Southside, Virginia, any warm, bright winter's day after Christmas, the stranger may be startled to see a dense column of smoke rising from the forest beyond. He anxiously inquires of the first person he meets—probably a negro—if the woods are on fire. Cuffee shows his white teeth in a grin that is half amusement, half contempt, as he answers: 'No, sar, deys jis burnin' a plant-patch.' For this is the first step in tobacco-culture.

"A sunny, sheltered spot on the southern slope of a hill is selected, one protected from northern winds by the surround-

"BURNING THE PATCH."

ing forest, but open to the sun in front, and here the hot-bed for the reception of the seed is prepared. All growth is felled within the area needed, large dead logs are dragged and heaped on the ground as for a holocaust, the whole ignited, and the fire kept up until nothing is left of the immense wood-heap but circles of the smoldering ashes. These are afterward carefully plowed in; the soil, fertilized still further, if need be, is harrowed and prepared as though for a garden-bed, and the small brown seed sown, from which is to spring the most widely-used of man's useless luxuries. Later, when the spring fairly opens, and the young plants in this primitive hot-bed are large and strong enough to bear transplanting, the Virginian draws them, as the New Englander does his cabbages, and plants them in like manner, in hills from two to four feet apart each way. Lucky is he whose plant-bed has escaped the fly, the first enemy of the precious weed. Its attacks are made upon it in the first stage of its existence, and are more fatal, because less easily prevented, than those of the tobacco-worm, that scourge, *par excellence*, of the tobacco crop. Farmers often lose their entire stock of plants, and are forced to send miles to beg or buy of a more fortunate planter. Freshly-cleared land—'new ground,' as the negroes call it—makes the best tobacco-field, and on this and the rich lowlands throughout Southside is raised the staple known through the world as James River tobacco.

"On this crop the planter lavishes his choicest fertilizers; for the ranker the growth, the longer and larger the leaf, the greater is the value thereof, though the manufacturers complain bitterly of the free use of guano, which, they say, destroys the resinous gum on which the value of the leaf depends. Once set, the young plant must contend, not only with the ordinary risk of transplanting, but the cut-worm is now to be dreaded. Working underground, it severs the stem just above the root, and the first intimation of its presence is the prone and drooping plant. For this there is no remedy, except to plant and replant, until the tobacco itself kills the worm. In one instance, which came under our observation, a single field was replanted six times before the planter succeeded in getting 'a good stand,' as they call it on the plantations; but this was an extreme case.

"When the plants are fairly started in their growth, the planter tops and primes them, processes performed, the first by pinching off the top bud, which would else run to seed, and the second by removing the lower leaves of each plant, leaving bare a space of some inches near the ground, and retaining from six to a dozen stout, well-formed leaves on each stem, according to the promise of the soil and season, and these leaves form the crop. The rejected lower leaves

STRINGING THE PRIMINGS.

or primings, in the days of slavery, formed one of the mistress' perquisites and were carefully collected by the 'house-gang,' as the force was styled, strung on small sharp sticks like exaggerated meat-skewers, and cured, first in the sun, afterwards in the barn, often placing a pretty penny in her private purse. Now when all labor must be paid for in money, they are not worth collecting, and, except when some thrifty freedman has a large family which he wishes to turn to account, are left to wither where they fall.

There is absolutely no rest on a large tobacco plantation, one step following another in the cultivation of the troublesome weed—the last year's crop is rarely shipped to market before the seed must be sown for the next—and planting and replanting, topping and priming, suckering and worming, crowd on each other through all the summer months. Withal the ground must be rigidly kept free from grass and weeds, and after the plants have attained any size this must be done by hoe; horse and plow would break and bruise the brittle leaves.

"'Suckering' is performed by removing every leaf-bud which the plant throws out after the priming (and topping), thus retaining all its sap and strength for the development of the leaves already formed, and this must be done again and again through the whole season. Worming is still more tedious and unremitting. In the animal kingdom there are three creatures, and three only, to whom tobacco is not poisonous—man, a goat found among the Andes, and the tobacco-worm. This last is a long, smooth-skinned worm, its body formed of successive knobs or rings, furnished each with a pair of legs, large prominent eyes, and is in color as green as the leaf upon which it feeds. It is found only on the under side of the leaves, every one of which must be carefully lifted and examined for its presence. Women make better wormers than men, probably because they are more patient and painstaking. When caught the worm is pulled apart between the thumb and finger, for crushing it in the soft mold of the carefully cultivated fields is impossible. Carelessness in worming was an unpardonable offence in the days of slavery, and was frequently punished with great severity. An occasional penalty on some plantations—very few, in justice to Virginia planters be it said—was to compel the delinquent wormer to bite in two the disgusting worm discovered in his or her row by the lynx-eyed overseer. Valuable coadjutors in this work are the housewife's flock of turkeys, which are allowed the range of the tobacco lots near the house, and which destroy the worms by scores. The moth, whose egg produces these larvæ, is a large white miller of unusual size and prolificness. Liberal and kind masters would frequently offer the negro children a reward

WORMING.

for every miller captured, and many were the pennies won in this way. One of these insects, placed one evening under an inverted tumbler, was found next morning to have deposited over two hundred eggs on the glass.

"As the plant matures the leaves grow heavy, and, thick with gum, droop gracefully over from the plant. Then as they ripen, one by one the plants are cut, some inches below the first leaves, with short stout knives,—scythe or reaper is useless here,—and hung, heads down, on scaffolds, in the open air, till ready to be taken to the barn. A Virginia tobacco-barn is totally unlike any other building under the sun. Square as to the ground plan, its height is usually twice its width and length. In the center of the bare earthen floor is the trench for firing; around the sides runs a raised platform for placing the leaves in bulk; and, commencing at a safe distance from the fire, up to the top of the tall building, reach beams stretching across for the reception of the tobacco-sticks, thick pine laths, from which are suspended the heavy plants. Safely housed and beyond all danger of the frost, whose slightest touch is sufficient to blacken and destroy it, the crop is now ready for firing, and through the late autumn days blue clouds of smoke hover over and around the steep roofs of the tall tobacco-barns. A stranger might suppose the buildings on fire, but not a blaze is within, the object here, as in bacon-curing, being *smoke*, not *fire*.

"For this the old field-pine is eschewed, and the planter draws on his stock of oak and hickory-trees. Many use sassafras and sweet gum in preference to all other woods for this purpose, under the impression that they improve the flavor of the tobacco-leaf. When the leaves, fully cured, have taken the rich brown hue of the tobacco of commerce, so unlike the deep green of the growing plant that a person familiar with the one would never recognize the other as the same plant, the planter must fold his hands and wait until they are in condition for what is technically known as striking, i. e., taking down from the rafters on which they are suspended. Touch the tobacco when too dry and it crumbles, disturb it when too high or damp, and its value for shipping is materially lessened, while if handled in too cold weather it becomes harsh. But there comes a mild damp spell, and the watchful planter seizing the right moment, since tobacco, like time and tide, waits for no man, musters all the force he can command for the work of stripping and stemming. This done, the leaves are sorted and tied in bundles, several being held in one hand, while around the stalk-end of the cluster is wrapped another leaf, the loose end of which is tucked through the center of the bundle. Great care is taken in this operation not to break the leaf, and oil or lard is freely used in the work. During this process the crop is divided into the various grades of commerce from 'long bright' leaf to 'lugs' the lowest grade known to manufacturers. These last are not packed into hogsheads, but are sent loose, and sold without the trouble of prizing, in the nearest market-town.

"Shades imperceptible to a novice, serve to determine the value of the leaf. As it varies in color, texture, and length, so fluctuates its market price, and at least half the battle lies in the manner in which the crop has been handled in curing. From the mountainous counties of South-western Virginia, Franklin, Henry, and Patrick, comes all the rarest and the most valuable tobacco, 'fancy wrappers' but these crops are smaller in proportion to those raised along the lowlands of the rivers. This tobacco is much lighter in color, much softer in texture, than the ordinary staple, and is frequently as soft and fine as silk. Some years ago a bonnet made of this tobacco was exhibited at the Border Agricultural Fair, and had somewhat the appearance of brown silk. Only one such plant have I ever seen grown in Southside, and that, a bright golden brown, and nearly two feet in length, was carefully preserved for show on the parlor-mantel of the planter who raised it.

"After tying, the bundles are placed in bulk, and when again 'in order,' are 'prized' or packed into the hogsheads,—no smoothly-planed and iron-hooped cask, by the way, but huge pine structures very roughly made. The old machine for prizing was a primitive affair, the upright beam through which ran another at right angles, turning slightly on a pivot, heavily weighted at one end, and used as a lever for compressing the brown mass into the hogsheads. Now, most well-to-do planters own a tobacco straightener and screw-press, inventions which materially lessen the manual labor of preparing the crop for market. Each hogshead is branded with the name of the owner, and thus shipped to his commission-merchant, when the hogshead is 'broken' by tearing off a stave, thus exposing the strata of the bulk to view. Of late years some planters have been guilty of 'nesting,' or placing prime leaf around the outer part and an inferior article in the center of the hogshead.

"At a tobacco mart in Southside, occurred perhaps the only instance of negro-selling since the establishment of the Freedman's Bureau. At every town is a huge platform scale for weighing wagon and load, deducting the weight of the former from the united weight of both to find the quantity of tobacco offered for sale. A small planter has brought a lot of loose tobacco to market, which, being sold, was weighed in this manner, and for which the purchaser was about to pay, when a bystander quietly remarked, 'You forgot to weigh the nigger.' An explanation followed, and the tobacco, re-weighed, was found short 158 lbs., or the exact weight of the colored driver, who had, unobserved, been standing on the scales behind the cart while the first weighing took place.

"Thirty years or more ago—before the Danville and Southside Railroads were built—the tobacco was principally carried to market on flat-boats, and the refrain to a favorite negro song was:—

"'Oh, I'm gwine down to Town!
An' I'm gwine down to Town!
I'm gwine down to Richmond Town
To cayr my 'bacca down!'

"Then all along the rivers, at every landing, was a tobacco warehouse, the ruins of some of which may still be seen.

With no crop has the Emancipation Act interfered so much as with this, and the old tobacco planters will tell you with a sigh that tobacco no longer yields them the profits it once did: the manufacturers are the only people who make fortunes on it now-a-days; $12 per hundred is the lowest price which pays for the raising, and few crops average that now. Still every farmer essays its culture, every freedman has his small tobacco patch by his cabin door, and the Indian weed is still the great staple of Eastern Virginia."

The first planters of tobacco at the West were the Ohioans, who began its culture about fifty years ago. From the first they have taken much interest in the plant, and as the result of many experiments not only produce seed leaf, but the finest cutting leaf grown in this country. The Ohio tobacco growers have shown a spirit of enterprise in this direction that is as commendable as it is rare. While they have not tested the great tropical varieties like their brother tobacco growers of Connecticut, they have succeeded in producing a leaf for cutting that is the admiration of the world. At first their experiments were unsuccessful, and the early growers were ridiculed for entertaining the belief that tobacco could be grown at the West. Yet despite all objections and seeming failures, the growers continued its cultivation until it has become one of the great products of the State. Of late the Ohio growers have demonstrated that their soil is better adapted for the finer grades of cutting leaf, than for seed leaf or even the more common "cinnamon blotch."

The soil is rich, and an experience of half a century has at length given them a thorough knowledge of the plant and the most successful modes of cultivation. In appearance an Ohio tobacco field resembles those of the Connecticut valley —the leaf is large, and though coarser, cures down a dark rich brown, like "cinnamon blotch," or a light yellow, the color of the famous "white tobacco." The Ohio growers have taken much pains with the Ohio broad leaf, and have produced a seed leaf tobacco that in many respects is a superior wrapper for cigars. While it does not possess the fine texture of Connecticut seed leaf it still has many good qualities, and with the careful culture given it will doubtless become still finer as a leaf tobacco, for wrapping cigars. But it is in the production of cutting leaf that the Ohio growers take rank, and ere long will supply the vast demand made upon them for their great cutting variety.

With a degree of pride peculiar to all tobacco growers, (when any new variety has originated,) they point with no little egotism to their fields of "white tobacco," and ask their fellow-growers of New England to rival this "great plant." So successful have they been of late with cutting leaf, that their fields yield them returns not inferior to many of the choicest tobacco farms on the Connecticut River. The Ohio growers have one advantage over earlier growers of the plant

OHIO TOBACCO FIELD.

—their land has not been cultivated as long as the famous tobacco lands of the Connecticut valley, and does not require that thorough fertilizing which is so necessary in New England. Still the tobacco field cannot be too thoroughly prepared for the growth of tobacco, whether in the tropics or in the more temperate regions.

In the curing of tobacco, the Ohio growers have but few equals, and no superiors. At first, the complaint made by the buyers of Ohio tobacco was, that "Ohio tobacco has the appearance of being too hard fired, indeed so much so as to have the flavor of being baked." The early culture of tobacco in the State attracted the attention of tobacco buyers, especially those who had dealt largely in Maryland leaf, and so much so, that one large firm issued a circular and sent to all the prominent growers in the tobacco growing section giving instructions in regard to its cultivation and management. We copy from one lying before us, and dated 1842. It reads as follows: "As tobacco is every year becoming a more prominent article in your State, we deem it of so much importance that we have had this circular printed on the subject of its Cultivation and Management, and take the liberty to address it to you. New ground produces the finest and highest priced tobacco. The plants should be set about 2 feet 9 inches or three feet apart, which will give them sufficient air and sun to ripen, and give the leaf a good body. It should be topped as soon as it buttons, kept clear of suckers, and cut as soon as it is ripe—if favorable weather, it will be fit for the house in 15 to twenty days after it is topped.

"When cut, let it remain until sufficiently lank to handle without breaking; but it should be housed before it is sun-killed, or much deadened, to prevent which, put it up in small heaps, say as much as a man can carry, with the heads to the sun, as soon as cut, and even then the top plants may be too much deadened, unless soon removed to the house. If sun-killed, it will not cure fine. The Maryland system is to fire without flues, and when the precaution is taken to lay planks or boards directly over the fire, accidents seldom occur.

" Slow fires are kept up for the first four or five days after the house is filled, so as to give it a moderate heat throughout, until the Tobacco is generally yellow, then the fires are raised or increased so as to kill the leaf and stem in forty-eight hours or less. When cured on the stock, as is done in Maryland, it can be better assorted, or the different qualities more readily separated than when stripped in the field and cured in the leaf. When stripping and tying up in bundles, it should be assorted according to the following classifications: 1st, Fine Yellow; 2d, Yellow; 3d, Spangled; 4th, Fine Red; 5th, Good Red; 6th, Brown and Common. It is often put up as if there were but two or three qualities, hence there is a great mixture of the several sorts, which is a very serious disadvantage in selling, as the purchaser generally values it at the price of the most inferior in the sample.

"The process of curing unfired, or air-dried tobacco, is similar to the above, except the firing; when so cured, it is more difficult to condition, so as to make it keep; but it generally sells quite as well. Planters should be very careful to have their Tobacco in good dry condition when they deliver it to the dealer or purchaser, as it is all-important to him to receive it free from dampness or moisture, which bruises it and injures its quality. We think such management as directed above would raise the value of Ohio tobacco as high as similar quality of Maryland."

As when first cultivated, the Ohio growers still select new land as the best adapted for tobacco, though not as easy of cultivation. When the tobacco growers are ready for preparing their "new ground" they invite in their friends and neighbors, and the field is "grubbed" in a short time. "Grubbing Day," with the young people, is an event of no common interest; the farmers gather from the adjoining farms and with mirth and muscle soon render the field fit for the "Indian herb." In the evening, the planter's home is filled with the young people, bent on having a right good time, and with "stripping the willow" and other games, close the day if not the night in the most enjoyable manner. Many of the country merchants take the tobacco of the growers when in condition to handle, paying them (or at least a portion of it,) in goods, or purchasing the tobacco as they do other merchandise. They have large warehouses where they receive and pack the tobacco until shipped to market. In the early Spring the growers take their tobacco to the workhouses, where it is packed by the merchants who frequently

TOBACCO WAREHOUSE.

have a claim on the crop for advances made on the same.

Having given a description of the Connecticut, Virginia and Ohio tobacco growers, we come now to the most extentensive cultivators of tobacco in America—the Kentuckians. With the exception of the Virginians they are the oldest growers of the plant in the United States,* and are confessedly among the most thorough cultivators of the plant in the world. The soil of Kentucky is admirably adapted for the great staple, and along the banks of the Green River may be seen the largest tobacco fields in the world. The plant attains a large size, and grows with a luxuriance common to all products grown in the famous "blue grass" region.

The system adopted by the Kentucky growers is similar to that adopted by all growers of cut tobacco, and the fine quality of Kentucky "selections" has deservedly gained the leaf a reputation that must place it in the front rank of American tobacco. The vast quantity grown in the state is an evidence not only of the good quality of Kentucky tobacco, but of the adaptation of the soil and of the method

KENTUCKY TOBACCO PLANTATION.

of cultivation in use. As a cut tobacco, Kentucky-leaf is held in the highest esteem, the exportation of the leaf to all parts of Europe gaining for it a reputation hardly equaled

---

* Kentucky was originally a part of Virginia.

by any Southern tobacco. The system of cultivation is similar to that pursued by the Virginian, and the same process of curing is also adopted.

The Kentucky growers generally succeed in getting a "good stand" and when once the plants have commenced to grow they come forward with a rapidity that is truly astonishing. The soil of Kentucky is well adapted for the production of the largest varieties of tobacco as well as the finest grades of cutting leaf. Much attention is paid to the selection of soil, that the light standard of Kentucky leaf may be further advanced. On the large plantations a vast amount of tobacco is grown, in some instances equaling the entire product of some of the tobacco-growing towns in the Connecticut Valley. The tobacco is packed in hogsheads, each one containing twelve hundred pounds, the same as in Virginia and Missouri.

The Kentucky planter prides himself on the superior quality of tobacco, as well as his famous blooded stock. If there is anything more remarkable than the high character of the latter, it must certainly be the renowned plant which has given the wealthy planters of Kentucky a national popularity among all cultivators of tobacco. The Kentuckians are thorough in all of their methods of cultivation, and with the first stock and tobacco farms in the country bid fair to achieve still further honors as "tillers of the soil." Possessed of the largest means, they have brought their farms up to a high state of cultivation, and produce in their famous valleys the very finest of Nature's products.

Kentucky planters are men of the largest endowments; Nature, in her gifts to them has been most lavish, and the princely fortunes which they have acquired shows how well they have benefited by her munificence. In manners affable, and in benevolence unsurpassed, the Kentucky planter gains the plaudits of all. He is polite to both friend and foe, and possessed with all of that polished manner which marks the true gentleman, and especially all growers of the " kingly plant." Easy of approach, he has still that reserve that bids all sycophants mark well their conduct and demeanor. On the plantation or at the race, the Kentuckian is ever in his best mood for recreation and enjoyment.

His attachment for the horse has developed qualities of patience and thoroughness that are shown elsewhere than on the "course." Benefiting by years of training and study, the success that follows his efforts shows at once that such talents are not confined to a single field of operations. In many respects like the Virginia planter, they differ somewhat in their taste in all that pertains to the turf and the field. But we would not lose sight, among his many noble traits of character, of that love of his State that pre-eminently characterizes the Kentuckian. He is justly proud of her soil and of her sons, and whether in the halls of Congress or on the field of carnage and blood, fears not to maintain the honor and safety of the one and the other.

THE KENTUCKY PLANTER.

It is surprising to one acquainted with the growth of tobacco and the value of the Southern States for its production that so small an area of land is devoted to its culture in Georgia, Florida and Louisiana. When owned by Spain, West Florida was noted for its tobacco, and produced large quantities which were exported to Spain and France. The soil of Florida is well adapted for tobacco, and the rich hummock lands produce an excellent quality for cigars, not unlike Havana leaf. Its cultivation has been tried in various parts of the State, but the result has not warranted its cultivation to any great extent excepting in Gadsden County where the plant flourishes as well as in Cuba.

The seed used in Havana and the plant resembles it so closely that even Cuban planters cannot distinguish it from that grown on the island. The mode of cultivation is nearly the same, and the soil is said to produce a leaf of tobacco similar to that of the celebrated Vuelta de Albajo. Formerly the product was sent to New Orleans, and the leaf was pronounced by some dealers to be bitter, but most of them considered it valuable. The planter selects the high lands or hummocks, the soil of which is light and rich for the tobacco field. The plants are carefully drawn from the bed, and transplanted afterwards. The mode of culture is to plow between the rows and hoe the plants carefully.

A Florida tobacco field in appearance is not unlike a *vega*, or Cuba tobacco field; the same luxuriant growth of the forest may be seen on every hand, and the "queen of herbs" grows underneath or near the fragrant Orange and the stately Magnolia. The soil of Gadsden County is in some respects unlike that of the rest of the State in that there is an entire absence of limestone, which is found elsewhere all through Florida. The climate of the State is well adapted for the growth of tobacco, and is less changeable on the Gulf side than along the Atlantic coast.

Formerly larger crops were raised than now. Under the old *regime* when on every plantation were a score or more of idle negro urchins, a large portion of the labor could be performed by them, such as worming, dropping the plants, and picking up the primings, while now the labor has to be paid for in money or its equivalent. At this time, the "wrapper leaf" was considered to be among the best for cigars, and brought high prices. In the days of slavery, tobacco was considered to be as profitable as the cotton crop, and good tobacco plantations were considered to be the most valuable in the State.

This peculiar tobacco region is without doubt capable, with proper management, of producing a superior article for cigars, both wrappers and fillers, and when grown on "new ground" the staple is exceedingly fine. The leaf cures as rapidly, and is of as good color as in Cuba, and in a favorable season and when harvested fully ripe, is destitute of that bitter taste formerly ascribed to it. The plants grow large, and have that smooth, shiny appearance peculiar to Havana to-

FLORIDA TOBACCO PLANTATION.

bacco, the leaves growing erect, and frequently covered with "specks" or "white rust," one of the best evidences of a fine flavored and a good-burning tobacco. A Florida tobacco-grower gives the following account of the plant:

"The Gadsden 'wrapper-leaf' was always in high repute, and extensively used in the manufacture of cigars, being in size, firmness, and texture fully equal to the best Cuba, and far superior to the Connecticut seed-leaf. Where the variety known as the Cuba filler has been tried, it has succeeded finely in this county, possessing that delicate and peculiar aroma so highly prized in the Havana cigars. We need but the capital to make the most profitable crop that is grown. It is a fact, that of all the counties of the State, many of them abounding in the very finest soil, Gadsden is the only one that has succeeded in making the Cuba tobacco a staple market-crop. Prior to 1860 it rivaled in net returns the great staple cotton, and from present indications, it is about to resume its former status among the great agricultural products of the country."

"Whether this success is attributable to any peculiarity in the elements of the soil, I am not able to determine, but this fact is worthy of note, that, except immediately on the banks of the Apalachicola River, which forms the Western boundary of the County, there is an entire absence of the rotten limestone which so largely pervades the other sections of the State. For the planter of limited means, there is no crop so well suited to his condition as the Cuba tobacco. To produce a given result there is a less area of land required than is demanded for the production of any other field crop. The cultivation, harvesting, and preparation for market is simple, and the labor so light that it may be participated in by every member of the family, male and female, over six years of age. The growth of the plant is so rapid, and its arrival at maturity so quick, that it never interferes with any of the provision crops, and rarely with a moderate cotton crop."

In Louisiana the tobacco plant flourishes well and grows as well and as luxuriantly as sugar cane. Even along the banks of the Mississippi the plants attain good size, and succeed as finely as in some of the other parishes in the interior of the State. The Perique and Louisiana tobacco are the principal varieties cultivated, and attain nearly the size of Connecticut seed leaf. In St. James parish the soil seems well adapted for Perique tobacco, and here it readily takes on that black hue that is one of the peculiar features of this singular variety. In Coddo parish tobacco is cultivated to some extent, but does not produce a leaf equal to that grown in St. James Parish. The tobacco grown in the Parishes of Bossier and Natchitoches is used chiefly by the growers of the parishes and is fitted for both smoking and snuff.

The Louisiana planters have adopted the method of the French in doing up their tobacco—twisting it in rolls, or as the French call them, "Carrots." The planters of St. James Parish annually put up from ten to fourteen thousand carrots of Perique, each carrot weighing about four pounds.

Mr. Perique, from whom the tobacco takes its name, made many improvements in the manner of preparing the tobacco for market, one of which consisted in taking up the twisted lumps (after remaining in press for six months), spreading them to fifteen or sixteen inches in length and having completed four pounds in weight, rolling it into a lump which retained its shape by means of a rope one-fourth inch in diameter, tightly twisted around it. The labor in pressing and twisting is entirely done by hand, and attended to with the most scrupulous care.

The Creole planters sometimes raise two, and even three crops on the same field, two of them being the growths of

LOUISIANA TOBACCO PLANTATION.

suckers or shoots from the parent stock or stump. The growers of Perique tobacco have tested Havana seed, but can see but little difference between the product and that from Virginia or Kentucky tobacco seed, while the growth is much smaller. In color Louisiana tobacco is very dark, entirely different from any other variety grown in the Mississippi valley.

Some few years since tobacco culture was introduced into California, and the belief then entertained by those who planted the consoling weed, that the state would soon become as famous for raising tobacco as she now is for producing wheat and gold seem likely to be realized. The soil and climate of California are admirably adapted for tobacco. In the valleys the land is a deep alluvial loam, easily worked, producing bountiful crops of the finest leaf tobacco. The planters have experimented with several varieties, including Havana, Florida, Latakia, Hungarian, Mexican, Virginia, Connecticut, Standard and White leaf. Large crops are grown, especially of Florida tobacco, which, with careful culture, produces two thousand five hundred pounds of merchantable leaf to the acre. The planters get their Havana seed from Cuba, preferring to do so rather than to risk the seed from their own plants. At first they used home-grown seed and could not see any serious deterioration or change in the quality of the tobacco, but a singular change in the form of the leaf took place. That from home-grown seed grew longer, and the veins or ribs, which in Havana tobacco stand out at right angles from the leaf stalks took an acute angle, and thus became longer and made up a greater part of the leaf. Of Florida tobacco the home-grown seed comes true.

Tobacco is now being tested in the several counties in the State and with every promise of success. Many of the ranches seem well adapted for the plant and the planters are confident by their new process of curing, of being able to produce an article equal to the best Havana brand. The plants attain a remarkable size, and grow up like many kinds of tropical vegetation, without much care being bestowed upon them, although the plants are regularly cultivated and hoed. The planters are not troubled with that foe of most tobacco fields, "the worm." They attribute this in part to the excellence of their soil and partly to the abundance of birds and yellow jackets. The planters do not always "top" the Havana and do very little "suckering." If the ground is rich, and free from weeds they let one of the suckers from that root grow, and thus become almost as large and heavy as the original plant. They believe that the soil is strong enough to bear the plants and suckers, and that they get a better leaf and finer quality without suckering.

In summer the roads are very dusty in California, and this dust is a disadvantage to the tobacco planter. On some of the plantations double rows of shade trees are planted along the main roads, and gravel is spread on the interior roads; and to protect the fields of tobacco from the high winds which sweep through the California valley, almonds and cottonwoods are planted for wind-breaks in the fields.

Some of the planters employ Chinese to cultivate the plants, who are very careful in hoeing and weeding the tobacco, living an apparently jolly life in shanties near the fields. A witty California correspondent of the *Tobacco Leaf* writes concerning the early cultivation of tobacco in that State:

"We are doing a great many other things in California now besides raising grain, fruit, wine, wool, and gold. We are doing a lively business in tobacco. Fifteen years ago I was down East on one occasion when they were gathering the tobacco crop—which goes to New York, and, by a process equal to wine making, becomes Havana tobacco. It struck me that this country was admirably adapted to its cultivation, and I brought back some seed, which I gave to a friend living on the bank of the Sacramento River, instructing him to plant it as per direction given me. We sat down and calculated the immense fortune we would make raising tobacco, if the experiment was a success. A week later my friend, who was an impatient sort of a fellow, wrote me just a line—'No results.' I replied, and asked him if he expected a crop of tobacco in seven days. A few weeks later he wrote, 'Here she comes;' two weeks later, 'How big is the stuff to be?' two weeks later, 'Not room for tobacco and me too. Who shall quit?' I heard no more for a month and thought I would go up and see it. I did so, and the steamboat landed me at my friend's ranch. I could not see the house, and hallooed. I heard an answer from the depths, and then following a path, I found my friend swinging in a hammock in the shade of a grove of tobacco trees. I desire to maintain my reputation for truth and veracity, so necessary to a correspondent, so I won't say how big or how high those tobacco plants were; but my friend's hammock was slung from them—and he was no feather-weight—the leaves completely embowered the cottage. I congratulated him on the results—such a grove and such a shade—and moreover I said, 'You will be permanently rid of mosquitoes.' 'Will I!' said he. 'Do you know that these gallinippers have learned to chew already, and the habit is spreading so that all the old he-fellows are coming down from Marysville to take a hand.' I inquired if my friend had cured any or smoked any. He pointed to a Manyanita pipe split open on the ground, and said. 'Before it was real strong, some three weeks ago, I tried a leaf in that pipe. Observe the result—busted it the second whiff, and knocked me off the log I was sitting on.' Such was the first experiment in tobacco raising in California. But now they have learned the trick. They have searched the State for the poorest and most barren soil, and, having found it are cultivating a splendid article of genuine Havana leaf tobacco, manufacturing cigars as good as you get one time in twenty even in Havana, making several brands of smoking tobacco, and, lastly, an article of Louisiana perique, ('peruke' proper,) that any old smoker would go into ecstasies over, fully equal, it is said to the genuine old-fashioned article, and that is saying a good deal. Now if we can supply the world with cigars and tobacco, we have got a dead sure thing for the future, even if gold gives out, grain fails and the pigs eat up all the fruit. Your people who have been paying fifteen cents apiece for genuine Havana cigars imported direct from—Connecticut, should rejoice and join in an earnest *hooray!*"

In Mexico the tobacco plantations exhibit a diversity of scenery not met with in other portions of America. The soil is well adapted for the crop, and on many of the plantations in the Gulf States the plant grows as finely as on any of the *vegas* of Cuba. The Mexicans are among the best cultivators of the plant in the world, and, like the Turks, prefer its culture to that of any product grown. The plant is a strong, vigorous grower, and ripens early, emitting an odor like that of Havana tobacco. The climate is so favorable that from one to three crops can be grown on the same field in one year, and yield a bountiful harvest without seemingly impoverishing the soil.* Transplanted in the summer or autumn, the plants grow through the winter months, and

---

*Shepard says of the cultivation of tobacco by the Indians:—"The tobacco which is raised on the Tehuantepec isthmus is said, by good judges, to rival that of Cuba, and commands, in the capital, equal prices with the far-famed Havana. It is cultivated by the Indians, whose fields, or '*milpas*,' according to Indian custom, are situated at some distance from their villages, often in the depths of the forest. Upon these little patches they bestow whatever labor is consistent with dislike for exertion, leaving the rich soil to accomplish the balance."

in spring are gathered and taken to the sheds. Sartorius, in his work on Mexico, says of its culture on the plantation:—

"Various kinds of tobacco are planted, mostly that with the short, dingy, yellow blossoms, which has a very large, strong leaf. But there is little doubt that the sorts would be more carefully selected, if the trade were not fettered by the monopoly. Most of the government planters enter into an arrangement with the small farmers and peasants who have to grow a certain number of plants, on condition of handing over the harvest at a low figure—six to eight dollars per crop. These *aviados* receive something in advance, and their chief profit consists in securing the sand leaf and the

MEXICAN TOBACCO PLANTATION.

greater part of the after-harvest, which they sell to the contrabandists. It is indeed allowed to export whatever remains; but it is attended with so many annoyances from the authorities, that it is never attempted. The many ships which enter the Mexican harbor of the east coast with European manufactures, find no return freight except gold and silver, cochineal, vanilla, a few drugs and goat skins, all of which take up very little room in the ships (money is usually sent off in the English government steamers); consequently they must either proceed to Laguna to buy log-wood, or they must take in sugar, coffee, or tobacco, in a Cuban or Haytian port. As soon as tobacco becomes an export article, its cultivation must increase immensely in the Coast States, the Mexican being very partial to this branch of agriculture, which occupies him part of the year only."

Mayer also alludes as follows to the same subject:—

"A large portion of the tobacco sold in the republic is contraband; for the ridiculous and greedy restrictions and exactions with which a plant of such universal consumption is surrounded, necessarily disposes the people to violate laws which they feel were only made to impair their rights of production and trade under a constitution professing to be free."

The government planters in the State of Vera Cruz have large, fine plantations, and the plants are carefully tended and cultivated as in all countries where tobacco is a government monopoly. On each plant a certain number of leaves are taken off, including the sand leaf, which is thrown away, and everything in the way of topping and suckering performed as carefully as on the tobacco farms in Cuba. The small farmers who raise only a few thousand plants are not as careful as the large planters, and are sometimes guilty of planting more than the number agreed upon, while the mountain passes towards the table-land are carefully guarded to prevent smuggling of the crop, which is far more remunerative than selling to the government.

We will now take the reader to the primitive tobacco plantations of America about the middle of the Sixteenth Century. The plantations were not located in Cuba as many have supposed but what has been variously named Hispaniola, Hayti, and St. Domingo. It was in this island that the Spaniards first began the cultivation of tobacco and inaugurated (under the guise of Christianity) that career of monstrous cruelty, with which their insatiable appetite for the burning of heretics and for the baiting of bulls so well accords. In 1509, Diego Columbus, the eldest son of the great discoverer, assumed in St. Domingo, or as it was then called,

Hispaniola, the vice-regal powers which had been intrusted to him. Diego as portrayed by the historian "was a man as noble as his father, and almost as gifted; and he had his father's fate. Like his father, he had to bear all that Spanish envy and Spanish malignity could inflict. In 1511, Diego Columbus sent Diego Velasquez to conquer Cuba." From historians Velasquez gets a better character than most of the *Conquistadores*, who in general were as ferocious as they were audacious and fortunate. No serious opposition was or could be offered. With the name of Velasquez the prosperity of Cuba is inseparably identified. As Governor of Cuba he was a vigorous colonizer and civilizer. He founded Havana, which he called the Key of the New World, and which is said to rank as the eighth place in the hierarchy of commercial cities. Havana, however had long been flourishing before the seat of Government had been transferred to it from Santiago. It was Velasquez who introduced slavery into Cuba; and it was during his vice-royalty and under his sanction that those memorable exploratory and conquering expeditions began, the most astonishing of which was that to Mexico, led by Cortez, the insubordinate lieutenant of Velasquez, whose death is said to have been hastened by the rebellious and ungrateful conduct of Cortez, and perhaps by the spectacle of such immense and rapid success. The agricultural, commercial, and general growth of the West India islands at this period would have been much more rapid if the Spaniards had not annihilated the native population, and if they had not been exposed to incessant piratical attacks. These were often of the most desolating kind. In 1688, the city of Puerto Principe was plundered and destroyed. From its strongly fortified position Havana set the buccaneers at defiance, and sometimes saved the whole island from ruin.

The exact period of the first cultivation of tobacco in St. Domingo is not known, but we find that as early as 1535 the negroes had habituated themselves to the use of it in the plantations of their master. Soon however its cultivation increased, and during the latter part of the Sixteenth Century the Spaniards shipped vast quantities to Europe, a very large amount of which found its way to England, where it brought fabulous prices. The Spaniards, by the application of the lash and other cruelties, extorted from the negroes an amount

ST. DOMINGO TOBACCO FIELD, 1535.

of labor never equaled by any other task masters in the world. Forcing these slaves to labor on the plantations from morning until night, with the fierce rays of a tropical sun shining full upon their uncovered backs, and goaded on to the performance of the severest toil, is it any wonder that the haughty cavaliers of Spain grew rich from their industry, and feasted on the products of the Indies. Cultivated on the rich soil of this fertile island, the tobacco of St. Domingo had no competitor, until the Spaniards began its culture a little later on the island of Trinidad, the product of which in time stood at the head of all the tobaccos of the Indies and of South America. The tobacco trade at this time was wholly controlled by the Spaniards, who, though successful in this direction, made but slow progress in colonization. Compared with the British colonies in the New World, the Spanish possessions were weak and incompetent, and for all their advantages in their great product, it was ultimately rivaled by the English Colonial tobacco. In the conquest of the New World, Spanish energy and enterprise seem to have exhausted themselves; and as Spain was declining, its colonies could not be expected rapidly to advance. The history of the Spanish conquest in America is a record of cruelty and of blood, while that of English colonization is marked by English rigor and enterprise, and is one of successful daring and ultimate triumph.

The West India plantations, however, were still worked, and for more than a century St. Domingo yielded a vast amount of tobacco, until the soil of Cuba was found to be better adapted for its production than any other of the West India islands, not excepting even the island of Trinidad.

Hazard, in his work on Cuba, describes the celebrated *vegas* or tobacco plantations, of the island as follows:

"The best properties known as *vegas*, or tobacco farms, are comprised in a narrow area in the south-west part of the island, about twenty-seven leagues broad. Near the western extremity of the Island of Cuba, on the southern coast, is found one of the finest tobaccos in the world. Within a space of seventy-three miles long and eighteen miles wide, grows the plant that stands as eminent among tobacco plants

A CUBAN *vega*.

as the lordly Johannisberger among the wines of the Rhine. Shut in on the north by mountains, and south-west by the ocean, Pinar del Rio being the principal point in the district. These *vegas* are found generally on the margins of rivers, or in low, moist localities, their ordinary size being not more than a *coballeria*, which amounts to about thirty-three acres of our measurement. The half of this is also most frequently devoted to the raising of the vegetable known as the *platano* (banana), which may be said to be the bread of the lower classes. A few other small vegetables are raised. The usual buildings upon such places are a dwelling house, a drying-house, a few sheds for cattle, and perhaps a small *bohio* (hut), or two, made in the rudest manner, for the shelter of the hands, who, upon some of the very largest places number twenty or thirty, though not always negroes—for this portion of the labor of the island seems to be performed by the lower classes of whites. Some of the places that are large have a mayoral, as he is called, a man whose business it is to look after the negroes, and direct the agricultural labors; but, as a general thing, the planter, who is not always the owner of the property, but simply the lessee, lives upon, directs, and governs the place.

"Guided by the results of a long experience transmitted from his ancestors (says a Spanish author), the farmer knows, without being able to explain himself, the means of augmenting or diminishing the strength or the mildness of the tobacco. His right hand, as if guided by an instinct, foresees what buds it is necessary to take off in order to put a limit to the increase or height, and what amount of trimming is necessary to give a chance to the proper quantity of leaves. But the principal care, and that which occupies him in his waking hours, is the extermination of the voracious insects that persecute the plant. One called *cachaga* domesticates itself at the foot of the leaves; the *verde*, on the under side of the leaves; the *rosquilla*, in the heart of the plant; all of them doing more or less damage. The planter passes entire nights, provided with lights, clearing the buds just opening, of these destructive insects. He has even to carry on a war with still worse enemies,—the *vivijagnas*, a species of large, native ants, that are to the tobacco what the locust is to the wheat. This plague is so great, at times, that prayers and special adoration are offered up to San Marcial to intercede against the plague of ants.

"The plant, whose original name was *cohiba*, seems to have been cultivated first by Europeans on the island in the vicinity of Havana. The island of Cuba is without doubt well adapted for the cultivation of tobacco—the soil, climate, and improved methods of culture all tend to the production of a leaf tobacco as celebrated as it is valuable.

"Between the 'Lower Valley,' in the Nicotian, not the

KILLING BUGS BY NIGHT.

geographical, sense of these words, lie the so-called *Partidos* which produce the tobacco that is sent to Europe as *Partido* or *Cabañas*. The leaf often surpasses that of the 'Lower Valley' in size and fineness, as well as in the beauty of the color; but it is inferior in quality. The tobacco farmers though stalwart fellows are not fond of work, and too often waste their time at the tavern. Many of them from thriftlessness are plunged into debt; and scarcely is the harvest ended when they borrow money from the tobacco merchant on the following harvest, who thereby obtains the right to interfere, it may be despotically, with the management of the crop. Continual embarrassments tempt the tobacco planters to be dishonest. To cheat their creditors, they often sell the best part of the crop in underhand fashion. Such of the tobacco farmers as wish to produce a great deal of tobacco, without regard to the excellence of the article, leave the plant to its natural growth, which is both scientifically and otherwise objectionable, for it is on a process of thinning and pruning a due diffusion of sap in the leaves depends, and consequently the quality of the tobacco."

SOIL AND CLIMATE.

The tobacco, after being being baled, is sent to the Havana market. The bales of tobacco are carried on the backs of mules or horses to the city or to the nearest railway station.

"In the long line or train of mules or horses, the head of one mule or horse is tied to the tail of the one before it.

GOING TO MARKET.

On the back of the foremost sits the driver. The hindmost carries a bell, which enables the driver to know whether any of the animals have broken loose."

From the description given by Hazard of Cuba, its soil, climate, and other resources, it will readily be seen by all acquainted with the tobacco plant that this famous island is well adapted for the production of a tobacco that for fineness and delicacy of flavor is hardly rivaled. With the peculiar composition of the soil, and with a climate well adapted for the perfection of all kinds of tropical plants and fruits, it can hardly be imagined that any finer variety of tobacco can be grown than that produced in Cuba and the adjoining islands. Doubtless the climate of Cuba is nearly the same as when

Columbus discovered the island, and wrote in such extravagant language its praise. The soil of Cuba is prolific, and the variety of tropical plants and fruits grown upon the island is quite remarkable. Nowhere is this seen to a greater extent than in the varieties of tobacco cultivated. Although there are several kinds and qualities grown on the island, the mode of culture upon all the *vegas* is nearly the same. These *vegas* or tobacco farms greatly outnumber the coffee and sugar estates, but are much smaller, and require a less number of hands to work them. Hazard estimates the number at ten thousand, while they are constantly increasing as new fields are being tried and new modes of culture introduced. Russell says of tobacco culture in Cuba:—

"In regard to climate, it is worthy of observation that tobacco is only cultivated during winter, when there is little rain. It grows most luxuriantly in summer with the increased heat and moisture; but the leaves grown in this season are devoid of those qualities for which the weed is esteemed. The conditions of growth are less powerful in winter, when the temperature is ten degrees lower, and the fall of rain small. At the same time, there is more sunshine to impart those aromatic qualities which are so much relished by smokers of tobacco. In Virginia the torrid heat and thunder showers during the summer months are by no means favorable for developing the mild aroma of a good smoking leaf. Such atmospheric conditions are better suited for cotton and Indian corn than tobacco, which must have dry weather and sunshine to produce it in perfection."

No country in Europe is more celebrated for its tobacco than Germany. The tobacco plant has been cultivated in some parts of Prussia for nearly two centuries. The tobacco of Germany is used for all purposes for which the leaf is designed—for cutting, cigars, and snuff. There are various kinds of German tobacco, the finest being grown in the Grand Duchy of Baden. The native tobacco of Germany, however, is not powerful in flavor, and may be smoked continuously to an extent which would be dangerous and disagreeable if American tobacco were used. Although it is cultivated in most of the States of Germany, and by a large number of growers, still the tobacco fields as a rule are small. The

Germans are among the most thorough cultivators of the plant in Europe, and every operation in the field is done at the proper time and in the right manner. After it is cured they prepare it nicely in rolls and carots, the latter for manufacturing into snuff. The tobacco fields are faithfully tended, and the utmost pains taken to secure large, well-formed leaves. The fields present a much more even appearance than similar fields in France, where the tobacco grown is small and uneven. The South German growers of tobacco are without doubt the most successful tobacco-growers in Europe, not excepting the Hollanders, who raise an excellent tobacco for snuff. The time of gathering the leaves is the occasion of quite a merry-making among the growers and villagers, and is considered an event of considerable importance. Fairholt says:—

"The time of harvesting the leaves is an interesting period for a stranger to visit the villages, which put on a new aspect as every house and barn is hung all over with the drying leaves."

German tobacco cures well, and some of the finer sorts make excellent cigar wrappers and are much esteemed

GERMAN TOBACCO FIELD.

throughout Europe. The following account of the cultivation and production of tobacco in the different German

States, will give some idea of the amount cultivated and used in Germany:—

"The aggregate area of land cultivated with tobacco in Prussia during the year 1871, amounted to 5.925 hectares (a hectare being equal to 2.47 English acres). It appears that the extent of tobacco-growing land has, during the last fifty years, been gradually diminishing in Prussia, and that accordingly the expectations entertained in the beginning of that period of a great future development of this branch of agriculture, have not been realized; for whilst the area of land planted with tobacco in the year 1825 was 12.374 hectares, it amounted in 1871 to less than one-half this amount. The reasons for this gradual decline are considered to be, on the one hand, the growing competition of the South German growers, and the increase in the importations of American tobacco; on the other, the fact that the cultivation of beet-root (for sugar manufacturing) and of potatoes (for the distilleries) has proved to be a more profitable business than the cultivation of tobacco. It has, moreover, been found by many years' experience, that whilst the quality of the tobacco cultivated in most parts of Prussia is not such as to enable the growers to compete successfully with the importers of foreign (particularly of North American) sorts, the labor attending its cultivation and its preparation for the market, as well as the uncertainty of only an average crop, are out of proportion, as a rule, to the average profits arising therefrom. The cultivation of the plant has, consequently, gradually become restricted, chiefly to those districts of the country where either the soil is peculiarly adapted for the purpose, or where it is carried on for the private use of the producer."

With regard to the various provinces of Prussia, it appears that "In East Prussia the extent of tobacco land is only a limited one, and is confined to the district around Tilsit, where about two-thirds of the entire cultivation is in the hands of peasants, who consume their own produce. In West Prussia (the western portion of the province of Prussia proper) the cultivation is rather more extensive, particularly near the town of Marienwerder; the tobacco, however, is very inferior. The most important districts of the province of Posen are those of Chodziesz and Meseritz. In Pomerania, next to Brandenburg the most important tobacco-growing province of the kingdom, the area of land cultivated is very large. The principal districts are those near Stettin. In Silesia the most important districts are those around

Breslau, Ratibor, and Oels. The principal tobacco-growing province of Prussia is Brandenburg, and here again, particularly the part of the government district of Potsdam, which contains the towns of Neustadt, Eberswalde and Prenzlau. Besides the districts mentioned, tobacco is grown largely in that of Frankfort-on-the Oder. In the province of Saxony the chief districts are those of Stendal, Salzwedel, Nordhausen, Burg, and Wittenburg. Hanover, like the other western provinces of the kingdom, produces a superior quality of tobacco to that raised in the eastern parts of Prussia—the most important district is that of Munden. The chief tobacco-growing districts of Hesse-Nossau are situated near the towns of Cassel and Hanau. In Rhenish Prussia the plant is cultivated, particularly in the neighborhood of Cleve, Emmerich, Coblenz, Creuznach, and Saarbruck; the districts first mentioned produce a very superior quality. The production of tobacco in Westphalia is extremely small, while in the province of Schleswig-Holstein the plant is not cultivated at all. In the account given it will be seen that the tobacco plant holds an important place among the products of Prussia, and although not as extensively cultivated as formerly, has not been entirely driven from the soil by other products which yield a larger profit to the producer. The plant is cultivated in other parts of Germany, especially in Bavaria, where large quantities of tobacco are grown, particularly so in the Bavarian Palatinate and in Franconia (viz., the districts around Nuremberg and Erlangen). In the Kingdom of Saxony but little tobacco is raised, as is also the case in Wurtemberg, although the soil and climate in parts of this state are said to be very favorable to the growth of the tobacco plant; the area of land cultivated is upon the whole, a very limited one, and in 1871 did not exceed 178.2 hectares. The Grand Duchy of Baden has at all times been the chief tobacco-growing part of Germany; as far back as the end of the Seventeenth Century, special laws for regulation of the cultivation, preparation, and warehousing of this article were in force. The most prominent tobacco-growing districts of Baden are those of Carlsruhe, Mannheim, Heidelburg, Badenburg, Schwetzingen, and Lahr; the quality of the plant grown in those parts being a very superior one (among the various kinds of German tobacco). The produce of the districts mentioned is therefore applied chiefly in the manufacture of cigar wrappers, and is exported in considerable quantities to Bremen, Hamburg, Switzerland, Holland,

and even to America for the use of cigar manufacture. The prices of the best kinds of Baden tobacco are consequently also, on an average, much higher than those realized by other German growers. In the Grand Duchy of Hesse the plant is cultivated, the chief district being that around the town of Darmstadt; in the Thuringian States, tobacco is grown; the most prominent among them as regards its production is the Duchy of Saxe-Meiningen. In Mecklenburg also some tobacco is raised, the most important district being that of Neu Brandenburg (in Mecklenburg-Strelitz). In Brunswick only a small extent of land is used for tobacco growing, the same being situated near the town of Helmstadt. In Alsace and Lorraine, the recently acquired provinces of Germany, the cultivation of tobacco has been extensively carried on for many years, more especially in the country around Strasburg, Mulhausen, Schirmeck, and Munster, and to a small extent near Metz and Thionville."

It is apparent from this account that the German tobacco fields produce a vast quantity of tobacco, some of which is of excellent texture and flavor, and well adapted to the taste of European smokers of the plant.

Ever since the introduction of tobacco into Holland, its cultivation and its use has been looked upon with favor by the "true-born Nederlander," who associates the plant with every social enjoyment. The Dutch, on the discovery of tobacco, were among the first to use it and encourage its cultivation. In the history of the Dutch colonies in the Indies it plays an important part. Tobacco began to be cultivated in Holland about Amersfoot in 1615, and from that time until now, its culture has increased until it has become one of the greatest of agricultural products of the country. The plant is grown in the Veluive (the valley of Guelderland), where the soil is particularly adapted for the rich snuff-leaf which is manufactured from Amersfoot tobacco. The Dutch, like the Germans, are excellent cultivators of tobacco, selecting the richest and the strongest land, and working the fields of as fine a tilth as possible. The plants do not grow as rapidly as in America, as they are transplanted into the fields in May, and are not harvested until the latter part of September or beginning of October. The plants attain good size—larger than most of the tobacco of Europe, and a tobacco field in Holland compares favorably with any in this country. The color of the plants while growing, is a dark rich green, and they are of a uniform size, maturing slowly but thoroughly. Connor says of Amersfoot tobacco: "This tobacco is much esteemed, the fineness of the leaf and its freedom from fibres fitting it for cigar-wrappers."

The Dutch planters of tobacco are among the happiest cultivators of the plant in Europe, if not in the world, and unlike the renowned Van Twiller never "have any doubts about the matter," and believe that tobacco is absolutely necessary to sustain life. After the evening meal the planter lights his pipe or calls upon the good dominie, to have a

DUTCH PLANTERS.

social chat, discoursing over their favorite beverage the virtues of two great luxuries. Oftener, however, he passes his evenings at the village inn, where, surrounded by other comrades, he discourses as follows of his favorite plant,—tabak:

"That the smoking of tobacco is of infinite benefit, no one who is impartial and unprejudiced can deny. In a country like Holland, where the atmosphere is always laden with heavy and hurtful particles, and where, while people breathe that atmosphere from above, they feel themselves not less affected from below by the cold, moist, swampy soil—the smoking and the chewing of tobacco are the wholesome prophylactics of which we can make use. To the Indians and the Negroes, tobacco is almost the only solace in this transient life. They learn, by means of it, to support nature, and to encounter valiantly, by its help, all the tribulations incidental to the human lot. If they are depressed, they smoke or chew tobacco, and gladden themselves therewith. If they are exhausted, and the sun and their hard and inhuman masters appear to conspire to destroy them, a little tobacco restores their strength, makes them forget their slavish life, and go vigorously to work again.

In the Thirty Years' War in Germany, the smoking and chewing of tobacco proved the salvation of many thousands of men, who by its aid guarded themselves against the deadly effects of deficient food and of bad meats and drinks. Nothing is so good, nothing so serviceable to human life, as the smoking of tobacco—which may well be called a kingly plant, seeing that the monarchs of the earth are not ashamed to use it. While tobacco cultivates sociality, and is of great avail in severe hunger and thirst, it strengthens the body and checks fluxions, and colds, and slimy humors. Nature has willed it that men should make use of plants like tobacco, which, by their heat and sharpness, draw the humors outward, and cause a slight salivation. Witness, as confirmation of what has been said, cloves and pepper, which hold sway nearly over the earth; betel, which to the Hindoos is the remedy for every disease; the onions and leeks of the Egyptians, who while building the pyramids and obelisks, spent their money eagerly on those dainties; and tobacco, which is adopted by the four quarters of the world.

The justly celebrated British physician, Cheyne, has remarked that both chewing and smoking of tobacco are exceedingly serviceable for those who suffer from rheumatic and catarrhal affections, have a sluggish digestion, or live a luxurious life. As tobacco has numerous slanderers, so there are many who know not how to turn tobacco to a good purpose. Excess and abuse may be found in the smoking and chewing of tobacco as in other things. Instead of using tobacco in moderation, there are persons who make themselves its slaves, and render themselves incapable of the immense benefit of the enlivening and stimulating effect they would otherwise owe to it. A little tobacco smoked or chewed three or four times a day cannot fail to be beneficial. But the adversaries of tobacco, in order to furnish themselves with an argument, make tobacco bear all the blame when some one who has given himself up to an intemperate and luxurious life, and who is besides a great smoker, becomes the victim of all kinds of discomforts and sickness. To condemn tobacco by saying those who begin to chew or smoke it nearly always suffer from malaise and nausea, is surely preposterous. May we not in fairness contend that tobacco is essentially wholesome, that it helps digestion, relieves the mind and cheers the spirits."

The following humorous account of "Thirsty Tobacco" is a most curious illustration of the superstitions which spontaneously grow up in the hearts of the people.

"Soon after the introduction of tobacco into Holland many of the Dutch were of the opinion that the tobacco plant drank in moisture greedily and required to be often and abundantly watered. From this insatiable thirst the belief arose that tobacco was the cause of rain, brought clouds to the heavens, and restored the general crops. Once, in the neighborhood of Amersfoot, the weather was very rainy, and the crops suffered accordingly. On the tobacco growing round the town the blame of the calamity was thrown; and it was resolved to punish tobacco, the sottish rain-drinker and wicked rain-bringer. A rabble, consisting chiefly of boys and youths, rushed to the tobacco fields, and scattered havoc with the ferocity of stupidity. The mad creatures pulled up the stalks, tore off the leaves, and trampled leaves and stalks under foot. Before they had done the work of destruction quite as completely as they desired, soldiers appeared on the scene. They sternly commanded the rioters to desist, but the rioters paid no heed either to entreaties or threats. Thereupon they drew their swords, as if by the mere flash of these to terrify the rioters, who laughed a laugh of contempt. Then effectually to frighten the rioters, the soldiers fired at them with blank cartridges. This harmless noise drove the mischief-makers to ignominious flight, and the tobacco plants which were still uninjured were left in peace."

At what exact time this destruction of "thirsty tobacco" took place we are left in doubt. It is doubtless a "good joke" got up by some "ponderous joker" for the amusement of Dutch smokers.

All admirers of tobacco like Holland and its people. It is emphatically the land of smoke. One is constantly in cloud-land, and whether in the house or on the street the incense of tobacco is perpetual, from the good natured dominie who puffs leisurely at many pipes to the humblest peasant who works modestly among the plants, all burn the fragrant weed and pay homage to its shrine. Ever since the Dutch looked upon the plant it has been more to them than king and courtier. The old Dutch burgomasters "who dozed away their lives and grew fat upon the bench of magistracy in Rotterdam; and who had comported themselves with such singular wisdom and propriety, that they were never either heard or talked of, owed all to the use and influence of the 'kingly plant.'" Not only are the Dutch prodigious smokers, but they use the pipe at all places and at all times. On the way to Church the pipe is lighted, and after service it is the solace of the evening hour.

In all public places the pipe plays an important part. The traveler is constantly reminded of the use of tobacco; for even the bridges have public notices affixed to them request-

SUCCESS TO VON TROMP.

ing all visitors to prevent the fall of tobacco-ashes on the gravel or grass; and not to knock out their pipes within bounds of the place. The old Dutch planters were fond of a "silent pipe," and after the labors of the day gathered together to drink and smoke to the success of Admiral Von Tromp, whose exploits in the British Channel carried terror to many a heart. Or, speculated upon the voyage of the "*Goede Vrouw*" (Good Woman), which had been fitted out to colonize the new country.

The progress of tobacco-culture in Oceanica, is shown in the following account which Connor gives of the tobacco plantations of Australia:

"The development of tobacco culture in Australia has been great and rapid. In these colonies, where only a few years ago the plant was not known, there are now hundreds of acres under tobacco. The local manufacture is also keeping pace with the production of the leaf, and the import of tobacco into the Australian colonies yearly diminishes in proportion to the increased consumption of locally grown and manufactured tobacco. Imported leaf is used in the manufacture of cigars, those made from colonial leaf being held in low esteem. Steady efforts are being made by the cultivators to improve the quality of the produce, and with every prospect of success, many places in the colonies being well adapted for the growth of the plant. Colonel De Coin says Australia is capable of producing very good qualities. Tobacco has hitherto been grown upon alluvial lands, but a preference is evinced for lands somewhat less rich but free from floods. Alluvial land gives a larger crop per acre, but the flavor is ranker. In 1872 there were 567 acres under tobacco in New South Wales. The average produce of the colonies is about 1,300 pounds to the acre. The amount of produce varied from 976 pounds to the acre in New South Wales to 2,016 in Tasmania, the climate of this island being moister and more favorable for tobacco than that of the other colonies. Manilla and Havana tobacco has been grown with great success for seed for many years at the Adelaide Botanic Gardens, and the seed raised has been largely distributed."

The Australian growers may demonstrate the fact that as good or better Manilla tobacco can be grown by them than in the Philippine islands. If the leaf will burn freely, and leave a white, firm ash, the product will no doubt prove a rival of the leaf grown in Luzon. From the composition of the soil, it is hardly probable that Havana tobacco can be grown to perfection; it may, however, resemble in some measure the Cuban leaf. The climate has much to do with the flavor of tobacco; more than with the size of the plants or the color of the leaf. Cuba in this respect has a decided advantage over Australia; and Havana tobacco will hardly find a rival in Australian leaf, though grown on the finest soil, and given the most thorough care.

So extensive is the cultivation of the tobacco plant, that even the Arab cultivates it in the burning desert. In Algiers it is an important product; and through the efforts and encouragement of the French government its cultivation is assuming large dimensions. Some portions of Algiers seem to be well adapted for tobacco, the finest of which is equal to

TOBACCO FIELD IN ALGIERS.

any obtained from America; but a large portion of the product from that province is of poor quality. It is a favorite plant with the Arab, and his attention seems to be about equally divided between his tobacco and his camels. The plant is light in color and of peculiar flavor, well suited to his taste, and in keeping with his idea of quality and excellence. The crop is usually bountiful, notwithstanding the heat of the summer and the absence of moisture in the soil.

TOBACCO IN AFRICA.

The tobacco plant is also cultivated in other parts of Africa besides Algiers. In Egypt and Nubia it is grown to a considerable extent, as well as by most of the native tribes of the South-west. Among some tribes it forms an important article of trade, and serves the purpose of money or its representative. The natives are partial to the plant, and devotedly attached to smoking. Little patches may be seen near their huts, on which they lavish their attention and care. In some parts of Africa tobacco grows to a very great height. Livingstone gives an account of a variety that attained an altitude much higher than the American plant. Several varieties are cultivated, some of them resembling the Shiraz and Latakia, while most of it is said to be similar to Virginia tobacco, only larger. With careful culture the plant would doubtless thrive in most parts of Africa, as the soil is light and the season usually favorable. Though the heat is extreme the plant flourishes even in the hottest part of the season, and attains a degree of perfection corresponding to the labor bestowed by the natives in cultivating.

TOBACCO FIELD IN AFRICA.

Their manner of curing is simply by drying the leaves, and is not suited to the taste of any besides themselves. In Egypt, Algiers, and Nubia, the plant is cultivated with more care, and a better system of curing is adopted than by the natives of the interior. Burton gives an account of the cultivation of tobacco by the natives of East Africa:—

"Tobacco grows plentifully in the more fertile regions of

East Africa. Planted at the end of the rains, it gains strength by sun and dew, and is harvested in October. It is prepared for sale in different forms. Everywhere, however, a simple sun-drying supplies the place of cocking and sweating, and the people are not so fastidious as to reject the lower or coarser leaves and those tainted by the earth. Usumbara produces what is considered at Zanzibar a superior article; it is kneaded into little circular cakes four inches in diameter by half an inch deep: rolls of these cakes are neatly packed in plantain-leaves for exportation. The next in order of excellence is that grown in Uhiao: it is exported in leaf or in the form called *kambari*, roll-tobacco, a circle of coils each about an inch in diameter. The people of Khutu and Usagara mould the pounded and wetted material into discs like cheeses, 8 or 9 inches across by 2 or 3 in depth, and weighing about 3 lbs.; they supply the Wagogo with tobacco, taking in exchange for it salt. The leaf in Unyamwezi generally is soft and perishable, that of Usukuma being the worst; it is sold in blunt cones, so shaped by the mortars in which they are pounded. At Karaguah, according to the Arabs, the tobacco, a superior variety, tastes like musk in the water-pipe. The produce of Ujiji is better than that of Unyamwezi; it is sold in leaf, and is called by the Arabs *hamumi*, after a well-known growth in Hazramaut. It is impossible to give an average price to tobacco in East Africa; it varies from 1 khete of coral beads per 6 oz. to 2 lbs."

Some of the most beautiful and fragrant tobacco fields in the world are to be found in Syria. Indeed it may truthfully be said that a field of Latakia tobacco is hardly inferior in beauty to the large and fragrant orchards of the olive and mulberry, or the wheat fields on the terraced sides of Mount Lebanon.

The tobacco plant is cultivated in various parts of Syria and particularly by the Druses on "The Lebanon," as it is usually called.

The cultivation of tobacco in Syria, has been a considerable industry, and the product has acquired a reputation in European markets that has demonstrated its real value, and a constant demand for this variety of the plant. Latakia tobacco resembles in flavor the yellow tobacco of Eastern Thibet and Western China, both of them grown from the same seed. Latakia tobacco is not sweated like most tobacco, but is first cured in the sun and then hung up in the peasants' huts to cure until ready for market. The plants ripen very fast and emit an aromatic odor, increasing in strength as the plants ripen. For smoking it has but few superiors. After curing, it is baled and sent to Europe, where it is manu-

TOBACCO FIELD IN SYRIA.

factured into smoking tobacco. The plants are well cultivated and watched against the ravages of birds, which seem to like the young and tender plants especially before they are transplanted. From the nature of the soil the plants are watered frequently, and when the leaves are about the size of a large cabbage leaf are ready to harvest. As the plants ripen the leaves gradually thicken and take on a lighter shade; the leaf when green is very thick, but after curing is quite thin and of a bright yellow or brown, according to the process employed in curing. The peasants take equal pains in its fumigation, using various kinds of wood according to the color of leaf they wish to obtain. They usually make two kinds of leaf, the finest being colored brown and known by the name of *abowri*. The tobacco is fumigated with two kinds of wood, *gozen* (pine wood) and *sindian* (oak), the tobacco fumigated with gozen having the best smell. The fumigation, however, is said not to be resorted to expressly for the tobacco, but the mountaineers of necessity burn much wood in their huts in the winter, and the smoke improves the tobacco in color, smell, and flavor. All the tobacco grown about Latakia derives its origin from the same seed, but the difference between the *abowri* and the other kinds is owing to the cultivation of the former about high mountains and with the use of pine wood in fumigating it. A field of Latakia tobacco presents a novel appearance, the short straight plants with their ovate leaves bearing yellow blossoms form a striking contrast to towering seed leaf rising fully two or three feet higher than the Syrian plant.

Fairholt says that "Latakia tobacco is a native of America but grows wild in other countries, and is a hardy annual in English gardens, flowering from midsummer to Michaelmas, so that by some botanists it has been termed 'common, or 'English tobacco.'" Burton's work on unexplored Syria is full of passages relating to tobacco and the custom of smoking.

"The tobacco which is grown on the slopes of the Libanus and the Anti-Libanus mountains appears to be one of the finest quality and most delicate flavor. The monks of the convents are famous for the production of a snuff, which for pungency, at least, is far superior to the snuffs of Europe. Personal experience of it convinces us that a great deal of the pungency of this snuff is due to the addition of some aromatic herb in addition to the natural acridity produced by the highly dried tobacco. The cultivation of tobacco in Syria, will probably increase in proportion to the improved condition of affairs in Syria, we have little doubt; and we trust that when agricultural science is better studied there, Englishmen will have the opportunity of testing the value and importance of Syrian tobacco products."

Connor says of the tobacco fields of India:—

"In the Bombay Presidency tobacco is largely produced, and its quality in such districts as Kaira and Khandesh is superior. In 1871 there were nearly 43,000 acres of land under tobacco in the presidency, the largest quantities being grown in Kaira, Khandesh, Belgaum Sattara, Shalopoor, and Poona. The trade is extensive. The exports of tobacco to foreign countries amount to several million pounds annually. Among foreign countries, Mauritius, Bourbon, and neighboring places, not reckoned as part of British India, take a large share of the exports. Bombay exports tobacco to other Indian presidencies. Small quantities of the fine Guzerat tobaccos find their way by rail into the North-western Provinces. Numerous endeavors have for many years past been made to improve the quality of Bombay tobacco. In 1831 the Resident in the Persian Gulf sent to the local Government a maund of seed of the 'very finest tobacco grown in Persia,' and with it he sent some observations on the mode of cultivating tobacco in the neighborhood of Shiraz. In 1867 fifteen small packets of genuine Shiraz tobacco were forwarded for trial in the Bombay Presidency. Of the seed sown in Kolhopoor, about eight or nine germinated, and the plants grew to a height of five feet two inches; of these only four survived. There were two varieties, one with oblong the other with circular leaves.

Of the seeds sent to Kandesh, only a few germinated. All the seed put down in the Victoria Gardens failed. That sent to Sind, though said to have been carefully sown, also failed to germinate. The Conservator of Forests had the seeds sent him sown in beds, and the plants, when a few inches in height, were transplanted into pots. They grew with the greatest luxuriance, and produced abundance of flowers and seed. Some of the seed was sent to the collector of Kaira, who forwarded a sample of the tobacco grown from it. The Conservator considered the produce very good, and the secretary of the Agri-Horticultural Society pronounced it 'of a superior kind.' The flavor was exceedingly fine, but it had not been allowed to come to maturity, hence it was thin and shriveled. It had also been spoilt

TOBACCO FIELD IN INDIA.

by rain, and consequently its market value could not be fairly tested. The experiment, it is clear, was not conducted with proper care by most of those to whom the seed was confided, but the Local Government considered that on the whole the result was satisfactory, as showing that there was every probability that Shiraz tobacco, with care and proper gardening, might be introduced into the Bombay Presidency.

"In August, 1869, the Bombay Government again distributed a small supply of seed of the Shiraz, Havana, and other varieties to the superintendents of cotton experiments, and to the collectors of Kaira, Khandesh, Dharwar, and Kurrachee, for experimental cultivation. The seeds did well in the hands of all the superintendents, who reported very favorably on the plants raised from them. In Sind only the soil in which the seed was sown proved unsuitable. In Dharwar all the five varieties germinated, though the Maryland failed to some extent, and a considerable quantity of seed of each variety was secured. Of Latakia, only twenty grains were sent to the superintendent; and the quantity in each case increased to one pound from the produce of the plants. These two varieties of tobacco, however, were not so much admired by the cultivators as Shiraz, Havana, and Maryland, to which they gave a decided preference. The only varieties of seed which were available for experiments at Broach and Veermgaum were Havana and Shiraz. In both places the plants came up well, and a large quantity of seed was obtained from them. That sent to Broach arrived a little too late in the season to admit of an extensive experiment being made; this indeed appears to have been the case at all the other places. The seed, however, was of good quality, germinated freely, and produced excellent plants in a very short time.

"The first transplanting was made out into a field in an open piece of land, where they commenced growing vigorously, but the rains being then over, swarms of small locusts made their appearance, and ate up the young plants before they had thoroughly established themselves in the ground. The second lot was transplanted into a more sheltered patch, where the progress was all that could be desired, both the varieties growing rapidly, the Havana especially producing some leaves of enormous size. The first cutting was entrusted to a potel, who managed it according to the native process of curing. The tobacco was so strong, however, that only old confirmed smokers could manage it. The most formidable difficulty which presented itself was the management of the midrib, which in the large leaves was extremely coarse and juicy. When the leaves were made up into hands for the purpose of fermentation before the midrib was thoroughly dry, the result was invariably mould and discoloration. On the other hand, when dried sufficiently to insure freedom from mould, the lamina of the leaf became so brittle that it was crushed to powder at the slightest touch, and so wrinkled and dry that the heaps did not ferment at all. Of the varieties supplied, the Shiraz, Havana, and Maryland attracted most attention and promised the best results. The great drawback was the curing part of the process. So far as the cultivation was concerned, there was every prospect of success; but not so with regard to the curing."

Robertson says of the curing of the leaf:—

"In my opinion, all efforts to produce good tobacco will be useless until the services of a competent curer are obtained."

He considers the fault of all Indian tobacco to lie in the curing. The leaf itself is good, and it is simply the art of curing that should be studied.

"I have cured tobacco of different varieties, some of which would hold a good place in the English market, but the fault generally found with the tobacco is that it is too full flavored. Further experiments were carried on in the same districts with varying results. In Sind the experiments and their results were insignificant. In Broach they were somewhat more successful, the superintendent thus summarising his experience:—' Havana, Shiraz, and other varieties of exotic tobacco will, with ordinary care and attention, yield fair and certain crops on ordinary black land, and presumably on every other kind to be met with in Guzerat. By the skillful application of manure, leaf of any desired quality or peculiarity of flavor and texture may be obtained. The quantity of produce is so great that, should it be found practicable to cure the leaf well enough to make it a salable article in the European market, a source of profit by no means insignificant would be opened up to the Guzerat ryot. For the native market the country plant is more suitable, and its cultivation consequently the more profitable.' In Dharwar the superintendent was enabled to distribute seed in sufficient quantities to those applying for it, but found the ryots would not cultivate it on a large scale, being apprehensive of loss. Native tobacco he considers less liable to injury than the exotic varieties during the squally weather prevalent about the time the leaf is approaching maturity."

Robertson, in replying to the assertion that the tobacco of India contains little if any nicotine, says:

"It appears to me that there must be some mistake as to the tobacco containing little or no nicotine. Very many have tried the tobacco, and pronounce it to be good, with, however, the fault of being exceedingly strong. Now, the strength of tobacco comes from its nicotine, and if the specimens I sent contain no nicotine, whence the strength? I believe that nothing destroys tobacco so much as moistening it. How, then, are acetic acid and chloride of soda to be used in the curing? If the process of desiccation had been carried on too quickly, the tobacco would have been of either a green or greenish-yellow color. If too slowly, it would have been black, like much of the country tobacco. I perceive that the amount of nicotine in a great measure depends on the extent to which the leaf is allowed to ripen. The riper the leaf the more the nicotine. The amount of nicotine does not appear to depend on the amount of curing. The soil the tobacco was grown in is a hardish red moorum soil, containing much iron; probably that may account for the red coloring matter being so much developed. I intend to have some of each description of the tobacco leaf analyzed, and also intend to submit the soil in which it was grown to the same process. I have had some of the cigars packed up for some months to test how far they are proof against insects. None have been attacked by insects. Some Manilla cigars, some Trichinopoly cheroots, all packed up at the same time, have, however, been entirely destroyed by insects.

"It is clear from the reports that both in Guzerat and Khandesh, Havana and Shiraz tobacco will flourish, and that they may be introduced without difficulty. The ryots, it is said, preferred the new kinds to their own, and desire their introduction, the foreign varieties commanding a higher price in the market. The chief drawback is the want of knowledge and appliances for the proper curing of the leaf. This, indeed, is the great drawback throughout India. In the district of Kaira the seed is always sown in nursery beds in the month of July, and transplanting commences about the end of August, the operation continuing for about two months. The tobacco planted on the dry soil called 'koormit' ripens and is fit for cutting in January and February; that which is grown on irrigated land during March and

April. In Canara, tobacco is generally grown in elevated situations. The seed is sown in August, and the seedlings are transplanted in November, the crop arriving at maturity in three or four months. North Canara derives its supply chiefly from Mysore, the leaf produced in that province being said to be less liable to affect the head than that of the Canara plant."

The Turk and his family love to cultivate tobacco as well as to smoke it; and give it their attention from seed-sowing until it is sold to the merchant. The Turk is very particular in cultivating it, as on its color depends in a great measure its value. He commences work on his plant-bed in March, sowing the seed about the same time as the Virginia planters. After the leaves are gathered the same scrupulous care is taken with them; especially in drying and baling, that the leaf may be in just the right condition to ferment properly, and be ready to be assorted by the "tobacco pickers." The Turk presses his whole family into the cultivation of the plants. The children are engaged in weeding while he waters the beds or prepares the tobacco field for the planting of the tobacco. In pruning and picking the leaves he removes only those that are small—the removal of which will still further advance the growth of the plants, and is careful to gather only those leaves that are turning yellow, giving evidence of their maturity. Says one in regard to the cultivation of tobacco in Turkey:

"The Turk and his family, it will seem, have now been occupied upon their tobacco crop for nearly a whole year. The leaf is just becoming a bright light yellow when it falls into the hands of the merchant, and it is during this period that the process of fermentation or heating generally occurs, before which the tobacco can not be shipped. The bales having been placed in the merchant's store, are left end up until a fermentation or baking has taken place, the ends being reversed every three or four days. In the course of a few weeks a bale is reduced to about two-thirds of its original size. It is then placed upon its sides to cool. When it is discovered to be cold it is broken open by the native tobacco-pickers, and every leaf sorted and classified. The patience with which this operation is carried out is truly astonishing.

There is a good deal of difference in their rate of work. One man may pick only fifty pounds weight a day, while another does twice that quantity. It is necessary to watch them closely, or they will put a dirty brown leaf with a pale yellow. They neither know nor care about the losses that may be incurred by the merchant, whose samples may be thus spoiled. A bale of leaf purchased at five piastres per *oque*, when dissected by the Greek for various markets will be found to contain varieties ranging in price from 5 to 60 piastres; of these some are dispatched to Odessa, some to Smyrna, others to Constantinople, Alexandria, and England —the mixed and common qualities generally to the latter country, the price there obtained being the least remunerative

TURKISH TOBACCO GOING TO MARKET.

to the Greek shippers. The bales are brought from the interior to the shipping ports upon mules, each animal carrying two bales; and it is a pretty sight to witness, say 150 mules at a time, crossing mountains and rugged paths with their burdens, followed by perhaps fifty camels laden with cotton, marching to the merry tinkle of the bells on their necks. When the tobacco reaches the shipping port the troubles of the exporter are intensified. The bales are first taken to the Custom House, and there weighed. The weights thus arrived at are compared with the quantity received from the interior, and if there be any material difference the shipper has to account for it. If any has been sold for consumption in Turkey, duty has to be paid upon the amount; and in order that no part of his shipment may be used in the country, he has to sign a bond that the tobacco shall not be landed in any other port of Turkey. On the arrival of the tobacco in England, the landing certificates are forwarded to Turkey. It is in this way that the trade is retained in the hands of a few Greeks, who naturally put every obstacle in the way of the foreigner, whose sole remedy is at last found to be the payment of the universal 'backshish,' to the comptroller of customs."

The merchant who buys the tobacco of the planter at a low price, and thereby takes the profit from him of cultivating it, is preyed upon in the same manner by the Greek buyers who have the sole monopoly of the trade. Like Shiraz tobacco, that of Turkey has to be handled frequently and pass through several stages of curing before it is ready to be manufactured. In this respect it is unlike most of the tobaccos of America, but its treatment is not unlike that of the varieties of the East.

The tobacco plant is cultivated with great success in many of the provinces of Japan, and is exported in large quantities to Europe. The leaf is excellent, and is in request by many buyers of Eastern tobaccos. Robertson gives the following interesting account of the Japan tobacco fields:—

JAPAN TOBACCO FIELD.

"According to a native account, tobacco was introduced into Japan in the year 1605, and was first planted at Nagasaki in Hizen. It is now very generally grown throughout the country. In the province of Awa, where a great deal of tobacco is grown, the seed is sown in early spring in fields well exposed to the sun and duly prepared for its reception. Well sifted stable manure is strewn over the field, and the seedlings appear after the lapse of about twenty days. The old manure is then swept away, and liquid manure applied from time to time.

If the plants are too dense they are thinned out. The larger plants are now planted out into fields well prepared for the purpose in rows, with about eight inches space between each plant, the furrows between each row being about two feet

TRANSPLANTING.

wide. They are again well sprinkled with liquid manure, also with the lees of oil at intervals of about seven days. A covering of wheat or millet bran is now laid over the furrows. The bitter taste of the leaf is in a measure an effectual safeguard against the ravages of insects, but the leaves are nevertheless carefully tended to prevent damage from such cause. If the reproduction from seed is not desired the flowers should be cut off and the stem pruned down, otherwise the leaves will lose in scent and flavor. In Osumi exceptional attention is paid to the cultivation of the tobacco plant. The lees of oil, if liberally used, and stable manure sparsely applied, have great effect on the plant, producing a small leaf with an excellent flavor; while, if the opposite course is followed, the leaves grow to an immense size, but are inferior in taste.

"When the flowers are in full bloom the 'sand' leaves are picked. After the lapse of twelve or fourteen days the leaves are gathered by twos. Any leaves that may remain are afterwards broken off along with the stalk. Any sand adhering to the leaves is removed with a brush; the stems having been cut off, the leaves are rolled round, firmly pressed down with a thin board, and cut exactly in the centre. The two halves are then placed one on the top of the other in such manner that the edges exactly correspond, and being in this position firmly compressed between two boards, they are cut into fine strips, the degree of fineness depending on the skill

GATHERING THE CROP. 373

of the cutter. A machine made of hard wood, but with the vital parts of iron, is used by some persons for this purpose. The machine was devised about sixty years ago by a skillful Yeddo mechanic, the idea being taken from those used in Osaka and Kiyoto for cutting thread used for weaving into silk embroidery. Since then numerous improvements have been made in it, and it is now extremely well adapted for the economization of labor. Another machine was invented about eight years since, also by a Yeddo mechanic. It is smaller than the first mentioned, but being very easily worked is much in use. Tobacco is sometimes cut in the following crude manner:—The leaves are piled one on top of the other, tightly compressed into the consistency of a board, and then cut into shavings by a carpenter's plane. This is, however,

CHINESE TOBACCO FIELD.

about the worst method, and even the best tobacco, if treated in such fashion, loses its flavor and valuable qualities."

In China* tobacco is cultivated in the western part of the

---

*I saw also great plantations of tobacco, which they call tharr, and which yield very considerable profit, as it is universally used in smoking, by persons of all ranks, of both sexes in China; and, besides great quantities are sent to the Mongolls, who prefer the Chinese manner of preparing it before any other. They make it into a gross powder, like saw-dust, which they keep in a small bag, and fill their little brass pipes out of it, without touching the tobacco with their fingers.—*Bell's Travels in Asia*, 1716, 1719, 1722.

empire, and grows almost as large as most American varieties. Chinese tobacco is usually light in color, of a thin, silky texture, and mixed with Turkey tobacco, is a considerable feature in the export trade of that country. The Chinese cultivate the plant like the Japanese, and give it as much care and attention as they do the tea plant. The leaves are gathered when ripe, and are dried and well-assorted before baling. The Chinese planter often raises large fields of the plants, and employs many hands to tend and cultivate them. We give a cut of a tobacco field and the planter looking at the field and noting the progress of the laborers.

In Persia tobacco is cultivated near Shiraz, which gives name to the variety. The soil is very fertile and richly cultivated. Not only does the tobacco plant flourish finely, but all kinds of vegetables grow to perfection. The Persians cultivate the plant principally for their own use. It is a fine smoking tobacco, and when cured properly is said to be equal to Latakia. Their mode of curing is unlike that adopted by any other cultivators of the weed but is very successful, and is no doubt the proper method of preparing the leaves for use. Their mode of pressing in large cakes is unlike that of

TOBACCO FIELD IN PERSIA.

any other growers—but doubtless adds to the aromatic quality of the leaf which makes it so popular in the East.

The tobacco field is trenched so as to retain water, while

the plants are set on the ridges where they flourish and mature until the buds and flowers are broken off. The harvest occurs in the autumn, when the singular process of curing begins.

Abbott says of the culture and commerce of tobacco in Persia:

"Jehrum, South Persia, is the principal mart for tobacco, which is brought here from all the surrounding districts, and disposed of to traders, who distribute it over the country far and near. These traders are numerous, and many established here are wealthy; they usually transact their business in their private houses, without resorting to the caravansaries of which there are six in the place. There are many grades and qualities of Shiraz tobacco but that produced at Tuffres (according to Forster), a town about one hundred miles to the south-west of Turshish, is esteemed the best in Persia.

"Of the many varieties of the tobacco plant grown in the East, that known as Manilla is among the most famous and the most extensively cultivated. It is grown in several of the Philippine islands, particularly in Luzon and the southern group, known as the Visayos. The Philippines are a large group of islands in the North Pacific Ocean, discovered by Magellan in 1521; they were afterwards taken posession of by the Spaniards, in the reign of Philip II., from whom they take their name.

"The islands are said to be eleven hundred in number, but some hundreds of them are very small, and all are nominally subject to the Spanish government at Manilla. The Philippines produce a great variety of tropical products such as rice, coffee, sugar, indigo, tobacco, cotton, cacao, abaca, or vegetable silk, pepper, gums, cocoa-nuts, dye-woods, timber of all descriptions for furniture and the buildings, rattans of various kinds, and all the agreeable fruits of the tropics. On the shores are found nacre, or mother of pearl, magnificent pearls, bird's-nests, shells of every description, an incredible quantity of excellent fish, and the *trépang*, or *balaté*, a sea-worm, or animal substance, found on the shores of the Philippine Islands, resembling a large pudding. The Chinese esteem it as a great delicacy and mix it with fowl and vegetables. The inhabitants practise various kinds of industry; they weave matting of extraordinary fineness and of the brightest colors, straw hats, cigar cases and brackets; they manufacture cloth and tissues of every sort from cotton, silk, and abaca;

they, from filaments taken from the leaves of the *etuana*, make cambric of a texture much finer than that of France; and they also manufacture coarse strong cloth for sails, and ropes and cables of all dimensions; they tan and dress leather and skins to perfection; they manufacture coarse earthen ware, and forge and polish arms of various kinds; they build ships of heavy tonnage, and also light and neat boats; and at Manilla they frame and finish-off beautiful carriages; they are also very clever workers in gold, silver, and copper; and the Indian women are specially expert in needlework, and in all kinds of embroidery.

"The island of Luzon is the largest of the Philippines, and extends from north to south for the length of about six degrees. It is divided throughout its whole extent by a chain of mountains, which in general owe their formation to volcanic eruptions. In the provinces of Laguna and Batangas there is the high mountain called Maijai, one of the loftiest in Luzon, which is beyond doubt an ancient crater; on the summit a little lake is found, the depth of which cannot be measured. At some period the lava that then flowed from the summit towards the base, in the neighborhood of the town of Nacarlan, covered up immense cavities, which are now recognizable by the sonorous noise of the ground for a great extent; and sometimes it happens that, in consequence of an inundation or an earthquake, this volcanic crust is in some places broken, and exposes to the view enormous caverns, which the Indians call 'the mouths of hell.' In the district about the town of San Pablo, which is situated on the mountain, are found great numbers of little circular lakes and immense heaps of rotten stones, basalt, and different descriptions of lava, which show that all these lakes are nothing else than the craters of old volcanoes. Altogether the soil to the southward, in the province of Albai, is completely volcanic, and the frequent eruptions of the volcano bearing that name may, as the natives say, be attributed to the same cause as the earthquakes so often felt in the island of Luzon. Over almost the whole of these mountains, where fire has played so conspicuous a part, there is a great depth of vegetable earth, and they are covered with a most splendid vegetation. Their declivities nourish immense forests and fine pastures in which grow gigantic trees—palm trees, rattans, and lianas of a thousand kinds, or gramineous plants of various sorts, particularly the wild sugar cane, which rises to the height of from nine to twelve feet from the ground;

in their interior are rich mines of copper, gold, iron, and coal.

"There are two distinct and strongly marked seasons in the island of Luzon, namely, the rainy or the wintry season, and the dry or summer season. For six months of the year —that is from June to December—the wind blows from the south-west to the north-east, and then the declivities of the mountains and all the western side of the island are in the season of the rains; in the six other months, the wind changes,

GROWING TOBACCO ON THE PHILIPPINE ISLANDS.

and blows from the north-east to the south-west, when all the eastern parts of the island have the season of winter. During the rainy season, the incessant fall of rain on the mountains causes the rivers, both large and small, to overflow and to become torrents, that rush down upon the plains, covering them with water, and depositing the broken earth and slime which they have gathered in their course. In the dry season, water is supplied for irrigation from reservoirs, which are carefully filled during the rains. From these causes it follows that without any manuring, and with scarcely any improvement from human industry, the soil of the Philippines is as fertile as any in the world; so that, without great labor, the cultivator has most abundant harvests."

The above description of the Philippines by Gironiere gives a faithful account of the vast resources of the islands. Of the products cultivated rice and tobacco are the most important. The finest tobacco plantations are situated in the northern parts of the island of Luzon, and furnish the finest quality of Manilla tobacco. That grown in the Visayos is of an inferior quality, and is sold to merchants holding a permit to purchase at the shipping ports and transport to Manilla for sale to the government. In the island of Luzon, the greatest quantity of tobacco is cultivated in the provinces of Nueva Ecija and Cagayan.

Tomlinson in an account of the tobacco of the Philippines says: "Manilla leaf comes from the three principal districts of the island of Luzon—Visayos, Ygarotes and Cagayan," The mode of cultivation does not differ in any great respect from that followed in other parts of the world. Great seed beds are made on the plantations where the plants are grown until ready to transplant in the tobacco ground. Unlike most land adapted for tobacco, large crops are grown without the aid of any fertilizer whatever. In cultivating the plants, buffaloes are used, yoked one after the other, going between the rows several times, and at the last ploughing leaving a trench in the middle of the rows, for letting off the water. The Indian plow used in cultivating is exceedingly simple: it is composed of four pieces of wood which the most unhandy

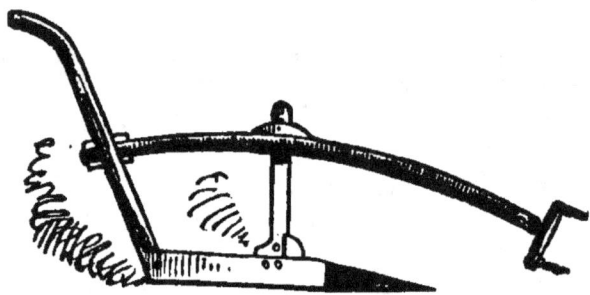

TOBACCO PLOW.

ploughman can put together, with the mould board and share, which are of cast iron. The lightness and simplicity of this plough render it easy to be used in every kind of cultivation, where the plantations are divided into rows, such as those of tobacco, maize and sugar cane. It is used with great advantage, not only for cutting down weeds, but also for giving to each row a ploughing, which is serviceable to the plantation, and which is less costly and quicker than simple weeding with the mattock.

When the leaves are ripe they are stripped from the stalks and separated into three classes, according to their size, and afterwards made into bunches of fifty or a hundred, by passing through them, near the foot, a little bamboo cane, as if it was a skewer, by which the bunches are afterwards hung up to dry in vast sheds, into which the sun's rays cannot enter, but in which the air circulates freely; they are left to hang there until they become quite dry, and for this, a greater or less time is required, according to the state of the weather. When the drying is effected the leaves are placed according to their quality, in bales of twenty-five pounds, and in that state they are handed over to the administration of the monopoly. Gironiere in describing the mode of culture on the tobacco plantations says:

"During the first two months after the transplanting it is indispensably necessary to give four ploughings to the ground between the rows of the plants, and every fifteen days to handpick, or even better, to root out with the mattock, all the weeds which cannot be touched by the plough. These four ploughings ought to be done in such a manner as to leave alternately a furrow in the middle of each line, and on the sides, and consequently, at the last ploughing, the earth covers the plants up to their first leaves, leaving a trench for carrying off all water that may accumulate during the heavy rains. As soon as each plant has gained a proper height, its head is lopped off to force the sap to turn into the leaves, and, in a few weeks afterwards, it is fit for being gathered."

The tobacco fields or plantations are very large, and together with the vast sheds for curing, the fields present a beautiful appearance; the long straight rows with their dark green leaves adding not a little to the beauty and variety of the landscape. The great growers of the plant are very careful in cultivating the fields and give the tobacco frequent hoeings, until ready to be gathered and taken to the sheds. The planters are obliged to take the utmost pains, as the product is obliged to be given up to the monopolizing government which is the sole purchaser, and which, in its great establishment at Binondoc, employs continually from 15,000 to 20,000 workmen and workwomen in manufacturing cigars for the consumption of the country and for exportation.

Manilla tobacco is much esteemed in the islands both by the Spaniards and the Chinese. The custom of smoking is universal among all classes and at all times. In the house, on the road and street, the aroma of a fragrant Manilla is ever borne on the breeze. The Spaniards are the principal owners of the tobacco fields, and, like their brother planters on the island of Cuba, are fond of the weed and its more potent companion. After a luxurious breakfast the planter

SPANISH PLANTERS.

elevates his feet for a quiet smoke, and lights either a cigar or cheroot, filling the room with smoke and with the most fragrant perfume.

Of all the various products cultivated, but few vie with the tobacco plant in beauty of form and general appearance. By its great variety of colors in leaves and flowers, it offers a striking contrast with the more sombre hues of most other plants. When left to grow until the plants have reached full size, the tobacco field has the appearance of a vast flower garden, the tiny blossoms exhaling their fragrance and the entire plant emitting odors as rare and as delicate as the most fragrant exotic. In the tropics the finest tobacco plantations are found, as nature is more lavish, not only in the richness of the soil, but in the variety of the vegetable products. Here the tobacco plant attains its finest form and most delicately flavored leaves. The hues of the flowers are brighter and their fragrance sweeter. In the tropics the tobacco field may be scented from afar, as its odors are wafted on the breeze. In its native home it flourishes and matures as readily as the more common kinds of vegetation, while it affords the planter a larger revenue than many of the more useful of nature's products.

CHAPTER XI.

VARIETIES.

THE tobacco plant almost vies with the palm in the number of varieties; botanists having enumerated as many as forty, which by no means includes the entire number now being cultivated. The plant shows also a great variety of forms, leaves, color of flowers, and texture. Each kind has some peculiar feature or quality not found in another; thus, one variety will have large leaves, while another will have small ones; one kind leaves flowers of a pink or yellow color, another white; one variety will produce a leaf black or brown, another yellow or dark red. The following list includes nearly all of the principal varieties now cultivated:—Connecticut seed leaf (broad and narrow leaf), New York seed leaf, Pennsylvania (Duck Island), Virginia and Maryland (Pryor and Frederick, James River, etc.), North Carolina (Yellow Orinoco, and Gooch or Pride of Granville, etc.), Ohio Seed leaf (broad leaf), Ohio leaf (Thick Set, Pear Tree, Burley, and White), Texas, Louisiana (Perique), Florida, Kentucky, Missouri, Wisconsin, Havana, Yara, Mexican, St. Domingo, Columbia (Columbian, Giron, Esmelraldia, Palmyra, Ambolima), Rio Grande, Brazil, Orinoco, Paraguay, Porto Rico, Arracan, Greek, Java, Sumatra, Japan, Hungarian, China, Manilla, Algerian, Turkey, Holland (Amersfoort), Syrian (Latakia), French (St. Omer), Russian, and Circassian. Many of these varieties are well known to commerce, and others are hardly known outside the limit of their cultivation.

All of these varieties may be divided into three classes,*

---
*Probably most writers would divide tobacco into but two classes, including tobacco used for the manufacture of snuff with cut tobacco.

viz.: cigar, snuff, and cut-leaf tobacco. The first class, cigar leaf, includes all those varieties of tobacco that are used in the manufacture of cigars, and embraces the finest quality of tobacco grown, including Connecticut seed leaf, Havana, Yara, Manilla, Giron, Paraguayan, Mexican, Brazilian, Sumatra, etc. The second class embraces all of the varieties used in the manufacture of snuff, such as Virginia, Holland (Amersfoort), Brazilian, French (St. Omer), etc. The third class includes all of those tobaccos used for smoking and chewing purposes, such as Virginia, Kentucky, Missouri, Ohio, Maryland, Latakia, Perique, Turkish, and others.

South American tobaccos are almost exclusively used for the manufacture of cigars. Although of various qualities, they possess the distinctive flavor which characterizes all tobacco used for this purpose. This is generally the case with most of the tobacco grown in the tropics—it seems to be especially adapted for the manufacture of cigars, rather than for cutting purposes. European tobaccos are milder in flavor, and are used extensively in the manufacture of snuff; while the tobacco of the East is well adapted for the pipe.

Tobacco to be used for cigars must not only be of good flavor, but must burn freely, without which it has no real value for this purpose. Non-burning tobaccos cannot be used, and are either employed in the manufacture of snuff or for cutting.

Of the many kinds of tobacco of both the Old and New World, doubtless the most curious of all is that kind known as

DWARF TOBACCO.

This plant is a native of Mexico, and was discovered by Houston, who found it growing near Vera Cruz. This is probably the smallest kind of tobacco known. The plant grows to the height of about eighteen inches, the leaves growing in tufts at the base of the plant. Some have supposed this tobacco to be what is known as Deer Tongue, which is used for flavoring, but it is quite probable that it is entirely different. The leaf is small and light green, and it is quite a showy plant when in blossom. As a curiosity it can hardly fail to attract attention from all those acquainted and interested in tobacco, but will hardly admit of cultivation, on account of the absence of leaves, with the exception of the few growing near the ground. Of all the tobaccos used for the manufacture of cigars, none have obtained an equal reputation (simply as a cigar wrapper)

MEXICAN DWARF TOBACCO.

with the famous and much sought for variety known as

CONNECTICUT SEED LEAF,

which in all respects towers far above the seed products of the other states. The varieties cultivated in the United States and known as "seed leaf" tobaccos, are grown in Connecticut, Massachusetts, Vermont, New Hampshire, New York, Pennsylvania, Ohio and Wisconsin. All of the seed leaf of these states is used exclusively in the manufacture of cigars. Connecticut seed leaf is justly celebrated as the finest known for cigar wrappers, from the superiority of its color and texture, and the good burning quality of the leaf. The plant grows to the height of about five feet, with leaves from two and one half to three feet in length and from fifteen to twenty inches broad, fitted preëminently by their large size for wrappers, which are obtained at such a distance from the stem of the leaf as to be free from large veins.

Connecticut seed leaf tobacco in color, is either dark or light cinnamon, two of the most fashionable colors to be found in American tobaccos. The plant is strong and vigorous, ripening in a few weeks, and when properly cultivated attaining a very large size. There are two principal varieties of Connecticut seed leaf, viz.—broad and narrow leaf: of these two, the broad leaf is considered the finest, cutting up to better advantage and ripening and curing fully as well.

CONNECTICUT SEED LEAF.

Connecticut seed leaf attains its finest form and perfection of leaf in the rich meadows of the Connecticut Valley, where it has been cultivated to a greater or less extent for nearly half a century.

The plant is one of the most showy of all the varieties of tobacco. The stalk is straight and large, while the leaf (especially the broad) is admirably proportioned, and the top is broad and graceful, rendering it far more symmetrical in appearance than many of the smaller varieties.

Before Connecticut tobacco became known as a wrapper, Maryland and Havana tobaccos were used for this purpose, and when Connecticut first came into use, it was only as a filler. This variety differs very materially from Havana in this respect—it has not that fine flavor of Cuba tobacco, but in texture is much superior. The lighter shades of it burn purely and freely, leaving a white or pearl colored ash, which is one of the best evidences of a good wrapper. The leaf also is very firm and strong, and sufficiently elastic to bear considerable manipulating in manufacture. The various shades also of the two colors, dark and light brown or cinnamon, are among the finest and most delicate of any to be found among the numerous kinds of tobacco used for cigars. The color of the wrapper, however, is merely a matter of taste; when first used for a wrapper the color in demand was a dark brown or cinnamon, now it is light cinnamon leaf that is the most fashionable, and leaf of this color is considered the finest and of the most delicate flavor. As a superior burning tobacco, seed leaf especially commends itself, and while all of the seed products of the various states producing this description of tobacco, are remarkable for their good burning qualities, none are more so than Connecticut seed leaf.

Thorough cultivation by the growers has made this quality of tobacco the most profitable of any grown in the United States. Some considerable controversy has arisen among tobacco-growers concerning the origin of this famous variety. One opinion sets forth that it sprung from plants or seeds brought from Virginia, while another is that tobacco seed from Cuba gave it origin. Most probably the former theory is correct, as the plant was cultivated in gardens in New England, during the reign of Charles I.

However this may be, the system of cultivation pursued has been successful in the production of a leaf tobacco that can hardly be improved, so far as the texture of the leaf is concerned. Some of the "selections" of seed leaf have that fine soft feeling peculiar to satin or silks, and we have seen specimens of such selections, that seemed almost destitute of veins, or anything that would naturally suggest that it was a leaf. In this respect it is quite remarkable, for while the leaf is very large the stem and veins are quite small, no larger than in many varieties with a much smaller leaf. From its first cultivation in the Connecticut valley, the quality has gradually improved until now, and it seems at last to possess almost every feature desirable in a good wrapper.

This famous variety of the tobacco plant is by common consent the finest flavored tobacco for cigars now being cultivated. Some, however, consider Paraguayian, Brazil, and Mexican coast tobacco its equals, while, according to Tomlinson, Macuba tobacco, grown on the island of Martinica, stands at the head of all varieties of the plant. These statements may, however, be regarded as mere opinions rather than acknowledged facts.

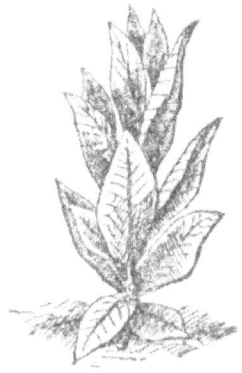

HAVANA TOBACCO.

Havana tobacco, according to Hazard, "grows to a height of from six to nine feet, as allowed, with oblong, spear-shaped leaves; the tobacco being stronger when few leaves are permitted to grow. The leaves when young are of a dark-green color and have rather a smooth appearance, changing at maturity into yellowish-green. The plant grows quickly, and by careful pruning a fine colored leaf is obtained, varying from a straw color to dark brown or black." The plant bears a pink blossom, which is succeeded by capsules not quite as large as those of seed-leaf tobacco. The finest is grown in the Vuelta de Abajo, which, for nearly a century, has been celebrated as a fine tobacco-producing district. When growing, a *vega* of Havana tobacco forms a most pleasing feature of the landscape. As the plants ripen, the dark, glossy green of the leaves is succeeded by a lighter shade and a thickening of the leaf. The plant ripens in from eight to ten weeks after being transplanted. The stalk and leaves are not as large as its great rival, Connecticut seed-leaf, but it far surpasses it in flavor. The plant emits a pleasant odor while growing, like most varieties of the plant grown in the tropics.

YARA TOBACCO.

This variety of tobacco, like Havana, is grown upon the island of Cuba, but is unlike it in flavor, as well as in the appearance of the plant. It is well known as an admirable tobacco for cigars, but is not sought after or grown to such an extent as Havana. The leaf when growing, is in color a fine green, and when cured is of considerable body and fine texture. A writer in alluding to Yara tobacco says:

"The most noted *vega* or tobacco plantation is situated near Santiago de Cuba and is called Yara. The choicest tobacco is that grown on the banks of rivers which are periodically overflowed. They are called Lo Rio, Rio Hondo, and Pinar del Rio, and the tobacco is distinguished from all other grown upon the island by a fine sand which is found in the creases of the leaves."

The flavor of Yara tobacco is so essentially different from Havana, that it is not cultivated as extensively, if indeed it could be. It is grown more particularly for home use and for exporting to Europe, where it is considered one of the finest of tobaccos. Of the other varieties grown in the West Indies such as St. Domingo, Jamaica, and Trinidad, much may be said both in praise and dispraise. St. Domingo and Trinidad have been cultivated for more than two hundred years. St. Domingo tobacco has a large leaf, but is of inferior flavor to most varieties of West India tobacco.

Virginia tobacco has acquired a reputation which has gradually strengthened for more than two hundred and fifty years. It was one of the first products to be cultivated by the English colony, and in less than a quarter of a century after the settlement of Virginia, had acquired a reputation hardly surpassed by its well known rivals, Trinidad, Brazil, St. Domingo, and Varinos tobaccos. The plant grows to the height of from five to seven feet; the leaves are long and broad, and when cured are of various colors, from a rich brown to a fine yellow. The finest of Virginia tobacco comes from the mountainous counties, but the amount is small in proportion to the vast quantities raised on the lowlands of

VIRGINIA TOBACCO.

the Dan and James rivers and their tributaries. The leaf grown in the higher counties of South-western Virginia is much lighter in color and much softer than the ordinary Virginia tobacco. Shades of color in Virginia tobacco (as well as in most others) serve to determine its use, while texture and length of leaf affect as well its market value. There are various grades of Virginia tobacco, especially in that grown in Southside, Virginia. "Long bright leaf" is considered the finest, while that known as "Luga" is the poorest and lowest grade of leaf.

The staple known as James River tobacco has acquired a world-wide reputation, and the same ground is cultivated and planted with tobacco now as in 1620. Virginia tobacco is known chiefly as a cut tobacco; "good, stout snuff leaf" is also obtained from it, which brings as much in European markets as "fine spinners." Missouri, Kentucky, and some parts of Ohio also produce large quantities for manufacturing into chewing and smoking tobacco.

OHIO TOBACCO.

The tobacco plant has been cultivated in this State for

OHIO WHITE TOBACCO.

nearly fifty years. Sullivan, in describing the kinds used for cutting, says:—

"Two kinds of seed are used, viz., the 'Thick Set' and the 'Pear Tree,' and of late years the 'Burley' has come into favor. Nearly all tobacco grown in Ohio is 'fired,' that is, cured by fires or flues; it is packed in hogsheads of about eight hundred pounds net."

Another writer says:—

"In some parts her soil produces a fine yellow article called 'Northern Ohio;' it is manufactured into the finest quality of smoking tobacco, and is extensively used by all epicures of the meerschaum, both in this country and in Europe. Ohio also produces another variety called Ohio seed leaf, or more familiarly, 'Seed.'"

While in another section she produces an excellent article of leaf for chewing. Ohio tobacco of all kinds is a large plant, and cures "down" to fine colors. One variety for cutting, known as "cinnamon blotch," is a leaf of good body and is considered an excellent tobacco for chewing. A few years since a variety originated in a very curious manner. We give the account as published by Prof. E. W. Smith:—

"This tobacco is known by the name of White tobacco. The seed was procured about three years ago, in a very singular way. There were a few hills of tobacco that looked very singular, situated near a thicket of bushes and trees. The rising morning sun sent its rays through this thicket, striking diagonally upon a few hills, and producing by some chemical law or daguerreotyping process the (white) tobacco. The tobacco was allowed to go to seed. This seed was sown the next year, and produced the same kind of tobacco. The tobacco, before the white tobacco was daguerreotyped, was a cinnamon blotch, so it may be seen by this freak of nature how it was changed from red to white."

PERIQUE TOBACCO.

There are many varieties of tobacco well adapted for smoking, of all colors and strengths. Of American tobaccos suitable for this purpose, none have acquired a wider reputation at home than Perique. It is cultivated only in small quantities in one or two parishes in Louisiana. Perique tobacco may be used not only for smoking, but for chewing and for snuff. The leaf when cured measures some eighteen inches in length by fourteen in width, is thick and substantial, has the appearance of a rich Kentucky tobacco, and when placed under press immediately after being cured becomes black without the aid of any artificial means. It is put up in rolls, or, as they are called, "carrots." This tobacco is raised mostly in the parish of St. James, La., and derives its name from an old Spanish navigator who settled in St. James parish in the year 1820. His first attempt at raising tobacco, for his own use, succeeded so well and gave him such a fine result, (the plant developing itself to a great extent and being very rich,) that he concluded to devote all his time to the culture of tobacco, in order to make a living out of it.

The seed first used by him was the Kentucky, but this was subsequently changed for the Virginia, which has been in use up to this time, being renewed every four or five years. The tobacco originally put up by Perique was twisted by hand and placed under press for three or four days, then taken out, untwisted, retwisted and replaced in the press for five or six days. After undergoing the same process three or four different times, it was finally left to remain under press for six months, and then taken out for use. Mr. Perique, however, soon made a capital improvement in the mode of putting up his tobacco; for, as early as the year 1824, we find the tobacco in beautiful rolls of four pounds, and as hard as a "Sancisson de Boulogne."

This tobacco, which has retained the name of its producer, is still manufactured in the same manner as it was fifty-four years ago, the work still being done entirely by hand. The plant is cultivated as the Virginia tobacco by about a dozen small planters in that part of the Parish called "Grande-Pointe," seven miles from the Mississippi river. A small quantity is also raised on the banks of the river in the same parish by a few planters. The growers of Perique tobacco have tried Virginia, Kentucky, and Havana seed, but prefer the former—Havana producing too small a plant without a much better flavor.

Tobacco is grown in other parishes of the State; it is

however of inferior quality, and is used only for smoking or snuff. Perique tobacco, when cut for smoking, is very black in appearance, exceedingly smooth, and of peculiar odor. It is probably the thinnest tobacco cultivated; and is strong, but of agreeable flavor.

PERUVIAN TOBACCO.

John Gerard gives the following description of the tobacco of Peru:

"Tobacco, or henbane of Peru, hath very great stalks of the bigness of a child's arme, growing in fertile and well-dunged ground of seven or eight feet high, dividing itself in sundry branches of great length; whereon are placed in most comely order very faire, long leaves, broad, smooth and sharp-pointed, soft and of a light green color; so fastened about the stalk that they seem to embrace and compass it about. The flowers grow at the top of the stalks in shape like a bell-flower, somewhat long and cornered; hollow within, of a light carnation color, tending to whiteness towards the rims. The seed is contained in long, sharp-pointed cods, or seed-vessels, like unto the seed of yellow henbane, but somewhat smaller, and browner of color. The root is great, thicke and of a wooddy substance, with some threddy strings annexed thereunto."

MEXICAN TOBACCO.

The tobacco plant seems to have been cultivated in Mexico from time immemorial. Francisco Lopez de Gomara, who was chaplain to Cortez, when he made conquest of Mexico, in 1519, alludes to the plant and the custom of smoking; and Diaz relates that the king Montezuma had his pipe brought with much ceremony by the chief ladies of his court, after he had dined and washed his mouth with scented water. The Spaniards encouraged its cultivation, and to this day it is grown in several of the coast states. Various kinds are cultivated, but chiefly a variety bearing yellow flowers, with a large leaf of fine flavor resembling the Havana. The plant is a favorite with the Mexicans, who prefer it to any other product grown. It is cultivated like most varieties of the tropics, and is hardly inferior to any grown in the West Indies, and is especially adapted for cigars and cigaritos. After the first harvest another, and sometimes a third crop is gathered by allowing one shoot to grow from the parent root, which oftentimes develops to a considerable size. The quality of leaf, however, is inferior; as is the case with all second and third crops grown in this manner.

ST. DOMINGO TOBACCO.

This well-known West India variety is inferior to most kinds grown on the neighboring islands. The plant attains a large size, cures dark, is coarse, and of inferior flavor. It is a favorite tobacco in Germany, and thousands of Ceroons are annually shipped to Hamburg. The West India islands produce many varieties of tobacco, which is owing more to the composition of the soil and climate than to the method of cultivation and curing.

The demand for St. Domingo tobacco is limited. It has no established reputation in this country, and on account of the high duties can not compete with our domestic tobaccos.

LATAKIA TOBACCO.

This variety of the tobacco plant is one of the most celebrated known to commerce. It attains its finest form and flavor in Syria, where it is cultivated to a considerable extent. For smoking it is among the best of the varieties of the East, and is used for the more delicate cut tobaccos and cigars. It grows to the height of three feet—each offshoot bearing flowers, the leaves of which are ovate in form, and are attached to the stalk by a long stem. The flowers are yellow, and number only a few in comparison with most varieties. When growing, the leaves are thick, but after curing are thin and elastic. The stalk is small, as are also the leaves. While growing, the plants emit a strong

LATAKIA TOBACCO.
(SYRIA).

aromatic odor not like that of Havana tobacco, but stronger and less agreeable.

The plant was introduced into this country by Bayard Taylor, and attains its full size in the Connecticut valley, where it has been tested by many growers. After curing, the leaf is a bright yellow of agreeable flavor, having the odor of ashes of roses. The flavor is similar to Turkish tobacco, but is said to be less delicate.

After harvesting, the plants cure rapidly and on account of their small size rarely sweat. Latakia tobacco, however, is not adapted to the taste of American smokers, most of whom prefer tobacco of home growth to even the finest of Turkish leaf. Latakia tobacco can be raised with less labor than most varieties. Its diminutive size and its unpopularity, however, prevent its general culture in this country.

RUSSIAN TOBACCO.

In no other country in Europe is the tobacco plant attracting as much as attention as in the empire of Russia. The varieties grown in America, Cuba, Turkey, and Persia, have been tried, renewing the seed once in two or three years. The tobacco of Russia is mild, and of inferior flavor, and brings from 40 to 80 kopecks per pood. A very good quality of tobacco is grown in the trans-Caucasian provinces; it also flourishes well in the Southern provinces.

The plants attain good size, but lack that fine flavor when cured that other tobaccos possess. A recent traveler through Russia, describing the tobacco, says:

"Russian tobacco is very mild and rather sweet flavored, though not equal in aroma to the Havana, or posessing that rich ripe taste so much prized in that well known tobacco."

COLOMBIA TOBACCO.

Colombia has long been celebrated for the quality and varieties of its tobacco. Its cultivation has been carried on for more than two hundred and fifty years, and Varinian tobacco had obtained a well established reputation in Europe long before Raleigh's "would-be-colonists" sailed for Virginia. The principal varieties grown are Colombian, Carmen, Ambalema, Palmyra, and Giron. Most of these tobaccos are used for cigar purposes, especially the latter. The leaf is fine, of good size, and marked with light yellow spots. Tanning says of the tobacco of Colombia:

"The Cumanacoa, Tobacco de la Cueva, de los Misones, de la Laguna de Valencia cura seca and Caraco, de la Lagunade Valencia cura negro, de Oriluca, de Varinos cura seca, de Casovare, de Baylodores, de Rio Negro en Andull, are equal to the tobacco of the Brazils. The tobacco of the Cueva, in the department of Cumana, is said to be grown from the excrements of certain birds deposited by them in a cavity, from which the natives extract it: it is considered the finest tobacco in Colombia. The birds are a species of the owl.

"The natives of Varinos, and in fact of the whole kingdom, chew a substance called chimo, which is made of a jelly, by boiling the Varinos tobacco, and afterwards mixed with an alkali called *hurado*, which is found in a lake near Merida. Both are an *estanco* of government, and produce a large annual income. The mode of cultivating the above tobacco by the natives is as follows:—They prepare a small bed, sifting the earth very fine, on which they sow the seed, and then cover it with plantain leaves for some days. As soon as the plants make their appearance, they raise the leaves about two feet, so as to give the plants free air, and to allow them sooner to grow strong. When they become large enough to transplant, they have the land prepared; and as soon as the rainy season sets in, they plant out their young plants, taking great care to protect them from the sun, and to keep them clean as they grow up, as well as to prevent the worms from destroying or eating the leaves. When the leaf is ripe, it gets yellow spots on it; and on bending the leaf it cracks. Then it is fit for pulling off, which is done, and the leaves are neatly packed in handsful, placed in a dry situation, and occasionally shifted from one place to another. When the leaves are well dried they are all packed closely, and well covered, to keep the flavor in.

"The leaf is left in this state for one or two months, and then made up for use. They never top their tobacco, and the leaves never ripen together. The mode adopted by the North American planters is somewhat different; they top their plants when they have eight full leaves, or they keep it suckered; and, by this means, the leaves are large and sappy.

"They cut off the stem at the ground, when ripe, and hang it on laths for one day and a night, with the leaves all hanging down; they then place it in their barns; and, when these are quite full, they smoke it for some days, and let it remain in that way until the stem, as well as the leaf, is quite dry; they then put it in a heap, and cover it up for market. They strip off the leaves, and pack them in hogsheads, and it is received in London."

SUMATRA TOBACCO.

Sumatra tobacco is one of the finest varieties cultivated, and commands in European markets the very highest prices. The plant is a vigorous grower, and produces large, fine leaves of most delicate odor. The leaf is of beautiful appearance, of almost a silky texture, and in color a rich brown. It is extensively used in the manufacture of cigars, and on the continent it frequently realizes as much as 5s. per pound for this purpose. It sells in London for from 3s. 6d. to 4s. per pound.

BRAZILIAN TOBACCO.

Brazil tobacco is grown chiefly in the valley of San Diego and San Francisco. The former being on the west side of the Brazilian mountains, and the latter on the east. The San Diego is the finest, and the following analysis of the San Diego of Brazil, and Vuelto de Abajo, will give one an idea of the soil of these famous tobacco lands:—

	Vuelta de Abajo, Cuba. Parts.	San Diego, Brazil. Parts.
Organic matter,	9.60	4.60
Silica,	86.40	90.60
Lime,		.40
Alumina,	.68	3.00
Oxide of Iron,	1.92	1.20
Loss by Evaporation,	1.40	.20
	100.00	100.00

The tobacco of Brazil is grown in the same manner as in other parts of South America. The planter raises two crops a year; curing for exportation as in Cuba or Venezuela. The plant grows to the height of about six feet, bearing leaves lanceolate in form, about thirty inches long, and from eight to twelve inches wide. The tobacco fields are very irregular. After it is cut it is placed on poles in the field, and afterwards carried to the drying sheds. It is gathered in the dry season in September. After curing, it is removed to the packing house and baled in packages, and then transported on mules to the coast for shipping. A large portion of the crop is shipped to Portugal. It is a dark maroon-colored leaf, and contains a large proportion of the nicotine oil. It is a high-flavored tobacco, and on this account is used for cigars and cutting.

Burton says of the tobacco of Brazil:

"The tobacco of the Rio de Pomba, especially the 'Fumo crespo,' is a dark strong leaf, well fitted for making 'Cavendish' or 'Honey-dew;' the weed flourishes throughout Minos Gerals. The soil will be much improved by compost; and the produce by being treated in Virginia style delicately dried in closed barns with fires."

VENEZUELAN TOBACCO.

The Orinoco tobacco grows from four to five feet high, bearing large ovate leaves, and is in all respects a fine quality of tobacco. The plant is grown during all seasons of the year. It is used chiefly for cigars, and is shipped to Northern Europe. It is packed in *carrottes*, and then baled. In color it is dark mahogany, and of good body and texture.

ORINOCO TOBACCO. (VENEZUELA)

The leaf is about eighteen inches long,

and about ten inches wide. The planters cure by air-drying in sheds, and afterwards it is tied up in hands and baled for export. For their own use, they have adopted the method of the Brazilians, sprinkling the leaf with water containing the juice of the poppy.

The flavor is rich and mellow; a little more oily than Havana leaf. It is used for the manufacture of cigars. Orinoco tobacco makes very fine flavored cigars, burning freely, and leaving a pearl-colored ash; it is considered by the Venezuelans to be much better than any variety grown in South America. In cultivating it the planters use no fertilizers whatever, taking up new land as the old wears out. The crop is gathered first in May, and then in September.

PERSIAN TOBACCO.

Shiraz tobacco is a native of Persia, and is one of the finest varieties for the pipe to be found in the East. The plant differs from most varieties in the color of the flowers and the form of the leaves. It is not adapted for cigars as it does not readily ignite, and this variety together with Manilla, are known as non-burning tobaccos. After curing, the color is a light yellow, the flavor mild and not unlike Latakia and Turkish tobacco. The color of the flowers like those of Guatemala tobacco, is white, but in other respects nearly similar to other kinds.

SHIRAZ TOBACCO, PERSIA.

AMERSFOORT TOBACCO.

This variety of tobacco is cultivated quite extensively in Holland, in the Veluwe (valley of Guelderland). The plant is of good size and averages 1.580 kilos to the hectare. The cultivation is very carefully conducted on the richest soil. The leaf is very fine and is free from large fibres, fitting it for cigars. Large quantities are also used in the manufacture of snuff. The tobacco plant has been cultivated in

Holland since its first introduction, with complete success, producing a variety for snuff unrivaled by any other tobacco grown in Europe.

In color Dutch tobacco is both dark and light; the former being used for snuff and the latter for cigars and cheroots.

ST. OMER TOBACCO.

Tobacco is an important product in France, and affords the government an immense revenue. In the north of France two varieties are cultivated, the Brazilian and the Mexican, but the tobacco is unlike that grown in those countries. Most of the tobacco of France is small and inferior to Havana and Manilla. In the South of France tobacco is cultivated to a considerable extent, but is of inferior quality, lacking the rich flavor of the tobacco of Cuba. The cultivation is permitted only in certain departments, and the cultivators must use only the seed supplied to them by the officers of the *regie*. This is selected with the greatest care, the kind and quantity depending upon the nature of the land, the soil being carefully analyzed, and cultivation prohibited in soils which do not possess the constituents necessary for the growth of good tobacco. These analyses also determine the quantities and sorts of manure required to bring the land into fit condition. Most of the seed used is the produce of seed imported at various times from North America and Cuba.

The cultivation is most carefully watched, and the statistics available concerning it are of the minutest kind. Not only is the area of each field of tobacco accurately measured, but each plant is noted down, and even each leaf on each plant is accounted for. St. Omer is used chiefly for snuff, sometimes used with other kinds and is much esteemed by the French who consider it among the best of tobaccos.

HUNGARIAN TOBACCO.

This variety is attracting considerable attention, from the fact that it is well adapted for the manufacture of cigars. Like Connecticut seed leaf, the leaves are large and well suited for cigar wrappers. A considerable portion is adapted for other uses, and it is in some respects a good cutting tobacco. When in fine condition, Hungarian leaf burns freely and leaves a clean, light-colored ash. No variety of tobacco grown in Europe is attracting more notice than this, and if good leaf tobacco suitable for cigars can be grown, American tobacco will diminish in proportion. Hungarian tobacco is a favorite with the Italians, and large quantities are sold to the Italian monopoly to be used both for cigars and cutting.

SPANISH TOBACCO.

For several years the growers of tobacco in the Connecticut valley have directed their attention towards the production of a tobacco possessing all of the excellencies of both wrapper and filler; in other words, if possible securing a leaf of light color and fine texture and good flavor, so as to combine all of the desirable features and qualities of tobacco in one variety. Some few years since the Department of Agriculture at Washington distributed a variety of tobacco seed among the Connecticut tobacco growers known by the name of Spanish tobacco.

SPANISH TOBACCO.

It has been tested by many of the largest tobacco growers in Connecticut, and found to be one of the best varieties of the plant ever cultivated in the valley. The plant grows to the height of eight feet, bearing leaves about two feet in length by one foot in width, is an erect, strong, growing tobacco with a small, hard stalk and stout, long roots. The plant, when growing, imparts a strong aromatic odor not unlike Havana tobacco, but is larger everyway, and of inferior flavor for cigars. By repeated trials its superiority has been demonstrated to a certainty, while the profit arising from its culture proves it worthy of attention from all cultivators of tobacco.

When cured the leaf is very fine and light of color, the stem and veins of the leaf are small, thus fitting it for a good wrapper as well as filler. If the tobacco growers in the Connecticut valley can succeed in raising this variety, they will produce a leaf tobacco much superior to the common variety known as seed leaf. Beyond all question a much finer flavored tobacco than Connecticut seed leaf can be grown, and still retain all of the excellencies of the latter, such as color, texture, and size of leaf.

TURKISH TOBACCO.

The tobacco of Turkey has been called by some enthusiastic smoker "the king of tobaccos," but whether it possesses this royal preëminence over all other varieties must be decided by other than ourselves. That it is a fine smoking tobacco, no one can doubt that ever "put breath" to the favored pipe that contains the yellow shreds, but we should prefer by far to part with it rather than with its great rival, Havana tobacco.

The plant is not as large as many varieties, but grows up strong and flourishes well on account of the care and attention given it by the Turk and his family, as it is in all respects a family plant, and the flower garden is generally the tobacco field. Turkey tobacco ranges in color from brown to light yellow, the latter being the most in demand. This variety is similar in flavor to Latakia and Shiraz, and these three tobaccos, Persian, Syrian, and Turkish, are considered the finest and best adapted of all tobaccos for the pipe. The work of cultivating a field of Turkish tobacco is very tedious, as large quantities of water have to be carried to sprinkle upon the plants. The finest colored, a pale yellow leaf, brings "inflated" prices, but more often by others than the poor Turk who grows it.

JAPAN TOBACCO.

Of the tobacco of Asia, the best known in Europe is the yellow leaf grown in Japan. In those provinces where a high degree of temperature prevails, the plant lives throughout the winter, but it is nevertheless customary to sow fresh seed in the early spring of each successive year. When fully grown, Japan tobacco attains an altitude of about six feet, bearing leaves long and pointed, completely enveloping the stalk. The leaves, however, differ in form in different provinces, some being round and wide, others narrow and pointed, and others thick and long.

JAPAN TOBACCO.

The mode of cultivating also varies in the different provinces. The sowing and transplanting are dependent on the temperature of the locality, and each place follows its own customs. In autumn a great number of flowers spring from the tip of the stalk. These are about an inch in length, and of a pale purple tint. To these succeed small round capsules, inside of which are three small chambers containing a great number of light red seeds. The method of cultivation is novel, the manuring of tobacco differing from that of other plants in that it is plentifully applied both to the roots and leaves.

GUATEMALA TOBACCO.

The tobacco of Central America, though possessing considerable excellence, has never become an important product, nor to any great extent an article of commerce. There are several varieties grown in Guatemala, Honduras, Nicaragua, and the other Central American states; some of which by proper cultivation might be valuable to both the user and the manufacturer. One variety bears white flowers like the tobacco of Persia, but in other respects it differs but little from South American varieties. Numerous other sorts occur, many of which are local, and differ principally, if not solely, in the size or form of the leaves.

The soil of Guatemala is well adapted for tobacco, and with careful cultivation it could hardly fail of becoming an important agricultural product. It is also probable that the soil of nearly all of Central America is adapted to the plant, and with the favorable climate, the varieties now grown would doubtless with proper care, become noted as tobacco well adapted for cigars.

MANILLA TOBACCO.

This variety is one of the most celebrated grown in the East.[*] It is used exclusively for the manufacture of cigars and cheroots, and supplies India and Spain with a vast quantity of the manufactured article. The plant is a strong, vigorous grower, bearing dark green leaves; coming forward rapidly under the careful culture bestowed upon the plants.[†] After curing, the leaves show a variety of colors ranging from dark brown to light yellow or straw color. The leaf when cured, has a peculiar appearance unlike that of any other tobacco. It is of good body but smooth, and has the appearance of tobacco that has been 'frost-bitten.' The leaf is not as porous as most other tobaccos, and therefore does not as readily ignite, and frequently 'chars' in burning—thus giving it the name of a non-burning tobacco.

The plants are 'set' wide apart, and during the first two

---

[*] Blanco thus describes the tobacco of the Philippines: "It is an annual, growing to the height of a fathom, and furnishes the tobacco for the *estancos* (licensed shops). General opinion prefers the tobacco of Gapan, but that of the Pasy districts, Laglag and Lambunao, in Iloilo, of Maasin or Leyte, is appreciated for its fine aroma; also that of Cagayan, after being kept for some years,—for this use like the tobacco of the island of Negros it burns the mouth."

[†] The seedlings are planted in January, and the greater part of the crop comes forward in May and June.

months are carefully cultivated, when the top is broken off and the leaves allowed to ripen. In some respects, Manilla tobacco is one of the best varieties of the plant cultivated, and were it not for its non-burning quality, it would have but few rivals among cigar tobaccos.* We have thus, at some length, described nearly half of the varieties of tobacco now being cultivated. There are, however, others as well known and of equal value and favor. Some of these are of superior quality and of world-wide repute. Of those described, the varieties grown in the tropics are the most celebrated and of the finest flavor. As when first discovered, the tobaccos of the tropics command the highest prices, and possess qualities not easily transmitted when grown in a temperate clime.

---

* "The soil of many of the islands especially of the Bisayas is favorable to the growth of tobacco. The island of Negros formerly produced some of very good quality."

CHAPTER XII.

TOBACCO HOUSES.

THE drying houses or sheds for the curing and storing of tobacco are among the most interesting objects to be seen on the tobacco plantation. These sheds vary in size from a small structure capable of holding only a few thousand plants to the immense sheds with sufficient capacity for hanging the products of several acres. In the Connecticut valley, the Southern States, at the West, and in the Philippine Islands these tobacco sheds are often several hundred feet in length, built in the most substantial manner and provided with suitable side doors and ventilators for the free passage of air, and the most perfect system of ventilation. The most substantial and finest tobacco sheds are to be found in the Connecticut valley, which are provided with every convenience for hanging and taking down or "striking" the crop. Many of them are painted and adorned with a cupola, which serves the double purpose of an ornament and a ventilator for the hot air to pass off from the curing and heated plants. Formerly, the tobacco being harvested was hung in barns and sheds, used for storing grain and hay, and better adapted to other purposes than to that of a tobacco shed, where thorough ventilation is necessary to avoid sweat and pole-rot, attending upon the curing of the plants. Of late, tobacco growers, throughout the world, have paid considerable attention to the method of curing, and to erecting more suitable buildings for the purpose. At the South and West, the log tobacco barns are giving way to the more substantial frame buildings, and better facilities are employed for "firing" the tobacco in the sheds. Formerly, the tobacco sheds at the South looked more like the rude huts of the herders on the pampas of South America, than buildings devoted to the curing of tobacco. Tobacco barns

OLD CONNECTICUT TOBACCO SHED.

and sheds are built of a great variety of material, and in various ways, according to the manner of building where the tobacco is grown. Thus in the Connecticut valley, such sheds or barns are large and commodious frame buildings; at the South and West, many of them are built of logs; in Cuba, of slabs covered with palm leaves or thatched. In Turkey, of stones covered with rough boards, and daubed with mud.

In selecting a site for the tobacco shed, not only should its proximity to the tobacco field be considered, but also the ground on which it is to be built. It should always be erected on dry ground, rather than upon moist, so that no dampness may arise and injure the leaves in curing. The tobacco shed should also be built on an elevated spot, so that a free circulation of air may be had, which is hardly possible if built on low ground or among trees or in the woods as at the South. This applies more particularly to sheds where the method of curing is by air-drying instead of by "firing" or by "flues." In New England the strongest timber, as oak, is used for building, as the weight of the plants before fully cured is immense. The shed is braced at every point and generally rests upon stone posts so as to allow a good circulation underneath the building. Poles are used for hanging, either round or sawed, when the plants are hung with twine; when hung on tobacco hooks, laths are used, the hooks attached to the lath; more frequently the plants are strung upon the laths without the aid of hooks, the lath

MODERN CONNECTICUT TOBACCO SHED.

passing through the center of the stalk an inch or two from the end. The doors lengthwise of the building are simply the outside boards hung on hinges, every second or third being chosen according to the ideas of curing entertained by the grower. Some planters are of the opinion that the plants need all the air that can be obtained, and keep the sheds open during both day and night, while others open the doors only now and then—closing during warm days, and during a storm. Sometimes the doors are hung on hinges at the top—opening but partially and not allowing as free circulation as when hung on the sides.

Another building of late has been built by the growers in the Connecticut valley, called a stripping house. This building is frequently attached to the shed or near by so that stripping may be performed during all kinds of weather, without danger of injuring the tobacco, or the health of the stripper. Such buildings however are needed only in tobacco sections where the cold is extreme during the winter, when most of the tobacco is to be stripped. The stripping room or house is provided with a stove, a long table, or elevated platform, in front of the windows, of which there should

STRIPPING ROOM.

be several to admit plenty of light, and a number of chairs to accommodate the strippers. On the stove a kettle of water is kept constantly boiling or heated, the ascending steam of which keeps the leaves of tobacco from drying and consequently from cracking or breaking. When in condition for "striking" or taking down, the plants are carried to the stripping-room, and covered with boards and blankets, when the operation called stripping commences. Many of the stripping-rooms are built large enough to contain the cases after the tobacco is packed, thus answering a double purpose.

In Virginia and the other tobacco-growing states of the South, the tobacco barn is built altogether different, as the method of curing is by fires or flues instead of air curing. The height of the building is usually twice its width and length. In the center of the smooth earthen floor, is the trench for "firing," while around the sides of the building runs an elevated platform for placing the tobacco leaves in bulk; and, commencing at a safe distance from the trench, up to the top of the building, reach beams stretching across for the reception of the pine laths, from which are suspended the tobacco plants. Many of the tobacco sheds at the South, are built like those of New England, but many log structures are still to be seen and many planters prefer them to those made like other frame buildings. The old Virginia planters of a hundred years ago, built rough log sheds for housing the plants, which afforded little protection from wind and rain, which, in consequence, injured much of the tobacco hanging around the sides of the building. Tatham gives the following description of the "Tobacco house and its variety" in his work on tobacco.

MODERN VIRGINIA SHED

"The barn which is appropriated to the use of receiving and curing this crop, is not, in the manner of other barns, connected with the farm yard, so that the whole occupation may be rendered snug and compact, and occasion little waste of time by inconsiderate and useless locomotion; but it is constructed to suit the particular occasion in point of size, and is generally erected in, or by the side of, each respective piece of tobacco ground; or sometimes in the woods, upon some hill or particular site which may be convenient to more than one field of tobacco. The sizes which are most generally built where this kind of culture prevails, are what are called forty-feet, and sixty-feet tobacco houses; that is, of these lengths respectively, and of a proportionate width; and the plate of the wall, or part which supports the eaves of the roof, is generally elevated from the groundsel about the pitch of twelve feet. About twelve feet pitch is indeed a good height for the larger crops; because this will allow four pitch each to three successive tiers of tobacco, besides those which are hung in the roof; and this distance admits a free circulation of air, and is a good space apart for the process of curing the plant. There are various methods in use in respect to the construction of tobacco houses, and various materials of which they are constructed; but such are generally found upon the premises as suffice for the occasion. And although these sizes are most prevalent, yet tobacco houses are in many instances built larger or smaller according to the circumstances of the proprietor, or the size of the spot of ground under cultivation.

"The most ordinary kinds consist of two square pens built out of logs of six or eight inches thick, and from sixteen to twenty feet long. Out of this material the two pens are formed by notching the logs near their extremities with an axe; so that they are alternately fitted one upon another, until they rise to a competent height; taking care to fit joists in at the respective tiers of four feet space, so that scaffolds may be formed by them similar to those heretofore

VIRGINIA SHED 150 YEARS AGO.

described to have been erected in the open field, for the purpose of hanging the sticks of tobacco upon, that they may be open to a free circulation of air during this stage of the process. These pens are placed on a line with each other, at the opposite extremes of an oblong square, formed of such a length as to admit of a space between the two pens wide enough for the reception of a cart or wagon. This space, together with the two pens, is covered over with one and the same roof, the frame of which is formed in the same way as the walls by notching the logs aforesaid, and narrowing up the gable ends to a point at the upper extremity of the house, termed the ridge pole. The remaining part of the fabric consists of a rough cover of thin slabs of wood, split first with a mall and wedges, and afterwards riven with an instrument or tool termed a froe. The only thing which then remains to be done, is to cut a door into each of the pens, which is done by putting blocks or wedges in betwixt the logs which are to be cut out, and securing the jambs with side pieces pinned on with an auger and wooden pins. The roof is secured by weighing it down with logs; so that neither hammer, nails, brick, or stone, is concerned in the structure; and locks and keys are very rarely deemed necessary.

"The second kind of tobacco houses differ somewhat from these, with a view to longer duration. The logs are to this end more choicely selected. The foundation consists of four well hewn groundsels, of about eight by ten inches, leveled and laid upon cross sawed blocks of a larger tree, or upon large stones. The corners are truly measured, and squared diamond-wise, by which means they are more nicely notched in upon each other; the roof is fitted with rafters, footed upon wall plates, and covered with clap-boards nailed upon the rafters in the manner of slating. In all other respects this is the same with the last mentioned method; and both are left open for the passage of the air between the logs.

"The third kind is laid upon a foundation similar to the second; but instead of logs, the walls are composed of posts and studs, tenoned into the sells, and braced; the top of these are mounted with a wall-plate and joists; upon these come the rafters; and the whole is covered with clap-boards and nails, so as to form one uninterrupted oblong square, with doors, etc., termed, as heretofore, a forty, sixty, or one hundred feet tobacco house, etc.

"The fourth species of these differs from the third only in the covering, which is generally of good sawed feather-edged plank; in the roof, which is now composed of shingles; and in the doors and finishing, which consist of good sawed plank, hinged, &c. Sometimes this kind are underpinned with a brick or stone wall beneath the groundsels; but they have no floors or windows, except a plank or two along the sides to raise upon hinges for sake of air, and occasional light: indeed, if these were constructed with sides similar to the brewery tops in London, I think it would be found advantageous. In respect to the inside framing of a tobacco house, one description may serve for every kind: they are so contrived as to admit poles in the nature of a scaffold through every part of them, ranging four feet from centre to centre, which is the length of the tobacco stick, as heretofore described; and the lower ties should be so contrived as to remove away occasionally, in order to pursue other employments at different stages in the process of curing the crop."

In Ohio, the tobacco barns are built in a manner similar to

those in Virginia; constructed of logs and provided with trenches for fires in curing the tobacco. The tobacco sheds for hanging the tobacco cured by air-drying, are built of the same material without trenches, as smoke is not employed in curing "seed-leaf" tobacco. The sheds for both kinds of curing tobacco are large structures, varying in size according to the area of tobacco planted. Sometimes the sheds are built near the woods where fuel can be procured, and in the immediate vicinity of the tobacco field. The tobacco houses are built in the strongest manner and of the most durable material, and are well fitted for the purpose designed. In the counties bordering the Ohio River, where a large quantity of tobacco is raised, the tobacco sheds are to be seen on every hand, the smoke issuing from the sides of the building, giving a stranger the idea of a burning building rather than the curing of a great staple.

OHIO TOBACCO SHED.

The following account of constructing tobacco barns in Missouri, is from a St. Louis paper:

"We believe in small barns for any kind of curing. A house built 16 feet inside and divided into four rooms and six tier high in the body is the preferable size for flue or coal curing. For flues they should be built on a very slightly sloping place; just enough to make the flues draw well. Flues four inches lower at the eye than the chimney will be slope enough. The door should always be between the flues and in the end of the house, to prevent the drip from falling before the door and the eye of the flues. The tiers should begin eight feet above the ground and be placed two feet above each other to the top. They should be placed across the house so that the roof tier can conveniently be placed above them. The door, three feet wide and six feet high, furnished with a good, close shutter. A barn of this size will cure 800 sticks of common size tobacco, which will weigh about 1200 lbs. The proper construction of flues is of great importance; they should be built of any stone that will stand fire without bursting. White sand-stone, bastard soap-stone, or any other that does not contain flint. The size of a flue for a sixteen foot barn, is generally about 12 inches wide by 14 inches high inside. Not much care need be taken to have them smooth on the outside. If stone can be had to make the inside smooth so as not to obstruct the putting on of wood, it is all that is necessary. They should be run just far enough from the house-side not to set the house on fire, and there is not as much danger of this as may be supposed. Run the walls of the house-side all around, running the stem out at the middle of the upper side. The stem should be run far enough above the wall of the house to avoid danger of sparks from the chimney. The height of the inside of the flue should be preserved its whole length. The width may be slightly decreased from the elbow to the chimney. The inner wall is carried all around. But too much explanation bewilders; we think we have said enough. As before said, we like small barns; where too much tobacco is together, it all can not receive the heat alike, which is our main objection to large barns. As to the number of barns necessary, we would say that there ought to be enough to receive all the crop without moving any. Say one sixteen-foot barn to every 8,000 hills of tobacco planted. As a general rule, plant one thousand hills for every hundred sticks house-room. That is, if you have three barns plant 24,000 hills, and if it is common tobacco, they will receive it. A much larger quantity may be saved in this number of barns by curing and moving out, but it is very troublesome."

In Kentucky and Tennessee the tobacco barns resemble those of Ohio and the other Western states, and are large, commodious structures, provided with every facility for curing the plants. In other tobacco-growing countries the tobacco barns and sheds differ but little from those in America, the only difference being in form and building material. In countries where tobacco is a government monopoly, large and comfortable buildings are provided for the crop with all the necessary accessories for the curing, packing, and storing of the tobacco. In South America many of the sheds are large and low, built on the plantation, and close to the tobacco field. In Cuba, the curing houses are located on the *vegas*, and as soon as the tobacco is cut it is placed on the poles to dry or cure. In Asia, a large quantity of the tobacco is cured in the peasants' huts, where the smoke is said to impart additional flavor to the already fragrant leaves. In the Philippines the largest tobacco sheds are found, described by Gironiere as "vast sheds," and of sufficient capacity to hold acres of the leaves. In Persia, where the celebrated Shiraz tobacco is grown, the sheds are simply covered buildings without any boards on the sides, the only protection afforded from the weather being supplied by light, thorny bushes, so that the plants may be exposed to the wind. After fully curing, the tobacco is removed to another drying-house and turned every day. The drying-houses in other tobacco-growing countries differ but little from those described, while the manner of curing is similar, the plants being "fired," sun-cured, or air-dried—the three modes now employed in drying the leaves. If the tobacco of the tropics is fragrant while growing, it is doubly so after being harvested and carried to the sheds.

PERSIAN TOBACCO SHED.

The odor from the well-filled barns is borne on the breeze alike to friend and foe of the plant. As the process of drying goes on, the plants gradually lose the strong perfume emitted during the earlier stages of curing, and by the time the leaves are "cured down" and the sheds closed, but little odor issues from the plants, and this continues to be the case until the leaves are entirely dried.

CHAPTER XIII.

TOBACCO CULTURE.

TOBACCO at the present time is one of the great products of the world. As an article of agriculture and of commerce, it holds an important place among the great staples, while as a luxury, its use has become as extensive as its culture. The tobacco plant is now cultivated in nearly all parts of the world with varying success, according to the system of cultivation adopted by its growers. Primarily cultivated by the aborigines of America in the rude manner common to uncivilized races, the plant has, by numerous experiments and careful culture, become one of the greatest of agricultural products. When first discovered by the Spanish and Portuguese, the plant was small, and in flavor "poor and weak and of a byting taste." As soon, however, as the Spaniards began its cultivation in the islands of St. Domingo and Trinidad, attention was paid to developing it, and in a few years the description we find of the latter variety is that it is "large, sharp, and growing two or three yards from the ground."

At the close of the sixteenth century the Portuguese began its cultivation in Portugal, the soil of which seemed well adapted to the plant, and still further increased the size and quality of the leaf. Tobacco is now cultivated through a wider range of temperature than any other tropical plant, and whether grown amid the sands of Arabia, the plains of South America, or in the rich valley of the Connecticut, develops its finest form and perfection of leaf. During the last half-century the plant has been developed to a greater extent than during the three hundred years succeeding its discovery. Now its cultivation has been reduced to almost an exact science, and the quality of the leaf is in a great measure within the control of the growers of the plant.

Formerly it was supposed that the varieties that grew in the tropics could not be successfully cultivated in the temperate regions, but recent and repeated experiments have demonstrated the fact that the tobacco of Cuba can be grown with success in the Connecticut valley. While the tobacco of the tropics is the finest in flavor, the more temperate regions produce the finest and best colored leaf. The tobacco of the tropics, as to the uses to which it is put, is limited, while the tobacco of the more temperate regions can be used for all the purposes for which the plant is designed.

The cultivation of the plant varies with the variety, the soil, and the use to be made of the leaf. Thus a tobacco designed for cutting purposes is cultivated somewhat differently from that designed for the manufacture of snuff or cigars. In the one case the plant is allowed to remain growing longer in the field, while in the other the work of topping the plants is performed at an earlier stage of their growth. Primarily but little attention was paid to the color and texture of the leaf, the principal object being the production of a leaf of large size, rather than one of good color, and of a silky texture. Now, however, these are most important considerations, and give value to the tobacco in proportion to the perfection of these qualities.

The soil, too, is carefully chosen and fitted in the most thorough manner, while the fertilizers used are selected with reference to the color of leaf desired. When first cultivated in the United States it was thought that tobacco designed for various uses could not be grown in the same state or section; now, however, tobacco for cigars and for cutting are grown nearly side by side. But in the fineness of the leaf, tobacco culture has made its greatest stride. By a careful selection of soil, and by the judicious application of proper fertilizers, the leaf tobaccos of Connecticut, Cuba, and

Virginia, resemble in texture the finest satins and silks. This result has been reached, not by the sacrifice of the strength of the leaf, but by the most careful culture and improved methods of curing.

The first labor to be performed in connection with the growth of a crop of tobacco, is the selection of a site for, and the making of, the "plant bed" or "plant patch." These beds for the early growth of the plants until large enough to transplant, are made in various ways and at different times, according to the method of tilling adopted and the climate. In California the tobacco bed is made in January, in the Southern States, Syria, Turkey, and Holland, in March. In New England in April. In Mexico and Java in June, and in Persia in December. In the Connecticut valley the manner of making the

PLANT BED,

as given by a Massachusetts tobacco-grower, is as follows:—
"No rigid rules can be given for any process in tobacco culture, which depends much upon weather and season, but certain advantages may be obtained by skillful adaptation of general principles to circumstances. This is especially true of raising tobacco plants, which occupy an extremely slight depth of ground for weeks after sowing, making it necessary to prepare the whole soil with reference to the state of this thin surface. Any slight mistake of treatment may make in the end a difference of several days; consequently each item is of importance. While every tobacco-raiser wants early plants, and appreciates the value of a good location for growing them, many naturally sheltered spots of ground, protected from northerly winds by buildings, trees, or hills, remain unappreciated. Tight board fences are no protection worth mentioning.

"A heavily manured crop of tobacco would fit such places for tobacco beds, and leave them freer from weeds than any other cultivation; and a subsequent use of some commercial fertilizer would avoid the introduction of weed seed. With these precautions, and a careful destruction of all neighboring weeds, a tolerably clean bed may be expected. To prepare the ground, plow or loosen deeply with a large cultivator;

harrow in two-thirds of the fertilizer to be used; rake the bed perfectly level, then rake in the other third; roll once, and another slight raking will fit the bed for sowing, after which it should be rolled down hard. If the soil is handled in drying weather, it should be done quickly, because damp ground, if prepared and rolled down before drying, will 'set' like mortar, and remain damp on the surface.

MAKING THE PLANT BED IN CONNECTICUT.

Moisture and darkness are essential to the germination of the seed, and these conditions can be secured only by making the surface compact while damp. The disintegration of the deeper lumps, and the decomposition of fertilizers, will cause the surface to grow gradually softer. The effect of plowing is to break the ground into lumps, which lie upon each other, giving free admission to the air between them. Harrowing makes finer the lumps near the surface, and mixes the fertilizer deeper than a rake can be used. The first raking is to pulverize and level, so that rains will neither collect in ponds, nor run off, but penetrate the soil evenly. The second raking is to mix the fertilizer equally through the soil, to the depth of an inch or less, and reduce the lumps to the size of peas, which is as fine as a medium loam can be made without danger of a tough crust. Too much working destroys the healthy grain of the soil, and reduces it to a paste, which the roots of the tobacco plants can penetrate but slowly.

"The bed should not be watered before nor after the plants come up. The ground will be cold enough without any extra evaporation, and if the place is suitable for tobacco plants, and rightly fitted, the surface will be damp in the morning, even in very dry weather. If the plants need stimulating, sow on them a coat of Peruvian guano or superphosphate at the commencement of a rain, regulating the quantity used by the amount of the water likely to fall. Superphosphate makes dark-colored, thick-leaved, stocky plants. Fish guano has about the same effect, but gives a lighter color and thinner leaf. Peruvian guano is more stimulating than either, and makes a light-colored, thin leaf. Great caution is necessary in the use of these powerful medicines to avoid an over-dose. A quantity that would be safe in a heavy rain, would in a light rain kill many or nearly all the plants.

"Old seed will sprout sooner than new. The seed should be measured while dry, and the same spoon used every year, so the effect of a given amount may be noted and the quantity regulated by experience. Level the seed in the spoon with a knife-blade, like measuring grain in a half-bushel. After sprouting again, allowing for the seed, increase in bulk for each rod separately. The amount of seed needed to the square rod varies with different seasons, soils, and seeds, but can be loosely a tablespoonful. There are many breeds of tablespoons. Too thick sowing will nearly spoil a bed by causing it to produce weak, yellow, spindling plants, while thin sowing will give good square ones. A bed should appear about half stocked till the plants are nearly ready to set, when they will suddenly spread and seem to multiply.

"Some growers sprout and some prefer dry seed. In favorable circumstances sprouting will give a gain of four to six days, but in many cases dry seed will be fully as early. A long sprout is liable to be broken off in sowing, or killed by cold, after it is in the ground. A sprout just showing will endure several nights' freezing if there is some warm sun in the day-time. One way to sprout is to spread the seed thinly on cotton cloth, and roll it up inside of woolen cloth, keep it in a warm place, and dip in warm water every day. In about four days the white spots will show. Sprouted no more than this, it will stand unfavorable weather as well as dry seed. A pint of meal and a pint of plaster to each rod, is a good mixture to sow in. Pouring from one dish to another many times will mix the plaster, meal, and seed perfectly if dry. If sprouted, it should be rubbed through the hands a few times with the mixture, to dry it and prevent any bunches of plants coming from seed stuck together. The plaster will show on the ground whether the sowing is being done evenly.

"Weeding should of course be done early and thoroughly. Weeds are stronger than the plants, and a little neglect will check them, making practically, perhaps, a difference of several days. A good way to prepare for weeding and taking up plants, is to make the bed about fifteen feet wide, and place round, straight poles across it about eleven feet apart. The poles should be three inches in diameter at the smallest end. They cost nothing and save moving blocks around with the weeding planks."

If the plants are tardy of growth, or the season is backward, wooden frames covered with cloth soaked in linseed oil may be placed over the beds, which is far better than to cover with pine boughs or glass even. The cloth soaked in oil draws the rays of the sun and keeps the earth dry and warm, causing a rapid growth of the plants, which at this stage need forcing in order to be forward enough for early transplanting. A Virginia planter gives the following description of making the

PLANT PATCH.

"Cut wood in September or October, so that it may season, to burn patches (beds) in winter or spring. For ten acres, or fifty thousand hills, burn and sow three patches each of seventy-five square yards. Say one (if the land be in good condition) the latter part of December, and if it be not in condition then, burn one hundred and fifty square yards the first good weather in January or February, and the other the first of March. Select a place on some small constant running stream, not liable to overflow, with a moist, sandy soil; cut down all trees close to the ground; get off all shrubbery, leaves, etc. The patch will then be ready for wooding. Commence by laying on skids ten or twelve feet long, four in diameter, three and a half feet apart; cover thickly with brush, then put on wood regular all over, and thick enough to burn dry an inch in depth. Commence your fires on the side, and continue to move after it has burnt hard enough. After it has burned, sweep off all coals, but not the ashes: then it will be ready for hoeing up, which can be done with good grub hoes; hoe deep, but do not turn over the soil; get off all large and small roots; chop over with hill hoes, and rake until the earth is thoroughly pulverized; then put on twenty-five bushels of good, fine, stable manure, without weed and grass seed, and twenty-five pounds of Peruvian guano, which should be put on regularly, hoed and raked in.

"For sowing, lay off beds four feet wide, so that the water from rains may run or drain off. For every bed four feet wide and twelve yards long, sow one chalk pipe bowl full of seed, after being mixed with ashes; tread with the feet or pat it over with weeding hoes, that it may be close and smooth; cover it with dog-wood, maple, or any fine brush, to the depth of twenty or twenty-four inches, to protect the young plants from cold or a drouth. After the plants have commenced coming up, re-sow the patches with half the quantity of seed first sown, which will not interfere with the plants first up, but make good re-planting plants. When the plants, or some of them, have grown to the size of a Spanish mill dollar, take off the brush, pick off all sticks, weeds, and grass, and keep them well picked until you have finished setting out.

"Should the plants not grow fast enough to suit, manure with Peruvian guano; have it fine, and sow over in the middle of the day when they are dry, or if it be raining briskly, it may then be sown over. Should the patches be suffering for rain, put five pounds of Peruvian guano in twenty gallons of water, and sprinkle it over with a watering-pot. To destroy the flea, bug, or fly, put dry leaves around the patch, and set fire to them at night, which will attract and destroy them if they are disturbed with a broom or leafy brush."

The old Virginia planters selected and made the plant patch as follows:—

"The quality of earth, and places which are universally chosen for this purpose, are newly cleared lands of the best possible light black soil, situated as near to a small stream of water as they can be conveniently found, due attention being paid to the dryness of the place.

"The beds, or patches, as they are called, differ in size, from the bigness of a small salad bed to a quarter of an acre, according to the magnitude of the crop proposed; and they are prepared for receiving the seed in March and the early part of April, as the season suits, first by burning upon them large heaps of brush wood, the stalks of the maize or Indian corn, straw, or other rubbish; and afterwards, by digging and raking them in the same manner of preparing ground for lettuce seed; which is generally sown mixed with the tobacco seed (the same process being suitable to both plants); and which answers the double purpose of feeding the laborer, and of protecting the young tobacco plant from the fly; for which intent a border of mustard seed round the plant patch is found to be an effectual remedy, as the fly prefers mustard, especially white mustard, to any other young plant; and will continue to feed upon that until the tobacco plant waxes strong, and becomes mature enough for transplantation."

A Tennessee planter gives the following description of making the plant bed as practised in his State. In some respects, especially in preventing the growth of weeds, it is the best process of making the "plant patch" that we have ever seen described. He says:—

"To make a good plant bed it requires good management and pretty hard work. It will hardly be done well without the presence of the farmer to attend to it. The time to make a bed is from the 15th of October to the first of April. The best beds are made in the Fall, for the reason that the ground is then very dry and therefore more easily burned, and besides there is more time for the ashes to rot before the hot weather. A bed turned in the Fall will hold moisture better than burned later. It takes less wood to burn well. The plants are more vigorous and tougher. The soil should be rich and light and never tilled before. The location should be as much exposed to the sun as possible. It is best for a bed to be surrounded by timber. The bugs are not so apt to find it. Low rich valleys will generally do better than ridges, though any good rich new ground will make good plants if well burned and prepared. The ground should be raked very clean of leaves before packing on the brush and wood. The fire must have a fair chance at the ground. The brush should be packed on straight and close, at least enough wood mixed with it to make it lie close. If the brush is green, endeavor to mix what dry stuff there is thorough, so the fire will burn through without trouble. It is very important that the fire should be as hot as possible while it is burning. The bed should not be rained upon after it is set before it is burned, as it will be doubtful whether the ground beneath the brush will get dry well.

"The ground should always be as dry as possible when it is burned. The bed should be set on fire in several places at once so as to have a very great heat on it at once. If the ground is well burned it will be a little crusty and whitish, and will pulverize beautifully. As soon as the ground is cool enough it may be loosened up and pulverized. This should be done well, and may be done with a good sharp harrow and then followed with hoes and grubbing hoes. Aim to keep the ashes and rich soil on the surface, and for this reason a bed is sometimes damaged by a too deep working. Rake carefully, getting off all the roots and trash. The bed should be drained by a little ditch around it on the upper side. If it is very early in the Fall, the seed should not be sown until the danger of very warm days has passed. After the last of November the sowing should be as soon as the bed is prepared. A little less than a heaping tablespoonful to ten steps square is about the quantity of seed. Cover the seed very lightly with the rake or tramping the ground with the feet. Cover the bed with a good layer of straight brush, not enough to keep the light rains from the bed, but at the same time enough to keep the ground in a moist condition even in hot weather. Make a low close brush fence around the bed to keep the leaves from being blown upon it. Re-sow whenever the plants are well up, so as to have two chances. Take off the brush cover when the plants are big enough to shade the ground themselves. If the plants are rather thin on the bed, do not uncover until you go there to draw the plants. If there is any danger of a scarcity of plants, always put the trash back after drawing."

In Cuba the

"SEMILLEROS"

or planting beds as a rule, lie higher than the rest of the farm. On the large *vegas* or tobacco plantations, numbers of planting beds are made under the supervision of the mayoral. Siecke gives the following account of making the beds or *semilleros*:

"On the island of Cuba any field selected for the cultivation of tobacco is divided into long beds (*Canteras*) twenty-five to twenty-eight feet long, and nineteen to twenty inches wide. The soil is then manured with a mixture of two parts of well rotten dung and one part of either sand or fine sandy earth. During the months of August, September, and even October, the beds are watered, and the seeds mingled with the nine-fold quantity of fine sand, are sown broad cast or through a fine sieve, and immediately after covered with a mixture of dung and triturated or molaxated earth, in such a manner that this mixture forms a covering layer of about 1-32 inches.

"The utmost care is taken to protect the seeds against the stifling heat of sunrays as well as heavy showers. To this end forked sticks about three inches high, are placed around the tobacco beds, opposite one another, and into

COVERING PLANT BED.

these forks thin twigs are laid, which are covered with palm-leaves in such a way as to form a slight roof."

In Syria the tobacco seed is sown in ground free from stones, well manured with goats dung, and strewn over with prickly bushes to protect the young plants from birds. The plants are watered daily till they reach the height of eight or ten inches, when they are transplanted. In Persia where the celebrated Shiraz tobacco is cultivated, the seed is planted in a dark soil slightly manured; the ground is covered with light thorny bushes to keep it warm, and these are removed when the plants are a few inches high. The ground is regularly watered if required, and when the plants are six to eight inches high are transplanted. In Turkey "the tobacco seed is sown early in the spring, in small beds carefully prepared for the early growth of the young plants. In a few weeks the plants appear thick; then begins the occupation of the farmer's wife, and their numerous children, whose little fingers are engaged day by day in thinning the beds, care being taken to leave the most healthy looking plants. The husband is engaged either in carrying water from the nearest well by the aid of his mule, or in preparing the land for the reception of the plants. The beds are well watered before sunrise and after sundown."

"The Hungarian peasantry always make their tobacco beds against the south ends of their houses. These beds are enclosed by hurdles two feet high, at the bottom of which stones are laid, and on the outside of these, thorns are thickly placed, to exclude the moles. They fill this enclosure to the height of eighteen inches with fresh, coarse manure, which they press closely by beating as they throw it on; covering with finely pulverized earth mixed with dung of the preceding year that had become soil. They do not regulate their time of sowing either by the moon, month, the season, but by the holy week of the passing year; it is on Good Friday that all of their beds are sown, and although this day may vary nearly one month in different years, they are faithful to their thermometer—their piety not permitting them to know any other. To the mysterious influence of the day, without regard to the season, they ascribe their success and they generally succeed." Bickinson gives an account of the manner of making the plant bed in the East Indian Archipelago. He says: "Not far from us is a hut inhabited by two natives, who are engaged in cultivating tobacco. Their *ladangs*, or gardens, are merely places of an acre or less, where the thick forest has been partially destroyed by fire, and the seed is sown in the regular spaces between the stumps."

After making the plant bed and tending through the weeding season, the next step to be taken is the

CHOICE OF GROUND

for the tobacco fields. Tobacco, unlike any other plant, readily adapts itself to soil and climate. The effect produced upon the plant may be seen in comparing the tobacco of Holland and France, the one raised upon low, damp ground, the other on a sandy loam. The early growers of the plant in Virginia, were very particular in the selection of soil for the plant. The lands which they found best adapted were the light red, or chocolate-colored mountain lands, the light black mountain soil in the coves of the mountains, and the richest low grounds.

Tatham says: "The condition of soil of which the planters make choice, is that in which nature presents it when it is first disrobed of the woods with which it is naturally clothed throughout every part of the country; hence in the parts where this culture prevails, this is termed new ground, which may be there considered as synonymous with tobacco ground. Thus the planter is continually cutting down new ground, and every successive spring presents an additional field, or opening of tobacco (for it is not necessary to put much fence round that kind of crop); and to procure this new ground you will observe him clearing the woods from the sides of the steepest hills, which afford a suitable soil; for a Virginian never thinks of reinstating or manuring his land with economy until he can find no more new land to exhaust, or wear out, as he calls it; and, besides, the tobacco which is produced from manured or cow-penned land, is only considered, in ordinary, to be a crop of second quality. It will hence be perceived, (and more particularly when it is known that the earth must be continually worked to make a good crop of tobacco, without even regarding the heat of the sun, or the torrent of sudden showers,) that, however lucrative this kind of culture may be in respect to the intermediate profits, there is a considerable drawback in the waste of soil." *

In the Connecticut valley where tobacco is grown for wrapping purposes, the selection of soil will depend upon the color of leaf in demand (as the soil as well as the fertilizers determine in a measure the color and texture of the tobacco). If the grower wishes to obtain dark colored tobacco then the soil selected should be a dark loam; on the other hand, if a light colored wrapper is desired he selects a light loam, and with the application of proper fertilizers the proper color will be obtained.

The tobacco plant flourishes well either on high or low ground, providing the soil be dry and free from stones, which are a source of annoyance during the cultivation of the plants and especially in harvesting. When grown on very low ground the plants should be "set" early, so as to harvest before early frosts. The plant may be cultivated on such soil in almost any part of the valley excepting only near the sound, or other body of salt water, the effect produced by planting tobacco too near the sea, more especially in Connecticut, being injurious to the leaf, which is apt to be thick and unfit

---

* Liancourt in his Travels in North America, says of tobacco culture in Virginia: "The nature of the country beyond the James River is much more variegated than on this side. At present they are preparing the lands for the planting of tobacco. After having worked the land it is thrown into small hillocks. * * The cultivation of tobacco, which has been very much neglected during several years, is more followed this year on account of the high price it bears in Europe; but the soil has been so long worked with this exhausting produce, and is so badly manured (for manure is absolutely necessary for tobacco when the soil is not newly broken up), that it is not capable of producing good crops."

for a cigar wrapper. In some countries, however, the leaf grown near salt water is equal in color and texture to any grown in the interior. But generally the plant obtains its finest form and quality of leaf—whether in the islands of the ocean, on the great prairies of the west, amid the sands of Arabia, on the mountains of Syria, or along the dykes of Holland—on lands bordering the largest rivers. This is true of the tobacco lands of Connecticut, Kentucky, Virginia, Florida, Brazil, Venezuela, and Paraguay, as well as of those in the islands of Cuba and St. Domingo, where the rivers flow to the southern coast from the mountains which lie to the north. It must not be imagined from this that tobacco can not be successfully cultivated at a distance from valleys enriched by large and overflowing rivers. Some of the finest tobacco grown in Connecticut is grown in counties some distance from the river that gives name to our state.

When possible, select that kind of soil for the tobacco field that will produce the color and texture of leaf desired. For Connecticut seed leaf a light moist loam is the proper soil. The same field can be used a number of seasons in succession; the result will be a much finer leaf than will come from selecting a new field each year. The early planters of tobacco in Virginia soon ruined their fields by failing to manure them. In Maryland the soil best adapted for the growth of tobacco is a light, friable soil, or what is commonly called a sandy loam, not too flat, but of a rolling, undulating surface, and not liable to overflow in excessive rains. New land is far better than old.

A Missouri tobacco grower gives the following account of the selection of soil for tobacco in that State:—

"Select upland, or black oak ridges and slopes, which comprise a large area of the tobacco lands of our county, and carefully clear off all the timber, and take out all the roots we can conveniently, and break up the ground as thoroughly as can be done by ploughing and harrowing until all the tufts and dirt are perfectly pulverized."

In Cuba the planters select the red soil as the best for fine tobacco. Some planters, however, prefer a soil mixed of $\frac{1}{4}$

sand and ⅛ to ¼ of decayed vegetable matter. In St. Domingo the soil is not uniform. The planters select a deep black loam or tenacious clay, or even loams mixed with sand. The most fertile places are on the banks of the Yuna, from Laxay to Jaigua, in the vicinity of Mocha, on the banks of the Camoo, and around La Vega. Around Santiago, clay and sand predominate, and the soil can not be highly praised. Most of the tobacco grown in the island is raised in the valley of the Vega.

Cnasree, in treating of this subject, says:—

"The quality of tobacco depends as much upon the nature of the soil as of the climate. The plant requires peculiarities of soil to develop certain of its qualities. And these peculiarities are such that art cannot furnish the conditions to produce them where they are naturally wanting. The sugar-cane grows chiefly on soils derived from calcareous formations; but few or none of these are fitted for tobacco, which is cultivated only on sandy loams. Both the Cuban and American planters concur in asserting that a large quantity of silicious matters in soils is essential for the growth of good tobacco.

"As already noticed, the rich clay loams on the banks of the James River, in Virginia, do not grow good tobacco; while the less fertile silicious soils in the county of Louisa produce it much superior in quality. Small patches of tobacco are everywhere seen growing over the sugar producing districts of Cuba; but I saw no tobacco plantations in the calcareous regions over which I traveled. The soils rest upon the primary formation. Even in the tobacco districts the planters know the spots in the different fields that produce the various qualities of leaf."

In

PREPARING THE SOIL

for the reception and growth of the plants, the fertilizing as well as the plowing of the fields should be performed in the most thorough manner. The first is essential for a large and vigorous growth, while the latter renders the cultivation of the plants much easier. The careful preparation of soil is so intimately connected with all that pertains to the plant, that it should be done well in order that the best results may follow. Tobacco of good body, color, and texture, cannot be grown on land devoid of fertility. The field selected for tobacco, if heavy sward, should be plowed early in the spring or the fall before, and later in the season if the turf is well rotted. After spreading on the manure, the field may be plowed again and harrowed frequently until all the lumps are made fine, and the surface mellow.

In the use of fertilizers select, if a light colored leaf is desired, either horse manure or tobacco stems. In the Connecticut valley nearly all kinds of Domestic, Commercial, and Special fertilizers are used. Of domestic fertilizers, horse manure is considered the best, as it produces the finest and lightest colored leaf of any known fertilizer. Of commercial fertilizers, Peruvian guano is doubtless one of the best— imparting both color and fineness to the leaf. Of special manures, tobacco stems are perhaps the best, at least the most frequently used. Of the other special fertilizers, such as cotton seed meal, castor pomace, ground bone, damaged grain, tobacco waste and saltpetre waste, much may be said both in praise and dispraise. Cotton seed meal, when used with domestic manure is an excellent and powerful manure.

If domestic manures are applied, use about twelve cords to the acre, composting before plowing under. As soon as spread, plow the field and see that all of the manure is covered. If tobacco stems are used, plow in from three to five tons to the acre, all of them at once, or a part in the fall and the remainder in the spring. If Peruvian guano is applied, sow on about three hundred pounds to the acre in connection with the domestic manure. Fish guano should be composted before sowing, either with loam or manure, and when used on light soil is a very good fertilizer, producing a light, thin leaf. After the tobacco field is harrowed it is ready for the ridger, which makes the hills and gathers together all of the loose manure on the surface, and collects it in the ridges. Where a ridger is not used, work off the rows from three and one half to four feet apart, or even wider than this. In the Connecticut valley the field is marked and hilled so as to give about 6000 hills to the acre. This will be a sufficient number if the growth is likely to be large. Where a ridger is used, manure can not be dropped in the hill and in many respects it is well not to do so, as the plants

A TOBACCO RIDGER.

are liable to be blown over during a storm—not standing as firmly in the hills as plants when no manure is used in the hills. If the hills are to be made with the hoe, avoid all stones, bits of turf and grass in making them, and select only the fresh earth—gently patting the top of the hill with the hoe. New made hills are better than old, but it will make but little difference unless the soil is very dry at the time of transplanting.

The following description of the manner of preparing the tobacco field in Virginia by the old planters is quite interesting, and gives some idea of the amount of labor to be performed on the tobacco plantation:—

"There are two distinct and separate methods of preparing the tobacco ground: the one is applicable to the preparation of new and uncultivated lands, such as are in a state of nature, and require to be cleared of the heavy timber and other productions with which Providence has stocked them; and the other method is designed to meliorate and revive lands of good foundation, which have been heretofore cultivated, and, in some measure, exhausted by the calls of agriculture and evaporation.

"The process of preparing new lands begins as early in the winter as the housing and managing the antecedent crop will permit, by grubbing the undergrowth with a mattock; felling the timber with a poll-axe;* lopping off the tops, and cutting the bodies into lengths of about eleven feet, which is about the customary length of an American fence rail, in what is called a worm or panel fence.† During this part of the process the negro women, boys, and weaker laborers, are employed in piling or throwing the brush-wood, roots, and small wood, into heaps to be burned; and after such logs or stocks are selected as are suitable to be malled into rails, make clapboards, or answer for other more particular occasions of the planter, the remaining logs are rolled into heaps by means of hand-spikes and skids; but the Pennsylvania and German farmers, who are more conversant with animal powers than the Virginians, save much of this labor by the use of a pair of horses with a half sledge, or a pair of truck wheels.

"The burning of this brush-wood, and the log piles, is a business for all hands after working hours; and as nightly revels are peculiar to the African constitution, this part of the labor proves often a very late employment, which affords many scenes of rustic mirth. When this process has cleared the land of its various natural incumbrances (to attain which end is very expensive and laborious), the next part of the process is that of the hoe; for the plough is an implement which is rarely used in new lands when they are either designed for tobacco or meadow. There are three kinds of the hoe which are applied to this tillage: the first is what is termed the sprouting hoe, which is a smaller species of mattock that serves to break up any particular hard part of the ground, to grub up any smaller sized grubs which the mattock or grubbing hoe may have omitted, to remove small stones and other partial impediments to the next process. The narrow or hilling hoe follows the operation of the sprouting hoe. It is generally from six to eight inches wide, and ten or twelve in the length of the blade, according to the strength of the person who is to use it; the blade is thin, and by means of a movable wedge which is driven into the eye of the hoe, it can be set more or less digging (as it is termed), that is, on a greater or less angle with the helve, at

---

* This is a short, thick, heavy-headed axe, of a somewhat oblong shape, with which the Americans make great dispatch. They treat the English poll-axe with great contempt, and always work it over again as old iron before they deem it fit for their use.
† The worm or panel fence, originally of Virginia, consists of logs or malled rails from about four to six or eight inches thick and eleven feet in length. A good fence consists of ten rails and a rider. It is called a worm fence from the zigzag manner of its construction.

pleasure. In this respect there are few instances where the American blacksmith is not employed to alter the eye of an English-made hoe before it is fit for use; the industrious and truly useful merchants of Glasgow have paid more minute attention to this circumstance.

"The use of this hoe is to break up the ground and throw it into shape; which is done by chopping the clods until they are sufficiently fine, and then drawing the earth round the foot until it forms a heap round the projected leg of the laborer like a mole hill, and nearly as high as the knee; he then draws out his foot, flattens the top of the hill by a dab with the flat part of the hoe, and advances forward to the next hill in the same manner, until the whole piece of ground is prepared. The center of these hills are in this manner guessed by the eye; and in most instances they approach near to lines of four feet one way, and three feet the other. The planter always endeavors to time this operation so as to tally with the growth of his plants, so that he may be certain by this means to pitch his crop within season.

DRAWING THE DIRT AROUND THE FOOT.

"The third kind of hoe is the broad or weeding hoe. This is made use of during the cultivation of the crop, to keep it clean from the weeds. It is wide upon the edge, say from ten inches to a foot, or more; of thinner substance than the hilling hoe, not near so deep in the blade, and the eye is formed more bent and shelving than the latter, so that it can be set upon a more acute angle upon the helve at pleasure, by removing the wedge."

The manner of preparing the soil in Virginia at the present time is thus described by a Virginia planter:—

"The crop usually grown in Virginia is divided into three classes, viz.:—Shipping, Sun-cured Fillers, and Bright Coal-cured Wrappers and Smokers. The first may be grown on any good soil, upland or alluvial: the latter two on dry, well-drained upland only. All require thorough preparation of the soil to insure good crops. The work first necessary for this crop is to burn a sufficiency of plant land. To prepare the land for transplanting, put the land in full tilth, then mark off with a shovel, plow furrows three feet to three feet four inches apart, and into these furrows sow the fertilizers; then with turning plows, bed the land on these furrows, and to facilitate the hilling, cross these beds three feet apart with furrows by a shovel plow, and the hills are made, except to pat them with hoes. Hilly lands will seldom admit of this cross-plowing, and the beds must be chopped into hills. On new ground apply the fertilizers broadcast. It acts well, and for fine yellow pays better on new grounds than any other lands. The culture is essentially the same for all classes of tobacco. Stir the land up as often as necessary to promote a rapid growth of the plants, and to keep down grass and weeds. 'Shipping' tobacco may be plowed later and worked longer than 'fine yellow.' For 'coal-curing' sacrifice pounds for color."

The next operation to be performed on the tobacco farm or plantation is

TRANSPLANTING.

As soon as four or five leaves on a plant about the size of a dollar have appeared, they are large enough to transplant.

TRANSPLANTING.

Take the plants up with care, sprinkling with water and keeping covered. In taking them up, the earth may be allowed to remain on the roots, or shaken off, at the option of the grower. As a general rule, however, the earth should remain rather than be shaken off. Remove to the field and drop one at each hill, and where the plants are small, two. A common custom is to "set" every tenth or twelfth hill with two plants. This is a good plan, as they are frequently needed during hoeing time to "fill in." If holes have not been made, insert the first two fingers, making a hole large enough for the roots to remain in an easy and natural position. Press the earth gently around the plant if the soil is moist, but if dry, more firmly. See that the plant stands in an upright position. If dry after "setting" the plants, water them, and if a protracted drought follows, cover them up with grass or hay dipped in water; remove, however, in a day or two.* Plaster may also be used to advantage, as it keeps

TRANSPLANTING.

the hill moist, besides fertilizing the plant; put a little just around the plants. In taking up from the bed select large ones, leaving the smaller ones to grow. Transplanting should commence as early as possible that this result may follow.

---

*Walker says of tobacco culture in Colombia (South America):—"It is advisable to cover the plant with a banana leaf, or something similar: by this means the tobacco is protected from the heat of the sun, and from the heavy rains, which would not prove less prejudicial."

Plants with large broad leaves are considered the best, while those that grow tall and "spindling" or "long shank" plants, as they are called at the South, are rejected and should not be set out when others that are more "stocky" can be obtained. Avoid, however, setting too large plants, as they are not as apt to live as smaller ones. Transplanting should be done as fast as possible, that the tobacco field may present an even appearance and be ready to harvest at one time. If the plants are to grow and ripen evenly, the transplanting should be finished in a week or two from the time of the first setting. This can generally be done unless plants are very scarce, when circumstances, beyond the growers' control, often make the field give apparent evidence of want of care, although the real trouble is a want of plants.

"It may be necessary to water the plants once or twice after transplanting; this in a measure will depend upon the season."

Tatham in his Essay on the Culture and Commerce of Tobacco, (London 1800,) gives an account of the manner of transplanting in Virginia at that period. Under the head of

"THE SEASON FOR PLANTING,"

he says:

"The term, 'season for planting,' signifies a shower of rain, of sufficient quantity to wet the earth to a degree of moisture which may render it safe to draw the young plants from the plant bed, and transplant them into the hills which are prepared for them in the field, as described under the last head; and these seasons generally commence in April, and terminate with what is termed the long season in May; which (to make use of an Irishism), very frequently happens in June; and is the opportunity which the planter finds himself necessitated to seize with eagerness for the pitching of his crop: a term which comprehends the ultimate opportunity which the spring will afford him, for planting a quantity equal to the capacity of the collective power of his laborers when applied in cultivation. By the time which these seasons approach, nature has so ordered vegetation, that the weather has generally enabled the plants, (if duly sheltered from the spring frosts, a circumstance to which a planter should always be attentive in selecting his plant patch,) to shoot forward in sufficient strength to bear the vicissitude of transplantation.

"They are supposed to be equal to meet the imposition of this task, when the leaves are about the size of a dollar; but this is more generally the minor magnitude of the leaves; and some will be of course about three or four times that medium dimension. Thus, when a good shower or season happens at this period of the year, and the field and plants are equally ready for the intended union, the planter hurries to the plant bed, disregarding the teeming element, which is doomed to wet his skin, from the view of a bountiful harvest, and having carefully drawn the largest sizable plants, he proceeds to the next operation, (that) of planting.

"The office of planting the tobacco, is performed by two or more persons, in the following manner: The first person bears, suspended upon one arm, a large basket full of the plants, which have just been drawn and brought from the plant bed to the field, without waiting for an intermission of the shower, although it should rain ever so heavily; such an opportunity indeed, instead of being shunned, is eagerly sought after, and is considered to be the sure and certain means of laying a good foundation, which cherishes the hope of a bounteous return. The person who bears the basket, proceeds thus by rows from hill to hill; and upon each hill he takes care to drop one of his plants. Those who follow make a hole in the center of each hill with their fingers, and having adjusted the tobacco plant in its natural position, they knead the earth round the root with their hands, until it is of a sufficient consistency to sustain the plant against wind and weather. In this condition they leave the field for a few days, until the plants shall have formed their radifications; and where any of them shall have casually perished, the ground is followed over again by successive replantings, until the crop is rendered complete."

In tropical regions, the plants are transplanted as well in summer and fall as in the spring, but more frequently in the early autumn. In Mexico, transplanting is performed from August till November. In Persia, the tobacco plants are "transplanted on the tops of ridges in a ground trenched so as to retain water. When the plants are thirty to forty inches high, the leaves vary from three to fifteen inches in length, when the buds are ready to be pinched off; the leaves increase in size until August and September, when they have attained their growth." In Turkey "when the young plants are about six inches in height they are removed from the small beds and planted in fields like cabbages in this country, and are then left to nature to develop them to a height of from three to four feet; three leaves, however, are removed from each plant to assist its growth."

A year or two since, a machine was invented and offered to the growers of the Connecticut valley, called a transplanter, of which we here give an engraving. The inventor claimed that the "American Transplanter" could do the work of several men and do it equally well. It rolls along the ridge something like a wheelbarrow, marking the hills with a sharp joint in the wheel and setting the plants as they are dropped into the receptacles at the top.

AMERICAN TRANSPLANTER.

The tobacco plant, like most of the vegetable products, has many and varied foes. Not only is it most easily affected and damaged by wind and hail, but it seems to be the especial favorite of the insect world, who, like man, love the taste of the plant. The first of them "puts in an appearance" immediately after transplanting, which necessitates the performance of what is known to all growers of the plant as

WORMING.

There are two kinds of worms that prey upon the plants; viz: the "cut worm"* and the green or "horn worm." The

---

* Hughes, in his History of Barbadoes, says that the common people call the worm kitifonia.

first commences its work of destruction in a few hours after transplanting in the field. During the night it begins by eating off the small or central leaves called by the grower the "chit," and often so effectually as to destroy the plant. The time chosen by the planter to find these pests of the tobacco

THE WORM.

field is early in the morning, when they can be found nearer the surface than later in the day. Remove the earth around the roots of the plants, where the worm will generally be found. Occasionally they are found farther from the hill. If they are numerous, the field should be "wormed" every morning, or at least every other day, which labor will be rewarded with a choice collection of primitive tobacco chewers. Sometimes the worms are very small and difficult to find, while at other times more are found than are required for the growth and development of the plants. As soon as they disappear they make way for the "horn worm" who now takes his turn at a "chaw." By some the cut worm is considered the most dangerous foe; as it often destroys the plant, while the other injures the leaf without endangering the plant. A little plaster sprinkled around the hill sometimes checks their progress, yet we have never found any remedy that would hinder their depredations very much. The plants should be kept growing as soon as transplanted, which will be found the better method, as they will soon be too large for the cut worm to injure them much, if at all.

The "horn worm" feeds upon the finest and largest leaves. They are not found as often on the top leaves—especially those growing on the very highest part of the stalk, as they prefer the ripe leaves and those lower on the plant.

WORMING TOBACCO.

The horn worm, if large, eats the leaves in the finest part of them, frequently destroying half of a leaf. They leave large holes which renders the leaf worthless for a cigar wrapper, leaving it fit only for fillers or seconds. In Cuba the tobacco plant is assailed by three different kinds of insects—one attacks the foot of the leaves; a second the under side; a third devours the heart of the plant. In Colombia the following are the great enemies of the tobacco plant: A grub, named *canne*, which devours the young buds; the *rosca-worm*, which commits its depredations in the night only, burrowing in the ground during the day; the grub of a butterfly, called by the Creoles *palometa*; a species of scarabæus called *arader*, which feeds on the root of the plant; and a species of caterpillar* which is called in the

---

* Wallace says of worming tobacco in Brazil: "The plants are much attacked by the caterpillar of a sphinx moth, which grows to a large size, and would completely devour the crop unless carefully picked off. Old men, and women, and children are therefore constantly employed going over a part of the field every day, and carefully examining the plants leaf by leaf till the insects are completely exterminated."

country the *horned-worm*, so voracious as to require one night only to devour an entire leaf of tobacco. At the South, and especially in Virginia, the housewife's flock of turkeys are allowed to range in the tobacco fields and devour many of these pests.

Almost as soon as the plants have been transplanted, the work of

CULTIVATING

should commence. As the tobacco plant grows and ripens in a few weeks from the time it is transplanted in the field, it is of the utmost importance that the plants get " a good start " as soon as possible. In a favorable season, and with ordinary culture, the plants will do to harvest or "cut" in from eight to ten weeks after transplanting. From the rapidity of its growth it will readily be seen that the plant should come forward at once, if large, fine leaves are desired. In a week from the time of transplanting a light cultivator should be run between the rows, stirring the soil lightly, after which the plants should be hoed carefully, drawing away from the hill and plant the old and "baked" earth and replacing it with fresh. If the hill is hard around the plant it should be loosened by striking the hoe carefully into the hill and gently lifting the earth, thus making the hill mellow. This is apt to be the case with stiff, clayey soil, which, if possible, should be avoided in selecting the tobacco field.

It is doubtless as true a saying as it is a common one with Connecticut tobacco-growers, that the plants will not "start much until they have been hoed." Where the first hoeing is delayed two or three weeks, the plants will to a certain extent become stunted and dwarfed, and will hardly make up for the delay in growing. In from two to three weeks, the field should be hoed again, and this time the cultivator should mellow the soil a little deeper than the first time, while the hoeing should be done in the most thorough manner. Draw the earth around the plant and cut up with the hoe all grass and weeds, and remove all stone and lumps of manure and any rubbish that will hinder easy cultivation, or retard the growth of the plants. At this period the most careful attention must be given to the plants, as they are (or ought to be) growing rapidly, and upon their early maturity will depend the color and texture of the leaf.

In a short time the plants may be hoed for the third and last time (as a fourth hoeing is but rarely necessary). At this time they have attained considerable size, (say two or three feet high) and are rapidly maturing, and ere long will be ready to harvest. At the last hoeing the plants should be "hilled up," that is, the earth should be drawn around the plant under the leaves, causing it to stand firmly in the hill, and keeping the roots well protected and covered. The tobacco plant requires constant cultivation, and the cultivator may be run through the rows after loosening the earth and turning up the manure towards the plants.

Some growers of tobacco in the early stages of its growth apply some kind of fertilizers to the backward plants; this will be found to be of advantage, and should be done just before a rain, when the plants will start in a manner almost surprising. A little phosphate or Peruvian guano may be used, but should be applied with care or the plants may be retarded instead of quickened in their growth.

There is much to be done in the tobacco field besides cultivating and hoeing the plants. In many hills there will be found two plants, which should be re-set at the second hoeing if needed, and if not, pulled up and destroyed, as it is better to have one large plant in the hill than two small ones. Again, after the last hoeing, the tobacco should be kept free from worms. If any have been overlooked they will have attained to a good size by this time, and will devour in a short time enough tobacco to make a "short six."

From this account of the cultivation of tobacco as practiced in the Connecticut valley, one will readily see that the labor performed during the growing of the plants should not be superficial. On their rapid growth depends the color and texture of the leaf. Plants that are slow in maturing never make fine wrapping leaves or show a good color. Where the growth is rapid the plants will be more brittle than if of slower growth, and must therefore be handled with care in passing through the rows to worm, top, and sucker the plants.

A century ago the Virginia planters cultivated their tobacco fields in the following manner:—

"Hoeing commences with the first growth of the tobacco after transplantation, and never ceases until the plant is nearly ripe, and ready to be laid by, as they term the last weeding with the hoe; for he who would have a good crop of tobacco, or of maize, must not be sparing of his labor, but must keep the ground constantly stirring during the whole growth of the crop. And it is a rare instance to see the plough introduced as an assistant, unless it be the slook plough, for the purpose of introducing a sowing of wheat for the following year, even while the present crop is growing; and this is frequently practiced in fields of maize, and sometimes in fields of tobacco, which may be ranked amongst the best fallow crops, as it leaves the ground perfectly clean and naked, permitting neither grass, weed, nor vegetable to remain standing in the space which it has occupied."

TOPPING.

The next operation to be performed in the tobacco field is known by the name of

TOPPING,

and is simply breaking or cutting off the top of the stalk, preventing the plant from running up to flower and seed. By so doing the growth of the leaves is secured, and they at once develop to the largest possible size.

TOPPING.

The leaves ripen sooner if the plant is topped, while the quality is much better. There are various methods of topping as well as different periods. Some growers top the plant as soon as the capsules appear, while others wait until the plants are in full blossom. If topped before the plants have come into blossom, the operation should be performed as soon as possible, as a longer time will be required for the leaves to grow and ripen than when topping is delayed until the plants are in blossom. In the Connecticut valley most growers wait until the blossoms appear before breaking off the top. Topping must not be delayed after the blossoming, in order that all danger from an untimely frost may be avoided. The top may be broken off with the hand or cut with a knife, the latter being the better as well as the safer way. Sometimes the rain soaks into the stalk, rotting it so that the leaves fall off, injuring them for wrappers. Top the plants at a regular height, leaving from nine to twelve leaves, so that the field will look even, and also make the number of leaves to a plant uniform. Late plants may be topped with the rest or not, at the option of the grower. This mode of topping refers more particularly to cigar rather than cutting leaf. Those varieties of tobacco adapted for cutting leaf should be topped as soon as the button appears; top low, thereby throwing the strength of the stalk into a few leaves, making them large and heavy. The number of leaves should not exceed fourteen. Let it stand from five to six weeks after it is topped. The object in letting it stand so long after topping is to have it thoroughly ripe. This gives it the bright, rich, golden color, entirely different from cigar leaf, but very desirable for chewing leaf. On account of the length of time it must stand after topping, it is desirable to take that which has been topped early, in order to have it ripen, and get it in before a freeze, although ripe tobacco is not injured by cold nights, and will sometimes stand even an ordinary frost.

The manner of topping in Virginia by the first planters in the colony, is thus described:—

"This operation, simply, is that of pinching off with the thumb nail* the leading stem or sprout of the plant, which would, if left alone, run up to flower and seed; but which, from the more substantial formation of the leaf by the help of the nutritive juices, which are thereby afforded to the lower parts of the plant, and thus absorbed through the ducts and fibres of the leaf, is rendered more weighty, thick, and fit for market."

Now the custom is to top for shipping from eight to ten leaves, for coal-curing from ten to twelve, according in both cases to strength of soil and time of doing the work.

In Mexico "as soon as the buds begin to show themselves the top is broken off. Not more than from eight to ten leaves are left on the plant, without counting the sand-leaf, which is thrown away," and destroyed in the same manner as the Dutch are said to do of spies. In some countries the plants are not topped at all, and the leaves are left upon the stalk until fully ripe, when they are picked.

The next labor following the topping of the plants is called

SUCKERING.

Immediately after topping the plants, shoots or sprouts make their appearance at the base of the leaves where they join the parent stalk. They are known by the name of suckers and the removal of them by breaking them off is called suckering. At first the suckers make their appearance at the top of the plants at the base of the upper leaves, and then gradually appear farther down on the stalk until they are found at the very root of the plant. The plants should be suckered before the shoots are tough, when they will be removed with difficulty, frequently clinging to both stalk and leaf, thereby injuring the latter, as the leaf very often comes off with the sucker if the latter is left growing too long. The plants should be kept clean of them and especially at the time of harvesting.

An old writer on tobacco says of Suckers and Suckering:—

"The sucker is a superfluous sprout which is wont to make its appearance and shoot forth from the stem or stalk, near to

---

*Many of the Virginians let the thumb nail grow long, and harden it in the candle, for this purpose: not for the use of gouging out people's eyes, as some have thought fit to insinuate.

the junction of the leaves with the stem, and about the root of the plant and if these suckers are permitted to grow, they

SUCKERING.

injure the marketable quality of the tobacco by compelling a division of its nutriment during the act of maturation. The planter is therefore careful to destroy these intruders with the thumb-nail as in the act of topping, and this process is termed suckering."

After this operation is performed the planter ascertains in regard to the

RIPENING OF THE PLANTS.

As soon as the plants are fully ripe they not only take on a different hue but give evidence of decay. The leaves as they ripen become rougher and thicker, assume a tint of yellowish green and are frequently mottled with yellow spots. The tobacco grower has two signs which he regards as "infallible" in this matter. One is that on pinching the under part of the leaf together, if ripe it will crack or break; the other is the growth of suckers to be found (if ripe) around the base of the stalk.

Tatham says:—

"Much practice is requisite to form a judicious discernment concerning the state and progress of the ripening leaf;

THE HARVEST.

yet care must be used to cut up the plant as soon as it is sufficiently ripe to promise a good curable condition, lest the approach of frost should tread upon the heels of the crop-master; for in this case, tobacco will be among the first plants that feel its influence, and the loss to be apprehended in this instance, is not a mere partial damage by nipping, but a total consumption by the destruction of every plant. I find it difficult to give to strangers a full idea of the ripening of the leaf: it is a point on which I would not trust my own experience without consulting some able crop-master in the neighborhood; and I believe this is not an uncustomary precaution among those who plant it. So far as I am able to convey an idea, which I find it easier to understand than to express, I should judge of the ripening of the leaf by its thickening sufficiently; by the change of its color to a more yellowish green; by a certain mellow appearance, and protrusion of the web of the leaf, which I suppose to be occasioned by a contraction of the fibres; and other appearances as I might conceive to indicate an ultimate suspension of the vegetative functions."

After the plants have ripened the operation of cutting or

HARVESTING

begins. The cutter passes from plant to plant cutting only

CUTTING THE PLANTS.

those plants that are ripe. In harvesting a light **hatchet or** saw may be used or a tobacco cutter which is the better and not as liable to injure the leaves. The plants may be cut either in the morning (after the dew is off) or just at night, providing there are no indications of frost. Lay the plants carefully on the sides to avoid breaking the leaves. If the plants are cut during a very warm day they should be examined from time to time as they are liable to "sun-burn," an injury much dreaded by the planter, as sun-burnt leaves are useless for cigar wrappers.

After the plants are wilted on one side they are turned so that the entire plant will be in good condition to handle without breaking. Harvesting should be performed in the most careful manner. At this time the leaves are very brittle and unless the cutter is an experienced hand much injury may be done to the leaves. The stem of each plant is severed as near as possible to the ground and afterwards if hung on lath they are divided longitudinally to admit the air and dry them sooner. When the plants are to be hung on lath they may be wilted before "stringing" or not, at the option of the grower. Most growers are of the opinion now that the plants should

PUTTING ON LATH.

be harvested without wilting at all, stringing on the lath as soon as cut and carrying them immediately to the shed.

When wilted in the field there is often much damage done to the leaves whether they are sun-burnt or not. Oftentimes the ground is hot and the plants in a few hours both on the under and upper sides become very warm and almost burnt by the rays of the sun. For this reason the manner of hanging on lath is the better way and in New England is fast displacing the old method of hanging with twine. When hung in this manner five or six plants to the lath are the usual

CARRYING TO THE SHED.

number unless they are very large. When placed or strung on the lath the plants are not as liable to sweat or pole rot, owing in part to the splitting of the stalk, which causes the rapid curing of the leaves as well as the stalk itself. A new method of hanging tobacco has been introduced of late in the Connecticut valley by means of tobacco hooks attached to the lath. This mode is considered by many growers the safest way, and by others as no better than the more common way of hanging simply on the lath.

In Virginia in "ye olden time," the following method of harvesting was adopted:—

"When the plant has remained long enough exposed to the sun, or open air, after cutting, to become sufficiently pliant to bear handling and removal with conveniency, it must be removed to the tobacco house, which is generally done by manual labor, unless the distance and quantity requires the assistance of a cart. If this part of the process were managed with horses carrying frames upon their backs for the conveniency of stowage, in a way similar to that in which grain is conveyed in Spain, it would be found a considerable saving of labor. It becomes necessary, in the next place, to see that suitable ladders and stages are provided, and that there be a sufficient quantity of tobacco sticks, such as have been described to answer the full demand of the tobacco house, whatsoever may be its size; time will be otherwise lost in make-shifts, or sending for a second supply.

"When everything is thus brought to a point at the tobacco house, the next stage of the process is that termed hanging the tobacco. This is done by hanging the plants in rows upon the tobacco sticks with the points down, letting them rest upon the stick by the stem of the lowest leaf, or by the split which is made in the stem when that happens to be divided. In this operation care must be taken to allow a sufficient space between each of the successive plants for the due circulation of air between: perhaps four or five inches apart, in proportion to the bulk of the plant. When they are thus threaded upon the sticks (either in the tobacco houses, or, sometimes, suspended upon a temporary scaffold near the door), they must be carefully handed up by means of ladders and planks to answer as stages or platforms, first to the upper tier or collar beams of the house, where the sticks are to be placed with their points refiting upon the beams transversely, and the plants hanging down between them. This process must be repeated tier after tier of the beams, downwards, until the house is filled; taking care to hang the sticks as close to each other as the consideration of admitting air will allow, and without crowding. In this position the plants remain until they are in condition to be taken down for the next process."

In Cuba about the beginning of January the tobacco is ready for cutting. If the harvest is good, all the leaves are taken from the plants at once. Tobacco consisting of those leaves is called Temprano, or "Early Pipe." If, on the contrary, the harvest is not good, the immature leaves are left to grow. Tobacco formed of these leaves has the name of Tardio, or "Late Pipe." In every respect, appearance included, the Temprano is much superior to the Tardio. In the purchase of tobacco, it is a principal thing to ascertain how much or how little Temprano a parcel contains. Moreover, there are what may be called bastard leaves, which grow after the leaves proper have been gathered.* Tobacco made from these bastard leaves is easily recognizable, the leaves being long and narrow, of a reddish color, and a bitter taste.

The mode of harvesting tobacco in Virginia at present is thus described by a Virginia planter:—

"In bringing to the barn place the tobacco on scaffolds near the barn-door, so that it can be readily housed in case of rain. As Bright Wrappers and Smokers pay so much better than dark tobaccos, it is advisable, whenever practicable, to coal-cure all that ripens of a uniform yellow color. The quality of the leaf will determine the hanging: 'Shipping' should be hung seven to nine plants to the stick four and a half feet long. To cure the plants properly requires some experience, great care, and much attention. The plants should not be 'cut' until fully ripe. Be careful in cutting to select plants of a uniform size, color, and quality, putting six or seven to the stick. Let the plants go from the cutter's hands on to sticks held in the hands of women or boys; and as soon as the sticks are full, place them carefully on wagons and carry them to the barn. Place the sticks on tiers about ten inches apart, and regulate the plants on the sticks.

"It is impossible to lay down any uniform system or give specific instructions. General principles will be suggested to guide the planter amid the changeableness of seasons and variableness of material to be operated upon."

In Turkey—

"The planters calculate always fifty-five days from May 12th, for their crops to be ready for gathering. When the leaves show the necessary yellow tips, they are carried to the house, and there threaded into long bunches by a large, flat needle, about a foot long, passed through the stalk of each."

In Ohio the process of harvesting tobacco for cutting is thus described by a grower:—

"When thoroughly ripe, having stood two or three weeks longer than is necessary for cigar leaf, it is ready to cut. This is done with a knife made for the purpose. It resembles a wide chisel, except that the handle and chisel are at right angles. Before cutting, the stalk is split down through the center. Being ripe, it splits before the knife, and following the grain the leaves escape unharmed. This splitting is

---

*Second crop, or Volunteer tobacco.

done in as little time as is necessary to cut the stalk off in the ordinary way. Split it to within about three or four inches of the ground, and cut it off in the ordinary way with the same knife. Cut it off and hang it over one of your sticks that you have driven slanting into the ground near you. Cut and put six stalks on the stick, and then lay it down on the ground to wilt, taking the usual care to prevent sun-burn. When it is sufficiently wilted, haul to the shed and hang it up."

In the East Indian Archipelago, "as soon as the leaves are fully grown they are plucked off, and the petiole and a midrib are cut away. Each leaf is then cut transversely into strips about a sixteenth of an inch wide, and these are dried in the sun until a mass of them looks like a bunch of oakum."

In Persia, when the plants are ripe they are cut off close to the root, and again stuck firmly in the ground. By exposure to the night dews the leaves change from green to yellow. When of the proper tint, they are gathered in the early morning while wet with dew, and heaped up in a shed, the sides of which are closed in with light thorny bushes, so as to be freely exposed to the wind.

In Japan, the leaves are gathered in the height of summer. When the flowers are of a light tint, two or three of the leaves nearest the root are gathered. These are called first leaves, but produce tobacco of second quality. After the lapse of a fortnight, the leaves are gathered by twos, and from these the best tobaccos are produced. Any remaining leaves are afterwards broken off along with the stem and dried. These form the lowest quality of tobacco. After gathering, the leaves are arranged in regular layers and covered with straw matting, which is removed in a couple of days. The leaves are now of a light yellow color. They are then fastened by the stem in twos and threes to a rope slung in a smoke room, and after being so left for fourteen or fifteen days, they are dried for two or three days in the sun, after which they are exposed for a couple of nights in order that they may be moistened with dew. They are then smoothed out and arranged in layers, the stems being fastened together, pressed down with boards, and packed away in a dark room.

D'Almirda says that in Java, the leaves are gathered and tied up in bundles of fifteen, twenty or thirty, and suspended from bamboo poles running across the interior of the shed, where they are left to dry for twenty days or more, according to the state of the atmosphere.

As soon as the plants have been hung in the shed the process of

CURING

begins. If fully ripe at the time of harvesting, the plants will "cure down" very fast and take on a better hue than when they cure less rapidly. During cool weather the doors and ventilators should be left open that the plants may have a free circulation of air and cure the faster. When, however, the weather is damp, they should be closed, to avoid sweating and pole rot. When a light leaf is desired, the tobacco shed should be provided with windows to let in plenty of sunlight, which has much to do with the color of the leaf. When a dark leaf is desired, all light should be excluded.

The time necessary for the curing of the plants will depend upon the ripeness of the plants as well as the weather during curing. There are three kinds or methods of curing, viz: air curing, sun curing and firing, or curing by flues. Air curing is the curing of the plants in sheds or barns. Sun curing is the process of curing in the open air, while "firing" is the process of curing by "smoke," the common method employed at the South and to some extent at the West. This is the common way of curing cutting leaf, while air curing is the manner of curing cigar leaf. Tatham, already quoted, gives the following account of the process as performed in Virginia of

"SMOKING THE CROP."

"From what has been said under the head of hanging the plant, it will be perceived that the air is the principal agent in curing it, but it must be also considered that a want of uniform temperature in the atmosphere calls for the constant care of the crop-master, who generally indeed becomes habitually weather-wise, from the sowing of his plants, until the delivery of his crop to the inspector. To regulate this effect upon the plants he must take care to be often among them, and when too much moisture is discovered, it is tempered by the help of smoke, which is generated by means of small smothered fires made of old bark, and of rotten wood, kindled about upon various parts of the floor where they may seem to be most needed.

"In this operation it is necessary that a careful hand should be always near: for the fires must not be permitted to blaze, and burn furiously; which might not only endanger the house, but which, by occasioning a sudden over-heat while the leaf is in a moist condition, might add to the malady of 'firing' which often occurs in the field."

In Virginia the manner of curing tobacco at the present time, is thus described by a planter. "For curing tobacco the simplest method is sun-curing or air curing and the one most likely to prove successful. The tobacco barn should be so constructed as to contain four, five or six rooms four feet wide, so that four and a half feet sticks may fit, all alike. Log barns are best for coal curing. All should be built high enough to contain four firing tiers under joists covered with shingles or boards and daubed close. Fire with hickory all rich, heavy, shipping tobacco.

"As soon as the barn is filled kindle small fires of coals or hickory wood, about twenty fires to a barn twenty feet square, four under each room. Coal is best, but hickory saplings, chopped about two feet long, make a good steaming heat. The successful coal-curer is an artist, and all engaged in the business are experimenters in nature's great laboratory." A North Carolina planter gives an interesting account of curing tobacco yellow. "Curing tobacco yellow, for which this section is so famous, is a very nice process and requires some experience, observation, and a thorough knowledge of the character and quality of the tobacco with which you have to deal, in order to insure uniform success. Much depends upon the character of the crop when taken from the hill. If it is of good size, well matured and of good yellowish color, there is necessarily but little difficulty in the operation. As soon as the tobacco is taken from the hill and housed, we commence with a low degree of heat, say 95° to 100° Fahr., 'the yellowing' or 'steaming' process. This is the first and simplest part of the whole process, and requires from fifteen to thirty-six hours, according to the size and quality of the tobacco, and this degree of heat should be continued until the leaf opens a lemon color, and is nearly free from any green hue. When this point is reached, the heat should be gradually raised to 105° in order to commence drying the leaf, and here lies the whole difficulty in curing (I mean in drying the leaf). The last degree of heat indicated, should be continued five or six hours, when it should again be gradually raised to 110°, when it should be maintained at this point, until the tail or points of the leaves begin to curl and dry. Indeed it will probably be safest for beginners to continue this degree of heat until one-third of the leaf is dried.

"The temperature may then be gradually increased to 115°, and kept for several hours at that point, until the leaf begins to rattle when shaken, then again raise the heat to 120°, at which point it should be continued until the leaf is dried, after which the temperature may be increased to 150° or 160° to dry the stem and stalks; the latter should be blackened by the heat before the curing is complete. Ordinarily it requires from two and a half to five days to cure a barn of tobacco, dependent entirely upon the size and quality. Put seven or eight plants on each stick and place them eight inches apart on tier poles. In the yellowing process the door of the barn should be kept closed to exclude the air. When this point is reached for drying the leaf, the door may be opened occasionally, and kept open for twenty or thirty minutes at a time, especially if the tobacco gets into a "sweat," as it is called, or becomes damp and clammy.

"The temperature is raised in the barn by cautiously adding coal from time to time to the fires, which should be placed in small piles on the floor, in rows, allowing about five feet between each pile, which should at first contain a double handful of coal. In adding coal, you will soon learn the quantity necessary to be applied by the effect produced. Avoid raising the heat hastily after the drying is commenced, lest the leaf should be scalded and reddened; on the other hand, it should not be raised too slowly for fear of 'raising the grain,' or the leaf becoming spongy and dingy. Both extremes are to be avoided, and the skill required is attained only by experience and observation. We usually cut tobacco the latter part of the week, house it and suffer it to remain until the first of next week, that we may not violate the fourth commandment."

In California tobacco is cured by the method known as

the "Culp process" from the name of its patentee. When the plant lies in the field, Mr. Culp's peculiar process begins which is described as follows:

"Tobacco had long been grown in California, even before Americans came. He had raised it as a crop for fifteen years; and before he perfected his new process, he was able usually to select the best of his crop for smoking tobacco, and sold the remainder for sheep wash. One year, two millions of pounds were raised in the State, and as it was mostly sold for sheep wash, it lasted several years, and discouraged the growers. Tobacco always grew readily, but it was too rank and strong. They used Eastern methods, topping and suckering, and as the plant had here a very long season to grow and mature, the leaf was thick and very strong. The main features of the Culp process are, he said, to let the tobacco, when cut, wilt on the field; then take it at once to the tobacco house and pile it down, letting it heat on the piles to 100° for Havana. It must, he thinks, come to 100°, but if it rises to 102° it is ruined. Piling, therefore, requires great judgment. The tobacco houses are kept at a temperature of about 70°; and late in the fall, to cure a late second or third crop they sometimes use a stove to maintain a proper heat in the house, for the tobacco must not lie in the pile without heating. When it has had its first sweat, it is hung up on racks; and here Mr. Culp's process is peculiar.

"He places the stalk between two battens, so that it sticks out horizontally from the frame; thus each leaf hangs independently from the stalk; and the racks or frames are so arranged that all the leaves on all the stalks have a separate access to the air. The tobacco houses are frame buildings, 100x60 feet, with usually four rows of racks, and two gangways for working. On the rack the surface moisture dries from the leaf; and at the proper time it is again piled, racked, and so on for three or even four times. The racks are of rough boards, and the floor of the houses is of earth. After piling and racking for three weeks, the leaves are stripped from the stalk and put into 'hands,' and they are then 'bulked' and lie thus about three months, when the tobacco is boxed. From the time of cutting, from four to six months are required to make the leaf ready for the manufacturer. "Piling" appears to be the most delicate part of the cure, and they have often to work all night to save tobacco that threatens to overheat."

In Mexico the leaves are hung up on bast* strings, dried in the shade and then sent to the chief depots, where, when they have undergone fermentation, they are sorted, and tied up in bundles. In Persia, the plants are carried to the shed and heaped, and in four or five days the desired pale yellow color is further developed. The stalks and center stem of each leaf are now removed and thrown away, while the leaves are heaped together in the drying house for another three or four days, when they are fit for packing.

In Turkey the bunches of leaves are exposed to the sun to

STRIPPING.

dry, and some months' exposure is necessary before they are sufficiently matured for baling. Rain sets in at a later period, and the tobacco becoming moist and fit for handling, is then

---
*The inner bark of the lime-tree.

removed from the threads, and made into bundles or "hands" of about sixty leaves each and tied around the stems.

After the leaves are thoroughly cured they are in condition for

STRIPPING.

The leaves of the tobacco are easily affected by the humidity of the atmosphere and during damp weather every opportunity is improved by the grower for taking down the tobacco preparatory to stripping. After taking down from the poles the plants should be packed in order to keep moist until stripped. The tobacco should not be removed from the poles when it drips or the juice exudes from either the stalk or the leaves. If stripped in this condition the leaves are apt to stain and thus become unfit for wrappers. The operation of stripping consists in taking the leaves from the stalk and tying them in bundles or hands with a leaf around the base of the hand.

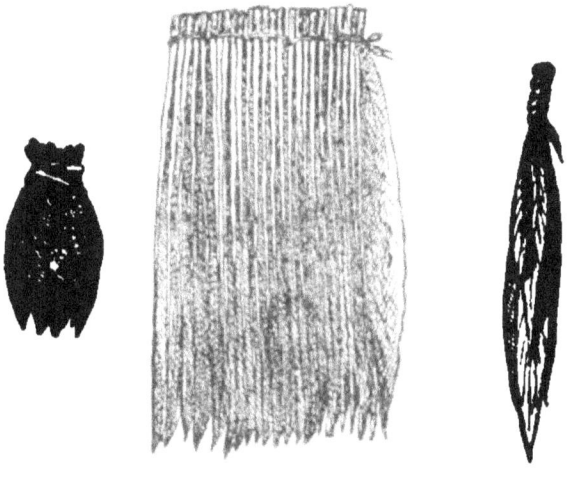

HANDS.

Each "hand" or bunch should contain at least eight leaves and from that number to twelve. If the plants are large the leaves of one stalk will form a hand; a poor leaf is used for binding as it can not be used for the same purpose as the leaves around which it is bound.

The old planters of tobacco in Virginia called this operation of taking off the leaves and tying them up "stripping and bundling" which is here described.

"When the plants of tobacco which are thus hanging upon the sticks in the house have gone through the several stages of process before the time of stripping, and are deemed to be in case for the next operation, a rainy day (which is the most suitable) is an opportunity which is generally taken advantage of when the hands cannot be so well employed out of doors. The sticks containing the tobacco which may be sufficiently cured, are then taken down and drawn out of the plants. They are then taken one by one respectively, and the leaves being stripped from the stalk of the plant are rolled round the butts or thick ends of the leaves with one of the smallest leaves as a bandage, and thus made up into little bundles fit for laying into the cask for final packing."

Hazard gives the following method of assorting and stripping tobacco in Cuba:—

"Among the Cubans, the leaves are divided into four classes: first, *desecho, desecho limpio*, which are those immediately at the top of the plant, and which constitute the best quality, from the fact that they get more equally the benefit of the sun's rays by day and the dew by night; second, *desechito*, which are the next to the above; third, the *libra*, the inferior or small leaves about the top of the plant; and fourth, the *injuriado*, or those nearest the root. Of the *injuriado* there are three qualities; the best is called *injuriado de reposo*, or 'the picked over,' and the other two, firsts and seconds (*primeros, segundos*).

"Tobacco of the classes *desechito* and *libra*, of which the leaves are not perfect, is called *injuriado bueno*, while all the rest, of whatever quality, that is broken in such a manner as to be unfit for wrappers are called *injuriado malo*. Amongst the trade in place of the above names, the different qualities are simply designated by numbers."

Meyer, a German writer who resided several years in Cuba, gives another classification, making ten classes altogether, while Hazard mentions only four general classes.

After the leaves are stripped from the stalk the process known as

ASSORTING

commences. Assorting tobacco is doing up in hands the various qualities and keeping them separate. In the Connecticut valley the growers make usually but two kinds or qualities excepting only when the crop is poor when three qualities are made, viz: Wrappers, Seconds, and Fillers. The Wrappers are the largest and finest leaves on the plant and should be free from holes and sweat as well as green and white veins. The leaves selected for this quality come from the middle and even the top leaves of the plant. The Seconds are made up of leaves not good enough for Wrappers and too good for Fillers. Such leaves sometimes are worm-eaten and of various colors on the same leaf — one part dark and another light. The fillers are the poorest quality of leaves to be found on the plants, and consist of the "sand" or ground leaves, one or two to each plant. Some of our largest growers in assorting the leaves keep each color by itself, an operation known as

SHADING.

This is a very delicate operation and requires a good eye for colors as well as a correct judgment in regard to the quality of the leaf. This mode of assorting colors in stripping is similar to that of shading cigars, in which the utmost care is taken to keep the various colors and shades by themselves. In shading the wrappers only are so assorted, and may be "run into" two or three shades depending on the number of shades or colors of the leaf. The better way is to make only two qualities of the wrappers in shading—viz., light and dark cinnamon "selections." Shading tobacco does not imply that it is carried to its fullest extent in point of color as in shading cigars, but simply keeping those general colors by themselves like light and dark brown leaves. Cutting tobaccos before being used are subjected to a process known as

STEMMING.

Tatham gives the following account of the process of stemming in Virginia a century ago:—

"Stemming tobacco is the act of separating the largest stems or fibres from the web of the leaf with adroitness and facility, so that the plant may be nevertheless capable of package, and fit for a foreign market. It is practised in cases where the malady termed the fire, or other casual misfortune during the growth of the plant, may have rendered it doubtful in the opinion of the planter whether something or other which he may have observed during the growth of his crop, or in the unfavorable temperature of the seasons by which it

STEMMING.

hath been matured does not hazard too much in packing the web with a stem which threatens to decay. To avoid the same species of risk, stemming is also practised in cases where the season when it becomes necessary to finish packing for a market is too unfavorable to put up the plant in leaf in the usual method; or when the crop may be partially out of case. Besides the operation of stemming in the hands of the cropmaster, there are instances where this partial process is repeated in the public warehouses; of which I shall treat under a subsequent head.

"The operation of stemming is performed by taking the leaf in one hand, and the end of the stem in the other, in such a way as to cleave it with the grain; and there is an expertness to be acquired by practice, which renders it as easy as to separate the bark of a willow, although those unaccustomed to it find it difficult to stem a single plant. When the web is thus separated from the stem, it is made up into bundles in the same way as in the leaf, and is laid in bulk for farther process. The stems have been generally thrown away, or burnt with refuse tobacco for the purpose of soap-ashes; but the introduction of snuff-mills has, within a few years back, found a more economical use for them."

As soon as the tobacco has been stripped it is ready for

PACKING.

It is necessary to pack the "hands" after stripping in order to keep it moist, or in nearly the same condition as when stripped. Select a cool place, not too dry or too damp, but one where if properly protected, the tobacco will remain moist. It should be packed loosely or compact, according as

PACKING.

the hands are moist or dry. It may be packed in the center of the floor so that it may be examined from either side, or against the sides of the packing house, as may be thought best. Hand the tobacco to the packer, who presses the hands firmly with his knees and hands, laying the tobacco in two tiers and keeping the pile at about the same height until all is packed. If possible pack all together, that is, each kind by itself, as it is better to have the wrappers or fillers all together rather than in several places, as the moisture is retained better than when it is packed in small piles or heaps. Use in packing a plank or board, placing it against the front of the tier and bring the ends of the hands up against it. This will make the tobacco look much better and also render the process of packing firmer.

The tobacco may be packed any height or length desired, according to the quantity, but usually from three to four feet high will be found to be convenient while the length may be proportioned to the height or not. Tobacco may be packed by the cord or half cord so as to be able to judge of the quantity—good large wrappers averaging a ton to the cord. Seconds and Fillers will not contain as many pounds to the cord as wrappers. After the tobacco is packed, cover first with boards—planed ones are preferable,—or even shingles—and press firmly, especially if the tobacco is dry, then cover with blankets or any kind of covering, adding plank or pieces of timber if additional pressure is needed. It can now remain packed until sold or cased, and will hardly need to be examined unless packed while very damp or kept packed until warm weather.

Wailes says of planting by the early planters of tobacco in Mississippi:—

"The larger planters packed it in the usual way in hogsheads. Much of it, however, was put up in carrets, as they were called, resembling in size and form two small sugarloafs united at the larger ends. The stemmed tobacco was laid smoothly together in that form coated with wrappers of the extended leaf, enveloped in a cloth, and then firmly compressed by a cord wrapped around the parcel, and which was suffered to remain until the carret acquired the necessary dryness and solidity, when together with the surrounding cloth, it was removed, and strips of lime-bark were bound around it at proper distances, in such a manner as to secure it from unwrapping and losing its proportions."

In Turkey, after the tobacco is made into bundles or hands, it is piled against the walls inside the dwelling rooms and a carefully graduated pressure put upon it until ready for baling. In Java, when the tobacco is ready to pack the leaf is examined, and if found quite brown, it is tightly pressed and packed up either in boxes or matting for exportation, or in the bark of the tree plantain, for immediate sale.

The next process on the tobacco plantation is that of

PRIZING, CASING, AND BALING.

The term prizing originated in Virginia, and as performed by the early planters, is thus described by an old writer on tobacco culture:—

"Prizing, in the sense in which it is to be taken here is, perhaps, a local word, which the Virginians may claim the credit of creating, or at least of adopting; it is at best technical, and must be defined to be the act of pressing or squeezing the article which is to be packed into any package, by means of certain levers, screws, or other mechanical powers; so that the size of the article may be reduced in stowage, and the air expressed so as to render it less pregnable by outward accident, or exterior injury, than it would be in its natural condition.

"The operation of prizing, however, requires the combination of judgment and experience; for the commodity may otherwise become bruised by the mechanic action, and this will have an effect similar to that of prizing in too high case, which signifies that degree of moisture which produces all the risks of fermentation, and subjects the plant to be shattered into rags. The ordinary apparatus for prizing consists of the prize beam, the platform, the blocks, and the cover. The prize beam is a lever formed of a young tree or sapling, of about ten inches diameter at the butt or thicker end, and about twenty or twenty-five feet in length; but in crops where many hands are employed, and a sufficient force always near for the occasional assistance of managing a more weighty leverage, this beam is often made of a larger tree, hewn on two of its sides to about six inches thick, and of the natural width, averaging twelve or fourteen inches. The thick end of this beam is so squared as to form a tenon, which is fitted into a mortise that is dug through some growing tree, or other, of those which generally abound convenient to the tobacco house, something more than five feet above the platform. Close to the root of this tree, and immediately under the most powerful point of the lever, a platform or floor of plank is constructed for the hogshead to stand upon during the operation of prizing. This must be laid upon a solid foundation, levelled, upon hewn pieces of

PRIZING IN OLDEN TIMES.

wood as sleepers; and so grooved and perforated that any wet or rain which may happen to fall upon the platform may run off without injuring the tobacco. Blocks of wood are prepared about two feet in length, and about three or four inches in diameter, with a few blocks of greater dimensions, for the purpose of raising the beam to a suitable purchase; and a movable roof constructed of clap-boards nailed upon pairs of light rafters, of sufficient size to shelter the platform and hogshead, is made ready to place astride of the beam, as a saddle is put upon a horse's back, in order to secure the tobacco from the weather while it is subjected to this tedious part of the process.

"That part of the apparatus which is designed to manage and give power to the lever is variously constructed: in some instances two beams of timber about six feet long, and squared to four by six inches, are prepared; through these, by means of an auger hole, a sapling of hickory or other tough wood, is respectively passed; and the root thereof being formed like the head of a pin to prevent its slipping through the hole, the sapling is bent like a bow, and the other end is passed through the same piece of wood in a reversed direction, in which position it is wedged. These two bows are in this manner hung by the sapling loops upon the end of the prize beam or lever; and loose planks or slabs of about five or six feet long being laid upon these suspended pieces of timber, a kind of hanging floor or platform is constructed, upon which weights are designed to act as in a scale. A pile of large stones are then carted to the place, and a sufficient number of these are occasionally placed upon this hanging platform, until the lever has obtained precisely the power which the crop master wishes to give it by this regulating medium.

"The prizing or packing by the old planters must have been a tedious affair, and far different from the quick work made by the screw-press now owned by all well to-do planters. The size of the hogsheads containing the tobacco was regulated by law to the standard of four feet six inches in length, but the shape of the cask varied according to the fancy of the cooper, or roughness of his work. At this period (a century ago), the tobacco hogshead was made most generally of white oak; but Spanish oak, and red oak, were sometimes used, when the usual kind could not be so readily commanded. Now the hogsheads are made of pine, but are nearly as rough as those made by the colonial growers.

"Tobacco, if well packed, and prized duly, will resist the water for a surprising length of time. An instance is recorded in strong proof of this, which occurred at Kingsland upon James river in Virginia, where tobacco, which had been carried off by the great land floods in 1771, was found in a large raft of drift wood in which it had lodged when the warehouses at Richmond were swept away by the overflowing of the freshets; an inundation which had happened about twenty years before this cask was found."

Tatham gives the following account of a similar instance:—

"On the sixth of October, 1782, I myself was one of a party who were shipwrecked upon the coast of New Jersey, in America, on board the brigantine Maria, Captain McAulay, from Richmond in Virginia, and laden with tobacco. Several hogsheads, which were saved from the wreck were brought round to Stillwill's landing upon Great Egg harbor; and amongst them some which had lost the headings of the cask, and the hoops and staves, were so much shattered by the beating of the surf, that it was not thought worth while to land them, and they were just tumbled out of the lighter upon the beach, and left to remain where the tide constantly flowed over them for several weeks, so that the outside was completely rotten, and they had the appesrance of heaps of manure. In this very bad condition, I still persisted in trying to save what I supposed might remain entire in the interior of the lump, and at last prevailed so far over the ignorance and prejudice by which I had been ridiculed, as to effect an overhauling and repacking of this damaged commodity and to save a proportion thereof very far beyond what I myself had expected. Some of the heart of this was so highly improved, that I have seldom seen tobacco equal to it for chewing, or for immediate manufacture; and what was repacked was sold to a tobacconist in Water Street, Philadelphia, at a price so little reduced below the ordinary market, that the man very frankly told me, that if he could have had the whole drowned tobacco in a short time after it was saved from the wreck, he would have made no difference in the price but would rather have preferred it for immediate manufacture, as it would have spared him some little labor in a part of the process."

Prizing tobacco applies to the packing of tobacco in hogsheads all such leaf being used for cutting purposes, cigar leaf being either cased or baled. In some sections about 800 pounds net is packed in one parcel, while in others 1000 pounds and sometimes even 1500 and 1800 pounds. "Seed leaf" tobacco in this country is all packed in cases instead of hogsheads, each case containing from 375 to 400 pounds net. It is necessary that all kinds of tobacco should be pressed in some kind of package before it is ready to be manufactured. There are exceptions, however, as in the case of Latakia tobacco, which is simply hung in the peasant's huts through the winter to be fumigated and to acquire the peculiar flavor this tobacco has. Tobacco in good condition to case must be damp enough to bear the pressure in casing without breaking and crumbling, while it must not be too moist or it will rot in the case. The number of pounds to the case will vary according to the size of the leaf, as well as the condition of the tobacco.

When ready to case the "hands" are packed in the case, laying them in two tiers. The case being nearly full the contents are then subjected to a strong pressure until it is reduced to one half its bulk, then another layer is placed in the case and again pressed, and succeeded by as many as are required to fill the case. The tobacco should be packed evenly in layers with the ends of the leaves touching one another or even crossing, and the whole mass presenting a smooth and even appearance. The "wrappers" should be cased by themselves and "the seconds" and "fillers" together or separate at the option of the packer. The tobacco should be cased

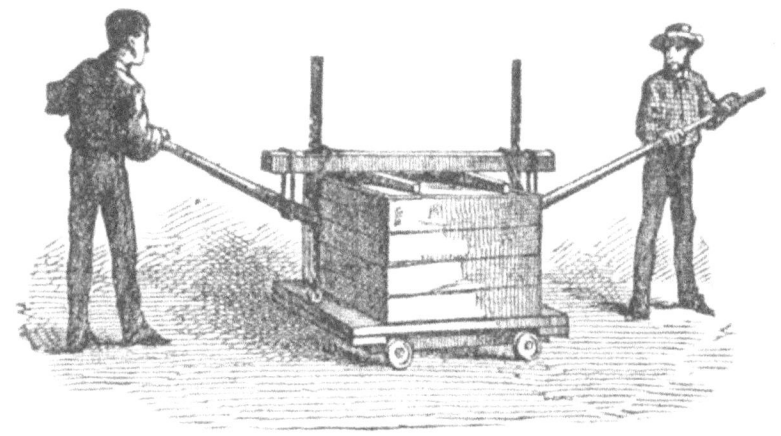

TOBACCO PRESS.

hard so that the mass will rise but little when the pressure is removed. As the fillers are usually dry they must be moistened before casing or subjected to a very strong pressure. After packing the cases should be turned on their sides, and the grower's name marked on each case, also the kind of tobacco, whether wrappers or fillers, together with the number of pounds and the weight of the case. This is necessary to ascertain the quality of leaf produced by each grower, as well as to protect the buyer against all fraud in packing and casing.

The cases may be piled one upon another, but should be kept from the rays of the sun and in a dry room, so that the sweating of the leaf may be sufficient to fit it for use. It is necessary that the season during sweating should be warm, in order to secure a good sweat. It will commence to "warm up" sometime in April or May, and will be ready to sample or uncase about the first of September. After "going through a sweat," the leaf takes on a darker color, and loses the rank flavor which it had before. It is better to let the tobacco dry off before being used or taken from the case. "Baling" is packing tobacco in small bundles or packages containing from one hundred to two hundred pounds, and is the manner of putting up tobacco for export in Cuba, Paraguay, Algiers, Hungary, Mexico, Syria, the Philippines, China, Sumatra, Japan, Java, Turkey, and in some other tobacco-growing countries. In Cuba after being formed into hands or "*gavillos*" and four of these tied together with strips of palm-leaf so as to constitute a "*manoja*," fifty or eighty of them are packed together, making what is called a "*tercio*" or bale, the average weight of which is two hundred pounds. Hazard says of the number of pounds produced on the *vegas*:

"A *caballeria* of thirty-three acres of ground produces about nine thousand pounds of tobacco, made up in about the following proportions: four hundred and fifty of *desecho*, or best; one thousand eight hundred pounds *desechito*, or seconds; two thousand two hundred and fifty pounds of *libra*, or thirds; and four thousand five hundred pounds of *injuriado*. From these figures, taking the bale at one hundred pounds, and the average price of the tobacco at twenty dollars per bale, (though this is a low estimate, for the crops of some of the vegas are sold as high, sometimes, as four hundred dollars per bale,) an approximate idea may be formed of the profit of a large plantation in a good year, when the crops are satisfactory."

In Mexico, after being baled, the tobacco is sent to the government factories, where it is not weighed until two months afterwards. The price is high, varying from twelve to twenty-eight dollars per crate; and is paid in ten monthly installments. In Persia, when the tobacco is fit for packing, the leaves are carefully spread on each other, and formed into cakes four or five feet round, and three to four inches thick, care being taken not to break or injure the leaves. Bags of strong cloth, thin and open at the sides, are provided, into which the cakes are pressed strongly down on each other. When the bags are filled they are placed in a separate drying house, and are turned every day. Water is then sprinkled on the cakes, if required, to prevent them from breaking. The leaf is valued for being thick, tough, of a uniform light yellow color, and of an agreeable aromatic smell.

In Turkey, the tobacco after remaining in the dwelling-room of the house a sufficient time, is ready for baling. The bales average in weight about forty *oques* (110 English pounds). The covering of the bales is a sort of netting made by the peasants from goat's hair; it is elastic and of great strength. Vamberry says of packing tobacco in European Turkey:

"The tobacco is packed in small packets (*bog tche*), and only after it has lain for years in the warehouses of the tobacco merchants, is it honored by the connoisseurs of Stamboul with the title of 'Aala Gobeck.' This sort of finely-cut tobacco resembling the finest silk, is held in equally high estimation in the palaces of the Grand Seignior, in the seraglio, and in the divan of the sublime Porte, where the privy council debate the most important affairs of the empire, under the soothing influence of its aromatic vapors."

In St. Domingo and the United States of Colombia, South America, the bales are called *Serous*, and in Holland and Germany, Packages. Tobacco is sent to market in bales of various sizes and made of various materials. In Cuba, the tobacco is bound with palm leaves. In South America it is packed in ox hides. From the East it comes in camel's hair sacks or "netting made from goat's hair," while from Persia, tobacco is exported in sacks of strong cloth. Manilla tobacco is shipped in bales containing four hundred pounds net. It is covered first with bass and then with sacking, made of Indian grass tied around with ratan. Each bale contains a printed statement, of which the following is a copy:

PROVINCIA DE CAGAYAN,

PARTIDO DE CITÁ. *Cosecha de* 186 .

Clas de conteine 40 manos de tabaco aforado por la junta de aforo y enfardelado por el que subscribe. Tuguegarao de de 186 .

El Gobernadorcillo caudillo. V.º B.º
Vicente Lasan. *El Interventor de aforo.*

The tobacco plant while growing is easily affected by a wet season, while it is also liable to injury by the opposite extreme of heat or drought. If a drought occurs soon after the plants are transplanted, their growth and development is greatly hindered. When, however, the plants are nearly grown, a severe drought affects the plants but little, the large palm-like leaves forming a kind of canopy and keeping the earth moist and cool. During a wet season, and sometimes when the plants have been set in damp soil, they are affected by "brown rust," or, as it is called at the South,

FIRING.

It is supposed to be caused by very damp weather, and is much dreaded by all growers of the weed, as it is sometimes quite common, and on low soil affects the crop to a considerable extent. It spots the leaf with hard brown spots that often fall out, producing holes fatal to the value of the crop.

FIRING.

The lower leaves on the plant are more likely to be injured than those higher on the stalk. The spots vary in size; sometimes they are as large as a three cent piece, but more frequently about the size of a small pearl button. At the South, rust or "firing" is much more common than in the Connecticut valley, and often whole fields are badly affected by the malady. Some seasons hardly any rust can be discovered on the leaves, and if any spots are found they are fixed and do not spread.

Small plants are more liable to be injured than large ones, and not unfrequently nearly every leaf is covered with the spots. Many theories have been advanced in regard to the cause of rust and how to prevent it. It usually occurs just before, or after, topping, and if the plants are ripe enough to harvest, they should be cut before the rust spreads to any great extent. It makes its appearance very suddenly, and if the weather be favorable (damp), spreads rapidly, often in a few days injuring the plants to a great extent. There are two varieties of rust or "firing," brown and white; and while the former is dreaded by the grower, as it injures the quality of the plant, the other is regarded with special favor, as it gives value to the leaf.

The white rust,* as it is termed, is a small white speck (often noticed on cigars), making its appearance on the leaves of the plant towards the latter part of its growth, and usually found on the top and middle leaves. It is usually found on the best, and more frequently on light than dark tobacco. Unlike the brown rust, the white does not fall out, but is as firm in its place as any part of the leaf; sometimes the spots are as white as chalk, and again they will be of a yellowish shade, though lighter in color than brown rust. The lighter the color the better their effect on the leaf upon which they are found. Leaves thus "spotted" make the finest of wrappers, and light-colored leaf thus affected brings the very highest price. It is well known to manufacturers of cigars that such leaves burn well, and almost invariably make a light ash. Good judges of cigars always pick for those thus affected, and watch with interest the ash of the cigar, noting the color as well as the flavor.

Some seasons this kind of rust is quite common, and it is supposed to be caused in the same way as the brown, although there are some growers who think that it is produced by altogether different causes. There is, however, a marked difference in the appearance of the leaves thus spotted; the white rust is not usually as thick upon the leaf, and is more generally found along the sides of the leaf, while the brown rust is found more in the center than along the sides. Tobacco of a light cinnamon color thus "marked" is considered

---

*Florida tobacco is noted for the white rust found on the leaves.

the most valuable, and could the planter obtain such a crop at option, he could realize the very highest price for it. Large growers who find much of their tobacco "spotted" in this manner, would do well to keep such leaves by themselves, and sell direct to the manufacturer. Both kinds of rust are more commonly seen on the plants during a wet than a dry season, and particularly if the plants have grown rapidly during the latter part of the time.

Formerly buyers of leaf tobacco were more interested in leaf of this description than now; and some of them, more anxious than others, made liberal offers to any grower of tobacco who could ascertain how such tobacco could be obtained. It is hardly probable that any method of culture could be devised so as to obtain such leaf; it seems to be a freak of nature, depending somewhat on the soil as well as the humidity of the atmosphere, and without doubt is beyond the control of the grower. Various theories propounded and experiments tried have not met with any success that we are aware of. Some growers are of the opinion that light manure spread on moist soil will tend to produce leaf affected with white rust, while others affirm that such leaf is common on high ground when manured with light fertilizers. It is a matter of doubt whether such leaf can be obtained by any preparation of soil, or any system of cultivation whatever.

SEED PLANTS.

The selection of large, well-formed plants for the maturing of the seeds, is of more importance than most growers are aware of.* Not only should the altitude of the plant be taken into account, but also the size and texture of the leaf.

If a variety foreign to the soil (on which it is cultivated) is grown, then particular pains should be taken to select seed plants resembling those cultivated in its native home.

In cultivating foreign varieties, even the first season plants may be seen that do not resemble the majority, but are

---

*Liancourt says of the selection of seed plants in Virginia:—"The seed for the next year is obtained from forty to fifty stalks per acre, which the cultivator lets run up as high as they will grow, without bruising their heads."

seemingly trying to accommodate themselves to the soil and climate, and in consequence resemble in a measure the variety commonly cultivated. Growers of Havana tobacco in the Connecticut valley can testify to this, and especially to the increased size of the plants. There are, however, growers of Havana tobacco, who claim that it will never deteriorate in quality, and that seed from Havana is not required in order to secure the delightful flavor of the *Vuelta de Abajo* leaf. Our experience is the reverse of this, and applies more directly to the flavor of the leaf than the size, color, or texture. In the Connecticut valley Havana leaf retains in a remarkable degree the texture and color of leaf, but not the flavor. Fresh or new seed is required from time to time. Sieckle says on the choice of seed:—

SPANISH SEED TOBACCO.

"The selection of seed is one of the principal conditions for raising good tobacco, especially when intended for the manufacture of cigars. In the United States now and then Havana seeds are planted. The tobacco raised therefrom generally resembles the real Havana in shape and color of leaves. But in order to reproduce approximately also the fine taste and flavor of genuine Havana tobacco, it would be required to impart to the soil exactly the components which constitute the famous tobacco-ground, viz.: the soil of the above-mentioned *Vuelta de Abajo* in Cuba. We say approximately, because the climate is a thing that can be neither transplanted nor fully equaled by artificial means. Havana seed propagated in the United States usually degenerates very soon, even in the course of two or three years. In other countries the experiment has been made to acclimate foreign seeds, for instance, Havana, by crossing, respectively changing the sexes and giving the male influence now to the foreign, then to the home plant."

In the Connecticut valley the cultivation of Havana tobacco is increasing year by year, and it promises to become the principal variety cultivated. All of the leading qualities of Connecticut seed leaf, such as color, strength, and texture, are preserved, while the flavor is as fine as that of much that is imported. The plants selected for seed should be allowed to fully ripen, when the leaves may be stripped from the stalks, that the capsules may receive all the strength of the growing and maturing plants. The seed plants should be left standing some six or eight weeks after the other plants have been harvested. If the nights are very cold and frosty, the top of the plants may be covered with a light cloth or paper to protect the seed buds.

When the capsules are of full size and brown in color, the top may be broken off and hung up in a dry, cool place to cure, after which the seeds should be taken from the capsules. To do this, the end of the seed buds may be cut, when most of the seeds will fall out if the buds are fully ripe and dry. A southern planter gives the following account of the curing and management of seed plants:—

"There are four classes of tobacco grown in Virginia and North Carolina, viz.: Shipping, filling, smoking, and wrapping; and it is important that planters desiring to raise either one of these should choose the kind of seed best adapted to each particular class. The Pryor makes the heaviest, richest shipping, and can only be grown to perfection on alluvial or heavily manured lands. The Frederick or Maryland grows larger, but is not so rich and waxy. The Oronoko is far preferable for fillers, smokers or wrappers, being sweeter in flavor, finer in fibre and texture, and more easily cured yellow. This is the kind best adapted to our gray soils, giving best returns. The product is not so large as on black or brown lands, yet with skill in curing and management, the difference in product is more than made up in quality.

"The Oronoko, therefore, is the only kind suited to our gray lands, and of this there are several varieties, the two most in favor being the yellow Oronoko, and the Gooch or Pride of Granville. The first is the kind that gave character to the Caswell (North Carolina) yellow tobacco more than twenty years ago, and is still preferred by a very large number of planters who grow the finest yellow smokers and wrappers. The latter is preferred in Granville county, North Carolina, that produces the finest yellow tobacco grown on this continent, or, perhaps, in the world. This latter is clearly an Oronoko tobacco, very much resembling the former, except that the leaf grows rather broader, and by some is considered sweeter. These two kinds have been grown with special reference to their adaptation to producing the finest quality of wrappers, smokers, and fillers. I am satisfied that the art of curing and management have not only been very far advanced toward scientific perfection, but that in perfecting the kinds of seed grown much improvement has been made. For instance, in the saving of seed, by adopting the plan of turning out the forwardest plants growing in the best soil, and afterwards observing to cut off all the heads of plants that ripen up coarse, narrow or ill-shaped, or of a green color on the hill, and saving only those heads that ripen yellow in color and of a smooth and fine texture, much has been done to improve the kind. Besides, the most important point in the saving of tobacco seed is to cut off all the lateral shoots, leaving only three crown shoots to perfect seed, thereby securing larger pods and more perfect seed that always ripen in good time, and are more reliable for seed beds and the production of early, vigorous plants.

"By following this mode of saving seed with special reference to the growth of a particular class of tobacco, in a few years the seed is not only greatly improved, but as like begets like in the vegetable as in the animal kingdom, becomes *sui generis*—the first of its species. The writer can bear testimony to the above facts and desires that others may profit thereby. Where any plant attains its highest perfection, there is the place to secure the best seed. The home of the tobacco plant is in Virginia and North Carolina, and the growth and perfection of the kinds here cultivated have reached a point unattained any where else. The West and South would do well to procure their seed from us, and then save and propagate after the instructions above given."

SECOND GROWTH.

The first account we find of raising a second crop of tobacco on the original field, is found in the early history of the Virginia colony; who, not satisfied with the vast amount cultivated in the usual manner, allowed a second growth to spring up from the parent stalk and thus obtained two crops from the same field in one year. The inferior quality of this growth at length caused its prohibition by law, as described elsewhere in this work. Of late, however, this "new departure" in tobacco culture seems to have attracted some attention, particularly in the Southern States, where numerous experiments have been made, and in some instances with complete success. In Mexico and also in Louisiana and California, two and even three crops are gathered, thus adding to the profit of the grower, but hardly to the fertility of the tobacco fields. Whatever the fertility of the tobacco field may be, or the care and attention given to the second crop by the planter, it can not equal the first crop, and must from the nature of the case be quite inferior in size, texture, and flavor of leaf.

Doubtless the varieties grown in the tropics will be much finer than the varieties grown in a more temperate region. There are many reasons why a second and third crop can not be equal to the first in the qualities necessary for fine leafy tobacco. In the first place, the soil will hardly produce a second crop of the size and texture of leaf that will compare with the first growth: the leaves will be small and resemble the top leaves of the original plant rather than the large, well-formed leaves of the center. Again, the season will hardly be favorable (unless in the tropics), for a second growth, which has much to do with the quality of the leaf and which alone ensures large, well-matured plants.

In the Connecticut valley but one crop can be grown of seed leaf, and even this when planted late is frequently overtaken by the "frost king" whose cold breath strikes a chill to the heart of the tobacco grower who has been so unfortunate as to have but a few plants; especially if his fields were "set" late in the season, or with "spindling" or "long shank plants" which come forward slowly and forbid all thought of a second growth, and sometimes give small hopes of even the first.

In Virginia and North Carolina the experiment has been tried of covering the stumps or trunk of the plants with straw, followed by plowing on both sides of the rows, thereby covering them to a depth of several inches, in which condition they are left until spring, when the covering is removed and the suckers or sprouts shoot forth and grow with great rapidity. This novel experiment may succeed so far as the growth and maturing of the plants is concerned, but will hardly add to the reputation of "Virginia's kingly plant" or to the profit of the growers, as the product must necessarily be small if the labor of transplanting is avoided.

Beyond all question, experiments with the growth and culture of the tobacco plant are among the most interesting and valuable, and afford the planter the most pleasure and instruction of all similar trials with the products of the vegetable kingdom. These experiments at once develop not only the rare qualities of the plant, but its various forms and habit of growth. They show as well as its adaptation to all countries and climes, and the preservation of its qualities when grown in regions far remote from its native home. The florist finds no more pleasure in the cultivation of the rarest exotic than the tobacco planter in testing some new variety of tobacco, and noting its varied qualities and adaptation to his fields. By trying new varieties, some of the finest qualities of the plant have been developed, and many other of its excellences still further advanced. In the United States numerous trials and experiments are constantly being made to still further perfect the various kinds already cultivated, as well as to test other varieties and note their qualities and adaptation to the soil. Already far advanced, the culture of the plant has not yet reached its highest point. The adaptation, soil, and fertilizers, are now attracting much attention, and further study of these elements promises to "bring out" qualities of leaf hitherto overlooked, or at least but partially developed.

CHAPTER XIV.

THE PRODUCTION, COMMERCE AND MANUFACTURE OF TOBACCO.

FEW comparatively of the users or even of the growers and manufacturers of tobacco, are aware of the vast amount cultivated, manufactured and used. Many suppose that its cultivation is confined to the United States and a few of the West India Islands, having no idea of the large quantities grown in Europe, Asia and Africa and the islands of the East India Archipelago. The Spaniards first began the cultivation of the plant on the Island of St. Domingo, afterwards extending it to Trinidad, the coast of South America, Mexico and the Philippine Islands. In Portugal the cultivation commenced about 1575–80, and continued some years. The Dutch a little later, began the production of tobacco in the East Indies, and in connection with the Spaniards and Portuguese were the only cultivators of tobacco until the English commenced its growth in Virginia in 1616.

The first production in St. Domingo by the Spaniards was sometime previous to 1535, and the island has continued to produce the great staple until now. In Trinidad, however, a finer article was yielded, and its cultivation became more general here until the Spaniards began to plant it in Cuba in 1580. From the West Indies, South America and the East Indies, Europe raised its supply of tobacco until the English colonists commenced its cultivation in Virginia. The Spaniards and Portuguese at first controlled the trade in tobacco, and extorted most fabulous prices for it. As soon, however, as the Dutch and English began to cultivate it and receive it from their colonies the price gradually fell while the demand and consumption for it increased in proportion to the falling off of prices. From the island of Trinidad, Europe received its finest tobacco, and it continued to maintain its reputation as such until that variety known as Varinas tobacco from South America appeared; this variety attracted the attention of European buyers and consumers, from its superiority in flavor and appearance which it has maintained for more than two hundred and fifty years.

In South America, the cultivation of tobacco took its rise in Venezuela, Brazil and Colombia. The varieties there produced had acquired an established reputation as early as 1600, together with St. Lucia, Philippine and Margarita tobaccos. Early in the Seventeenth Century, the Dutch became the great producers and importers into Europe, and the growths of their colonies continued to furnish a large proportion of the quantity used until English colonial tobacco made its appearance from Virginia.

The Plymouth and London companies from its first appearance in their markets, saw its vast importance as an article of agriculture and commerce, and in twenty years after the first planting of it, began to reap rich returns from its sale and production. From this time forward, not only in America, but in Europe and Asia, its cultivation spread among other nations until at length it has become one of the great sources of revenue of almost every country, and a leading product of nearly every clime. The islands of St. Domingo, Trinidad, St. Lucia and Martinique, do not produce as large quantities of tobacco as formerly; its cultivation in the West Indies being now confined chiefly to the island of Cuba.

This island produces at the present time the finest cigar leaf of the West Indies, which is considered by many as the best grown. The value of the annual product of Cuba is estimated at $20,000,000, nearly as much as that of the entire United States. Brazil, Colombia, Venezuela, and Paraguay, which are the tobacco-producing countries of South America, furnish Europe with a large amount of leaf tobacco. In

Brazil according to Scully it "occupies the fourth place in the exports" and is extensively cultivated in various parts of the empire. In Venezuela it is an important article of agriculture, and the product is of fine quality and in good repute in Europe. Colombia has long been noted for the amount and excellence of its tobacco; its various growths are fine in all respects and are among the finest cigar tobaccos grown. In Paraguay large quantities of excellent cigar tobacco are raised, much of which is used in various parts of South America, the remainder going to Europe.

All of the tobacco of South America is unrivaled in flavor and is well adapted for the manufacture of cigars. In Mexico, tobacco is raised to some extent, particularly in the Gulf States, where it develops remarkably and is of excellent quality both in texture and flavor. Mexico is doubtless as well adapted for tobacco as any country in the world, and if certain restrictions* were removed, its culture would increase and the demand would cause its extensive production. In the Central American States, some tobacco is cultivated, but not to the extent that is warranted by the demand or the adaptation of the soil. Some parts of the States, especially of Honduras, are well adapted for the production of the very finest leaf. As it is but little is grown; hardly any being exported to Europe. America is the native home of the tobacco plant, and in the United States vast quantities are produced of all qualities and suited for all purposes.

In New England from 20,000 to 30,000 acres are cultivated annually, estimated to yield on an average from 1500 to 1700 pounds to the acre. The annual product in cases is from 50,000 to 170,000.† Of the Middle States, New York and Pennsylvania furnish a large amount of "seed leaf" as it is called. In 1872 the latter state reported 38,010 cases, mostly grown in three counties. A fine quality of tobacco is raised in the immediate vicinity of the old William Penn mansion, and is known to all dealers as superior leaf. In New York

---

* Tobacco is not allowed to pass from one state into another without paying a certain duty.
† The amount in 1872, was 172,000.

the crop is usually good, and along the valleys are found some excellent lands for its culture.

As we go South, we reach the great tobacco-growing states, Maryland, Virginia, Kentucky, and others. Maryland has long been noted for its tobacco, and annually exports thousands of hogsheads to European markets. Virginia, as we have seen, is the oldest tobacco-producing state in the Union, and still continues to raise thousands of acres of the "weed" for home use and for export. In 1622, six years after its cultivation began, she produced 60,000 pounds of leaf tobacco. North Carolina also raises a fine article of smoking tobacco— of fine color and superior flavor. This state has long been noted for its superior leaf tobacco, and ever since the first settlement of the state has produced large quantities of it. In 1753 100 hogsheads were exported, the number constantly increasing until the present. In Georgia some tobacco is grown. Havana tobacco was first cultivated in this state by Col. McIntosh, and succeeded finely in some of the counties along the coast.

In Florida, Havana tobacco is cultivated altogether. It differs somewhat in flavor, however, so that it is called Florida tobacco, not because it is grown in that state, but because it is a little bitter, unlike that grown in Cuba. Kentucky is the great tobacco-producing state of the Union. Two-fifths of the entire amount grown in the country comes from this state. In 1871 nearly 150,000 acres were devoted to it in the state—producing 103,500,000 pounds of leaf tobacco. In Ohio and Missouri large quantities of tobacco are grown, the former state furnishing both cutting and seed leaf tobaccos. The other Western states including Illinois, Indiana, and Wisconsin, are engaged largely in its production, and furnish a good article of leaf.

California for the last few years has given the culture of tobacco some attention, and promises to become a great tobacco-producing state. The United States have cultivated in some seasons 350,769 acres of tobacco, valued at $25,901,-769. The average yield per acre is greater in Connecticut than in any other state, being 1,700 pounds, while the smallest yield is in Georgia, 350 pounds. The average price per pound in Connecticut is 25cts; in Kentucky 7 7-10cts; in Georgia 21 4-10cts; in Ohio 9 1-10cts; and in Pennsylvania 15 2-10 cts. In 1855 there was exported from this country 150,213 hogsheads and 13,366 cases of tobacco.

In Europe large quantities of tobacco are grown, excepting in England, Spain, and Portugal, where its culture is prohibited by law to benefit the colonial growers of the plant. Austria is the great tobacco-producing country of Europe, and yields an annual product of 45,000,000 pounds of tobacco; the leaf is of good quality, and is used for cigars. France also raises about 30,000,000 pounds of tobacco besides importing large quantities from the United States. In Russia the annual tobacco crop is about 25,000,000 pounds. In Holland about as much tobacco is grown as in the state of Connecticut— 6,000,000 pounds and the product is adapted for both cigar and snuff-leaf. Large quantities of tobacco are also imported, from 30,000,000 to 35,000,000 pounds. The tobacco factories in the country are stated to give employment to one million operatives. Belgium produces considerable tobacco, about 3,000,000 pounds annually. Switzerland also raises from 1,000,000 to 1,200,000 pounds of leaf. In Greece tobacco is an important product and the quality of leaf is very fine; her product has been as high as 5,500,000 pounds.

In Asia tobacco has long been cultivated, and is one of the greatest products of the country. In both Asiatic and European Turkey the annual production is about 43,000,000 pounds. In China and Japan large quantities are grown, as well as in Persia, Thibet, and other portions of Asia. In the Philippine Islands its cultivation is carried on by the Spaniards, as it has been for upwards of 250 years. Bowring says of its culture:—

"The money value of the tobacco grown in the Philippines is estimated at from 4,000,000 to 5,000,000 of dollars, say 1,000,000l. sterling. Of this nearly one half is consumed in the island, one quarter is exported in the form of cheroots (which is the Oriental word for cigars), and the remainder sent to Spain in leaves and cigars, being estimated as an annual average contribution exceeding 800,000 dollars. The sale of tobacco is a strict government monopoly, but the impossibility of keeping up any sufficient machinery for the protection of that monopoly is obvious even to the least observant. The cultivator, who is bound to deliver all his produce to the government, first takes care of himself and his neighbors, and secures the best of his growth for his own benefit. From functionaries able to obtain the best which the government brings to market, a present is often volunteered, which shows that they avail themselves of something better than the best. And in discussing the matter with the most intelligent of the empleados, they agreed that the emancipation of the producer, the manufacturer and the seller, and the establishment of a simple duty, would be more productive to the revenue than the present vexatious and inefficient system of privileges.

"In 1810 the deliveries were 50,000 bales (of two arrobas), of which Gapan furnished 47,000 and Cayayan 2,000. In 1841 Cayayan furnished 170,000 bales; Gapan, 84,000; and New Biscay, 34,000. But the produce is enormously increased; and so large is the native consumption, of which a large proportion pays no duty, that it would not be easy to make even an approximative estimate of the extent and value of the whole tobacco harvest. Where the fiscal authorities are so scattered and so corrupt;—where communications are so imperfect and sometimes wholly interrupted; where large tracts of territory are in the possession of tribes unsubdued or in a state of imperfect subjection; where even among the more civilized Indians the rights of property are rudely defied, and civil authority imperfectly maintained; where smuggling, though it may be attended with some risk, is scarcely deemed by any body an offense, and the very highest functionaries themselves smoke and offer to their guests contraband cigars on account of their superior quality,—it may well be supposed that lax laws, lax morals and lax practices, harmonize with each other, and that such a state of things as exists in the Philippines must be the necessary, the inevitable result.

"I am informed by the alcalde mayor of Cayayan that he sent in 1858 to Manilla from that province tobacco for no less a value than 2,000,000 dollars. The quality is the best of the Philippines; it is all forwarded in leaf to the capital. The tobacco used by the natives is not subject to the *estanco*, and on my inquiring as to the cost of a cigar in Cagayan, the answer was

'Casinada' (Almost nothing). They are not so well rolled as those of the government, but undoubtedly the raw material is of the very best."

In Sumatra some of the finest tobacco in the world is produced which has an established reputation in European markets.

In Africa tobacco is grown to some extent in Egypt, Algiers and Tripoli as well as by the natives of Central and South, Western Africa. The French have paid particular attention to its culture in Algiers and have succeeded in producing tobacco of good flavor and texture. In Australia the plant does remarkably well and promises to become as celebrated as that of other portions or islands of the East India Archipelago.

It readily appears from the extensive cultivation of tobacco that it can hardly fail of becoming an important article of commerce. The Spaniards and Portuguese found it to be an important source of revenue, and from South America and the West Indies exported large quantities to Europe. As soon as it began to be cultivated in Virginia its commercial value began to be apparent and attracted many navigators who came thither to barter for tobacco and furs, and other articles of inferior value. Most of the tobacco exported from the United States is shipped to Europe and from there is reshipped to Asia and Africa. Of foreign tobacco but little finds its way to this country, the duties* preventing many varieties of excellent quality competing with our domestic tobacco. Cuba, St Domingo and Manilla tobacco are the only varieties that are imported from other countries. West India tobacco, more particularly that of Cuba — is shipped to all parts of the world, especially to Spain, Great Britain, Russia, France and the United States.

The tobacco of South America is exported almost entirely to Europe. England receives a large quantity of South American tobacco as well as Spain and Portugal. The varieties cultivated in Asia and Africa for export are shipped mostly to Europe. Great Britain, Spain, France and Germany are the great tobacco-consuming countries of the

---

* Thirty-five cents a pound, gold.

world, or at least of Europe. In Great Britain, Spain and Portugal, no tobacco is cultivated, and these countries are therefore dependent upon their colonies for a supply of the great product. The commerce in the plant is extensive and reaches to every part of the globe. No nation, state, or empire now ignores the revenue to be derived from its import or culture, and many a government receives more from this plant alone than from any other source.

While some nations prohibit its culture at home, their colonies are allowed to grow it, and thus the article and the revenue are both secured. But while the production of the plant and the commerce depending on it are extensive, they are not more so than the manufacture of the leaf into the various preparations for use. The government work-shops of Seville and Manilla, as well as those of Havana and Paris are of enormous proportions and employ thousands of operatives in the manufacture of cigars and cigarettes. In this country and in England, large quantities of cigars are made both from domestic and foreign tobaccos.

In South America also many are made, but more for home use than for export. Cutting leaf is largely manufactured in this country, especially near the great leaf growing sections. Most of this is used here, the leaf for manufacture abroad being exported in hogsheads for cutting in any form desired. Snuff leaf is exported largely from this country to Great Britain and France, where are the largest manufacturers of snuff in the world. At the present time the demand seems greater for cutting than for cigar leaf. The growths of the West Indies and South America furnish a large quantity of fine tobacco for cigars, but comparatively little for cutting purposes. European tobaccos are adapted for both cutting and for cigars, and are used extensively at home though not considered equal to American varieties, being of a milder flavor. As an article of production and commerce, tobacco must be considered as important as any of the great products or staples, since the demand is constant and continually increasing. Year by year its cultivation extends into new sections, where it becomes a permanent production if the soil and climate prove congenial. From time to time new varieties become known, and are cultivated in various countries with success varying according to the soil and climate and the knowledge of the planter. Nowhere is the plant receiving more attention both in its cultivation and manufacture, than in this country. The varieties grown in the tropics have been tested with more or less success, and bid fair ere long to become the leading kinds in some sections. But not alone in this country is the plant attracting the attention of the great commercial nations. In Europe and Asia as well as in Africa, its production is assuming the large proportions due to its vast importance to Agriculture and Commerce.

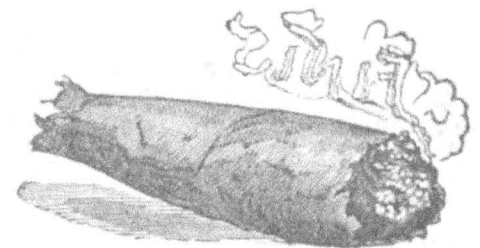

SMOKING IN THE SEVENTEENTH CENTURY.

ENGLISH GALLANTS.

www.ingramcontent.com/pod-product-compliance
Lightning Source LLC
Chambersburg PA
CBHW081017240526
45471CB00017B/3154